AF574365

Intelligence for Embedded Systems

Cesare Alippi

Intelligence for Embedded Systems

A Methodological Approach

Springer

Cesare Alippi
Dipartimento di Elettronica Informazione e
Bioingegneria
Politecnico di Milano
Milan
Italy

ISBN 978-3-319-05277-9 ISBN 978-3-319-05278-6 (eBook)
DOI 10.1007/978-3-319-05278-6
Springer Cham Heidelberg New York Dordrecht London

Library of Congress Control Number: 2014932818

Printed on acid-free paper

Springer is part of Springer Science+Business Media (www.springer.com)

To Alessandro Maria, my Family
and those who came before

Preface

This book was written having in mind researchers, practitioners, and students willingly to *learn, understand, or perfect the fundamental mechanisms behind intelligence* and how they can be used *to design the future generation of embedded systems and embedded applications.*

Adaptation strategies, active and passive learning abilities, robustness capabilities, design of embedded, and distributed cognitive fault diagnosis systems, techniques for assessing the performance and constraints satisfaction in embedded applications are some fundamental aspects intelligent embedded systems and embedded applications need to face to deal with those uncertain, nonstationary, and evolving environments the real world is proposing.

The approach is *methodological*; as such, the presented methods are technology independent and can be suitably adapted to software, hardware, or both implementations, depending on the application constraints. Although it is not the focus of the book, presented methodologies can also be fruitfully used to guide the hardware/software co-design phase to define which parts of the application are more suitably implemented on a dedicated hardware and which ones found in software the most adequate implementation.

For its nature the book is crossing several disciplines, from measurements and metrology to machine learning, from computer science to probability and system identification. As such, *it is a book designed to build bridges among those fundamental areas for electronic engineers, computer scientists, and physicists.*

As it will be immediately clear the book was neither written with the goal to propose a tutorial for all topics covering a specific issue nor detailing and listing all papers and methodologies related to a given argument. Instead, the focus is on the formalization of a given problem, proposal of the most relevant strategies for addressing it, and discussions about "what is behind" the theory, the method, the approach. I consider the book successful if the reader grasps, after reading the main strategies, ideas and challenges behind intelligence and how intelligent methods can be—and should be—used to boost the next generation of embedded applications.

The book has been designed with the goal to fill a gap existing among disciplines a computer scientist, particularly the one designing embedded applications, will face in his/her working life.

It is the author's view that *many sections of the book should constitute teaching material within computer science and electronic engineer curricula* possibly, but not necessarily, within an embedded systems or a machine learning course. The material should be provided at the graduated or Ph.D. student level to optimally benefit from skills and knowledge gained in the undergraduate program.

The book is mostly self-contained. It is expected the reader to be familiar with the basics of mathematics (integrals, linear algebra, gradients, and partial derivatives) and hold the principles of probability and statistics (mean, variance, distributions) and operational research (function optimization). The reader must also be acquainted with the basics of computer sciences and the very basics of electronics, even though deep knowledge is not requested at this level. Having that in mind, also undergraduate students can take advantage of many of the presented topics. For instance, Chaps. 2, 3, 8, and 10 would constitute appropriate material to be taught within an undergraduate course on embedded systems or computer science, with the other chapters' contents used to give the flavor of what is behind. Material presented in the book chapters would constitute instead a full course on advanced embedded systems.

If *multidisciplinary* is the most appropriate adjective characterizing the book from the content point of view, the process behind its birth and writing is surely *globalized*. The book was conceived and moved its first steps in Paris, at the École Supérieure de Physique et de Chimie Industrielles (ESPCI), France, following the very welcome invitation of Prof. Gérard Dreyfus. Then, it grew in Italy, at the Politecnico di Milano, both at the Milan and the Lecco campuses, Italy. The first draft took body in Beijing, China, at the Chinese Academy of Sciences, Institute of Automation (CASIA), following a research experience carried out within the group of Prof. Dongbin Zhao. Refinements were carried out at the National Library in Florence, Italy, and the Institute for Infocomm Research (I^2R), A*STAR, Singapore, thanks to Dr. Huajin Tang who invited me for a short visit. Deep discussions and presentations where given, among others, at the Advanced Learning and Research Institute of the Università della Svizzera italiana, Switzerland, the Tsinghua University and the Peking University, China, the National University of Singapore, Singapore, the National Taiwan University in Taipei, Taiwan and the Los Alamos Labs, USA.

I am beholden to my family for having continuously supported me in this time-consuming editorial challenge.

I also acknowledge the great contribution received by my collaborators in reviewing some chapters and run most of the experiments behind the introduced examples. At first I need to mention the precious collaboration of Dr. Manuel Roveri, then that of Dr. Maurizio Bocca, Dr. Giacomo Boracchi, Dr. Antonio Marullo, Mrs. Ouejdane Mejri, and Mr. Francesco Trovò, all from Politecnico di Milano, Italy.

I take the opportunity to thank those friends—before my colleagues—who contributed to review some chapters. In particular, a great thank you must be sent to Mariagiovanna Sami (Politecnico di Milano, Italy, Università della Svizzera italiana, Switzerland) for having always supported me over my research career.

Then, Prof. Ali Minai (University of Cincinnati, USA), whose competences were fundamental to critically assess the chapter on cognition, as well as Prof. Roberto Ottoboni (Politecnico di Milano, Italy) for the proofread of the chapter on metrology and measurements.

Finally, I acknowledge the support of those Agencies that contributed in different ways to advance both basic and applied research. In particular, this work was partly supported by the FP7 EU Project-i-SENSE Making Sense of Nonsense, Contract No: INSFO-ICT-270428, the visiting professorship of the Chinese Academy of Sciences, the INTERREG EU project M.I.A.R.I.A.: An adaptive hydrogeological monitoring supporting the alpine integrated risk plan and the KIOS Cyprus-funded project.

Milan, January 2014 Cesare Alippi

Contents

Acronyms

2cp	2's Complement
ACC	Anterior Cingulate Cortex
ADC	Analog to Digital Converter
AOA	Angle of Arrival
AR	Auto Regressive
ARL	Average Run Length
ARMAX	AutoRegressive Moving Average eXternal
ARX	AutoRegressive eXternal
BLB	Bag of Little Bootstraps
cdf	Cumulative Density Function
CDT	Change Detection Test
CI-CUSUM	Computational Intelligence CUSUM
CLT	Central Limit Theorem
CPM	Change Point Method
CPTM	Controlled Power Transfer Module
CUSUM	CUmulative SUM Control Chart
DCS	Discrete Controller Synthesis
DPR	Dynamic Partial Reconfiguration -FPGA
DV/FS	Dynamic Voltage/Frequency Scaling
ECM	Electronic Control Module
EM	ElectroMagnetic
FDS	Fault Diagnosis Systems
FFT	Fast Fourier Transform
FIR	Finite Impulse Response
FOA	Frequency of Arrival
FPGA	Field-Programmable Gate Array
FSM	Finite State Machine
FTSP	Flooding Time Synchronization Protocol
GPS	Global Positioning System
GPU	Graphics Processing Units
H-CDT	Hierarchical CDT
HMM	Hidden Markov Model
ICI	Intersection of Confidence Intervals
JIT	Just in Time

k-NN	k-Nearest Neighbors
LOO	Leave-One-Out
LPAC	Lateral Prefrontal and the Association Cortices
LTI	Linear Time Invariant
LTS	Lightweight Tree-based Synchronization
LUT	Look Up Table
M2M	Machine-to-Machine
MAPE(K)	Monitoring, Analysis, Planning, Execution, (Knowledge)
MEMS	Micro Electro-Mechanical Systems
MIPS	Million Instructions Per Second
MLE	Maximum Likelihood Estimation
MPPT	Maximum Power Point Tracker
MSE	Mean Squared Error
NTP	Network Time Protocol
OFC	Orbital preFrontal Cortices
PACC	Probably Approximately Correct Computation
pdf	Probability Density Function
ppm	Parts Per Million
QoS	Quality of Service
RBS	Reference Broadcast Synchronization
RSS	Received Signal Strength
RSSI	Received Signal Strength Indicator
SE	Squared Error
SEPIC	Single Ended Primary Inductor Converter
SNR	Signal to Noise Ratio
SPU	Sensing and Processing Units
TDOA	Time Difference of Arrival
TOA	Time of Arrival
UCEM	Uniform Convergence of Empirical Mean
UWB	Ultra-Wide Band
VM-PFC	Ventral-Medial PreFrontal Cortices
w.p.1	With Probability One

Symbols

$\mathbb{N}^d$	Space of natural d-dimensional vectors
$\mathbb{Z}^d$	Space of integer d-dimensional vectors
$\mathbb{Q}^d$	Space of rational d-dimensional vectors
$\mathbb{R}^d$	Space of real d-dimensional vectors
E	Expected value w.r.t. all random variables
E_x	Expected value w.r.t. x
$Var(x)$	Variance of x
$\lceil x \rceil$	Minimum integer greater or equal to x
$\Vert\cdot\Vert$	Norm operator
δx	Perturbation affecting x
$\mathcal{N}$	Gaussian distribution
$\circ$	Point-wise multiplication operator
$\bar{V}$	Structural risk
V_N	Empirical risk
ln	Natural logarithm
$O(\cdot)$	Big Oh notation
$trace(A)$	Trace of matrix A
$erf(\cdot)$	Error function
$\frac{\partial f(\theta,x)}{\partial \theta}\vert_{\theta^o}$	Function $\frac{\partial f(\theta,x)}{\partial \theta}$ is evaluated in θ^o
$f(x)\vert x$	$f(x)$ such that x

Chapter 1
Introduction

The emergence of nontrivial embedded sensor units (e.g., those embedded in smartphones and other everyday appliances), networked embedded systems, and sensor/actuator networks (e.g., those associated with services running on mobiles and wireless sensor networks) has made possible the design and implementation of several sophisticated applications where large amounts of real-time data are collected to constitute a *big data* picture as time passes. Acquired data are then processed at local, cluster-of-units or server level to take the appropriate actions or make the most appropriate decision.

Acquired data may be influenced by uncontrolled variables as well as external environmental factors, which impair the availability, validity, and usability of data. Sensors are affected by uncertainty and faults, either transient or permanent, negatively impacting on the subsequent decision-making process. Moreover, finite precision representation, algorithm pruning at the numerical computational level so as to satisfy execution time and memory constraints, and the use of algorithms and parameters learned from data, introduce additional levels of uncertainty affecting the accuracy of the decision-making algorithm.

On the technological side, advances in embedded processor have made available embedded systems with a computational capacity ready to support sophisticated intelligent mechanisms and the ability to host a plethora of sensors, hence providing a new degree of freedom to applications. For instance, a smartphone might host an ambient light sensor for adjusting the display brightness (which in turn saves battery power), a proximity sensor (e.g., to switch off the display), a MEMS accelerometer (e.g., to commute a picture visualization from landscape to portrait), a compass (to identify the magnetic pole and orientate maps to the north), and GPS (to provide the local coordinates). It should be emphasized that sensors can even be virtual, e.g., a sensor that detects a change in inclination by processing the output of an existing tri-axial MEMS accelerometer, a sensor providing the profiling of a web user, or one proposing email features for subsequent spam detection.

Interestingly, the adjective *intelligent*, when associated with a sensing system, can be inflected differently, depending on the reference community. As such, it may

C. Alippi, *Intelligence for Embedded Systems*, DOI: 10.1007/978-3-319-05278-6_1,

in some way imply: the ability to make decisions, the capability of learning from external stimuli, the promptness in adapting to changes, or the possibility of executing computationally intelligent algorithms. All the above definitions, explicitly or implicitly, rely on a computational paradigm or application which receives and processes incoming acquisitions to accomplish the requested task. Under this framework, the literature generally assumes that sensors are fault free, that data are stationary, time invariant, available, and ready to be used, and that the application is capable of providing outputs and taking decisions. Unfortunately, assumptions about the quality and validity of data are so implicitly taken as valid by scholars that, most of the time, even their existence as assumptions is forgotten.

From the perspective of the designer and the user, we expect that the coded application is well performing and fully satisfies technological and application constraints, e.g., power consumption, cost, execution time, and accuracy performance. But how do we assess the quality of the algorithm given the different degrees of freedom available in the design space of an unknown solution to a given problem? We recall that numerical embedded applications operate in an uncertainty-affected world: do we really need to waste resources to represent and process uncertainty? We should balance complexity with accuracy.

Moreover, we are interested in designing an application characterized by robust features, so that perturbations affecting its computation and the presence of uncertainty will be tolerated well and their effects mitigated. How can we evaluate an effective index for the application robustness allowing us to select/drive the design phase toward the most appropriate solution? Owing to the different forms of uncertainty affecting the numerical computation, the code execution provides an output that is approximated. If we investigate this concept a bit further, we discover that the natural way of carrying out a numerical embedded computation is to provide a result that approximates the correct one with high probability. That is what the Probably Approximately Correct Computation (PACC) is about.

Adaptation is another main feature any intelligent system should possess. It represents the lowest form of intelligence associated with passive, uncontrolled intelligent mechanisms, like those of emotional processing in the human brain. Some of the adaptation mechanisms are well known and operate at the hardware level to keep control of the power consumption, e.g., by means of voltage and clock frequency scaling. However, adaptation must be intended in a broader sense within an embedded framework, with learning mechanisms representing the main tools to keep track of the environmental evolution. Although adaptation plays a relevant role in managing evolving applications, it may not be sufficient in those applications requiring more sophisticated responses to meet a higher Quality of Service (QoS) and performance. Here, cognitive mechanisms must be envisaged, which act as controlled conscious processes. The basics of cognition, as well as some implications on existing mechanisms specified in the book, will be addressed since it is strongly perceived that the next generation of embedded systems will be based on cognitive-based approaches.

Another issue is that of learning in a nonstationary environment. For most applications, we assume that the data stream to be processed is time invariant and, hence, a time invariant-based application is designed to solve the problem. Although the time invariant hypothesis might hold for a short period of time, it probably does not in the mid term and never in the long run. We should not be surprised, our body evolves over time, so embedded systems and the way they interact thanks to sensors and actuators with the external environment do. This is a consequence of aging effects at the sensor level, faults or changes affecting the environment, and/or the interaction between the embedded system and the environment. Such mechanisms induce changes in the structure of the process generating the data, that evolves, with the consequence that the application being executed rapidly becomes obsolete unless learning strategies are taken into account. Learning in an evolving environment aims at addressing this aspect so that the information carried by incoming data is not only used for decision making but, also, to keep track of the changes and react accordingly. When the change is of a fault-type, we need to intervene with suitable fault diagnosis procedures whose effectiveness depends on a priori information about the environment, the nature of the process generating measurements to be acquired by sensors, and last, but not least, the type of expected faults.

Most of the above aspects need forms of intelligence, either basic or advanced, to learn in a time variant, evolving environment and diagnose fault occurrences. These aspects will also be tackled in next chapters.

1.1 How the Book is Organized

We briefly summarize the main topics addressed in the book to get a quick snapshot of both organization and content. The functional dependency among chapters in terms of content is given in Fig. 1.1.

1.1.1 From Metrology to Digital Data

Most embedded systems take advantage of a sensor platform to carry out the due task. However, the mounted sensors may not only be those requested by the application as seen by the user, since extra sensors are generally envisaged to improve the QoS needed by the application or mitigate technological problems that might impair overall performance. As an example of the former class we have the Received Signal Strength Indicator (RSSI) sensor that measures, in wireless communications, the power of the received radio signal. By exploiting the information provided by the RSSI, we can derive the quality of the communication link and identify suitable actions for maximizing it. As an instance of the latter class we have the temperature sensor internal to the sensor device (not to be confused with the temperature sensor mounted on the acquisition board), which measures the temperature of the sensor for thermal effect compensation. In fact, if a tiltmeter is deployed to inspect the structural health of a building, then both the transduction mechanism and the analog to digital conversion are parasitically affected by the temperature. The temperature

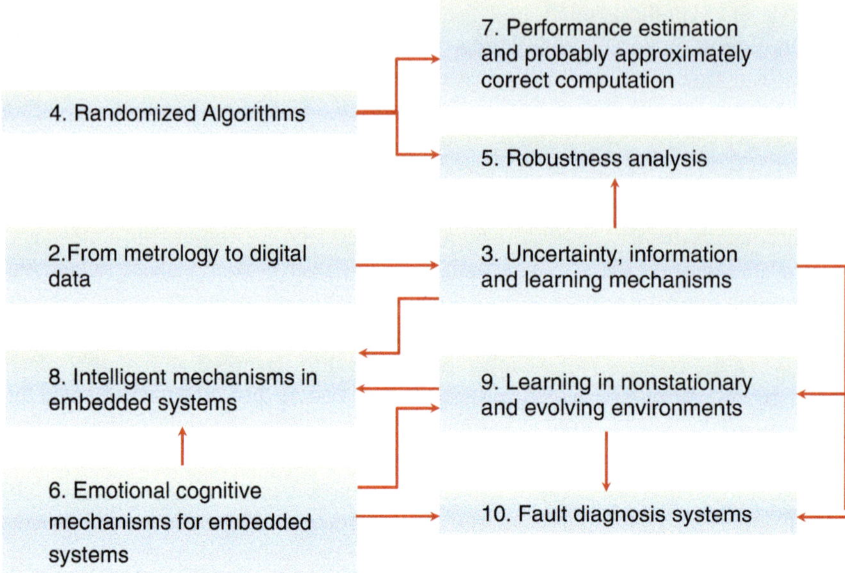

Fig. 1.1 The functional dependency among the chapters composing the book. An oriented arc from chapter i to j implies a strong functional dependency between the two chapters with material in i relevant to fully understand that in jth

sensor allows us for introducing measurement compensation actions. As a result, the data flow is affected by uncertainties from the transduction mechanism to the digital instance.

The chapter introduces the basic concepts behind measure and measurements, e.g., accuracy, resolution, and precision and sheds light on the elements composing the measurement chain (transducer, conditioning stage, analog to digital converter, estimation module). Since measurements are affected by uncertainty, we need to investigate how uncertainty corrupts the final acquired data. This analysis sets the basis for the subsequent propagation of uncertainty within the computational chain as well as introduces constraints on the final embedded solution.

1.1.2 Uncertainty, Information, and Learning Mechanisms

Uncertainty associated with available measurements is not the unique form of information corruption mechanism. In fact, in digital embedded systems, finite precision representations introduce an additional form of uncertainty that combines in a nonlinear way with that of measurements and propagates along the computational flow. The outcome is that the information content of the output is corrupted, hence affecting its validity in a subsequent decision-making process.

The chapter introduces and formalizes the most important forms of uncertainty that embedded systems have to deal with. In particular, in addition to measurement uncertainty, we will encounter uncertainty affecting the data representation level in a digital device, characterize it, and see how it propagates within a computational flow.

Another interesting form of uncertainty arises when the computational code in execution on the embedded system contains parametric models, with parameters estimated from available data or parameter-free models where the model is configured directly from available measurements. Machine learning-based solutions represent a relevant example of such mechanisms.

Since machine learning solutions play a main role in intelligent systems, the chapter presents a statistical formalization of the theory of learning by detailing those key points that combine model complexity with uncertainty and accuracy. In particular, we will see that different forms of uncertainty reside behind the learning mechanisms, depending on the noise affecting the data, the effectiveness of the learning algorithm, the number of available instances used by the learning algorithm, and the suitability of the model family envisaged to model the data.

Finally, we introduce uncertainty at the application level either when building a model from incoming data or designing an application solution, which, for its nature, is mostly unknown. Uncertainty at application level means that we can design different solutions for an application, possibly equivalent in terms of performance according to a given figure of merit (i.e., a quantity considered to assess the performance of a system solution or method). The problem of deciding which solution should be preferred among the set of feasible solutions is left to the performance estimation and probably approximately correct computation chapter. Here, differently, we are interested in discovering that a high level source of uncertainty exists and, in many cases, dominates over any attempt to improve performance.

As a final comment, all these sources of uncertainty combine in a nonlinear way and influence the result of the computation in execution on the embedded system. Results are no more uncertainty-free but uncertainty-affected since the embedded system does no more provide a correct deterministic outcome but an uncertainty-affected one.

1.1.3 Randomized Algorithms

The chapter introduces an intuitive key mechanism every engineer should be aware of: randomization. The idea behind the method is very intuitive. Every time we cannot solve a complex problem, either because it is too complex or computationally hard, we explore how the problem behaves on a number of instances. We take an input instance and feed it to the algorithm associated with the problem that provides a result. Then, we repeat the procedure by sampling other instances.

We find it natural to believe that, by having many instances or samples, we should be able to say something about the original problem. This is what the Monte Carlo method is about. However, randomization does much more than a blind sampling from an instance space. In fact, by merging the Monte Carlo method with the probabilistic theory of learning we derive the number of samples needed to solve a large class of hard problems within a probabilistic framework.

The proposed method is general and addresses the very large class of applications and figures of merit that are Lebesgue measurable (i.e., all those involved in a numerical computation associated with both physics and engineering problems). Examples of problems we wish to address in the embedded system world are the assessment of the performance level attained by the embedded application, the evaluation of the energy consumed by an algorithm, the estimation of the time needed to execute a task, the determination of the average latency in providing a result, and the satisfaction of the real-time execution constraint for the application.

After having introduced the main results coming from the randomization theory and those granting convergence of sampled quantities to the exact ones, the chapter will present some general methods based on randomized algorithms for solving a large class of performance assessment problems. Results, independent from the dimension of the sampling space and the probability density function induced on the sampling space, mostly unknown, hold in probability at arbitrary accuracy and confidence levels function of the envisaged number of samples.

1.1.4 Robustness Analysis

Robustness analysis deals with the problem of investigating whether a given solution (e.g., the computational flow in execution on the embedded system) is able to tolerate the presence of uncertainty/perturbations affecting it or not. Many embedded applications rely on subsystems implemented with analog solutions. Which is the impact of the production process on our components and, ultimately, the performance of the system?

If we design a computational solution on a high resolution platform, e.g., data are represented in floating point or double precision and, then, we wish to move the application to an embedded system characterized by a fixed point notation, can we be reassured that performance is within an application-tolerable loss or not? The answer to the questions above is that, a priori, nothing can be asserted unless a robustness analysis is carried out to evaluate/estimate the loss in performance associated with the introduced perturbation effect. Those of us senior enough to have worked on artificial neural networks and having tried to port families of neural models from a high precision platform to a lower precision embedded one are well aware of the drastic performance loss in accuracy we shall expect. Clearly, the problem is well beyond that of a specific application.

Most available results on robustness analysis assume the small perturbation hypothesis to make the mathematics amenable enough so as to derive the closed form relationship between perturbations affecting the computational flow/variables and the induced loss in performance. However, such results prove to be of limited use when envisaging the porting of an algorithm to an embedded system, since the small perturbation hypothesis is mostly violated. The chapter solves this problem by introducing a perturbation in the large analysis that, based on randomized algorithms, allow us to derive an estimate of the robustness index possessed by the application.

1.1.5 Emotional Cognitive Mechanisms for Embedded Systems

The chapter introduces the basis of emotional cognition since it is strongly believed that the next generation of embedded systems will natively integrate such mechanisms either in hardware or in software to allow the device/application to expose sophisticated intelligent behaviors. The focus is on the fundamental cognitive mechanisms obtained by modeling the functionalities of the human brain w.r.t its emotional processing ability. All methods presented in the book following this chapter inherit or expose levels of intelligence that can be associated with a same form of intelligence, either automatic or conscious. For instance, adaptation represents the lowest level of intelligence, finding in the Amygdala its neurophysiology counterpart. Reaction to the input stimuli is ruled by an uncontrolled process that grants an immediate reduced-latency action, modelable as an emotional response to perception. However, in many applications, the use of adaptation mechanisms is not enough, mostly due to the generation of erroneous actions (reactions are based on conservative principles), that must be subsequently validated by higher cognitive levels where conscious controlled processes are activated. For instance, this role is played in our brain by the vertical-medial orbital cortices. The same processes will play a major role in the learning in a nonstationary and evolving environment chapter.

1.1.6 Performance Estimation and Probably Approximately Correct Computation

The PACC theory formalizes the way a computation is carried out within an uncertainty affected environment. As such, it represents the natural characterization of those numerical algorithms for embedded systems or parts of the algorithm affected by those forms of uncertainty presented in Chap. 2. We comment that a deterministic computation, mostly requested to satisfy the worst-case scenario, is generally unacceptable since the cost necessary to grant a deterministic outcome is not justified by the high complexity requested by the solution.

It is shown that, by relaxing the request for determinism, which imposes the application output to be deterministically correct, we can formalize a simpler dual probabilistic framework requesting the computation to be correct in probability. The probabilistic problem is characterized by a lower complexity compared with the deterministic one.

The idea behind PACC—but not its formulation—comes from the robust control community where it has been pointed out that designing a deterministic controller introduces an unnecessary complexity compared to a probabilistic design. This additional complexity is not counterbalanced in most real applications by the gained determinism.

We now recall that a probabilistic computation is the natural way an embedded system processes numerical information since the different forms of uncertainty affecting the computational flow natively provide an output that is correct in probability under some mild hypotheses. Let us provide a simple example. Assume we have a scalar function $f(x)$ whose deterministic output y for each input, spanning its input dominion X, is

$$y = f(x), \forall x \in X.$$

If $f(x)$ needs to be executed on an embedded system, given the presence of uncertainty corrupting its evolution, we are satisfied if

$$\Pr\left(y \simeq f(x)\right) \geq \eta, \forall x \in X$$

where η is a confidence term we expect to be close to 1 and $\simeq$ is the approximate operator. In other words, we are satisfied if our embedded system is providing an output approximating the correct one according to a suitable figure of merit; however, the statement has to hold with high probability to be sure that the device behaves as expected, at least in probability. This is exactly what embedded systems, e.g., those designed for domestic appliances, do.

The framework should not be confused with fuzzy logic and fuzzy algorithms that can be cast in the PACC framework but do not naturally provide the confidence level η unless PACC is activated. Randomized algorithms are here used to solve the complex problem associated with the characterization of the PACC level of function $f(x)$.

Another strictly related problem is that of performance estimation and assessment. How can we assess the performance, e.g., in terms of accuracy, of the computation being executed in the embedded system? How can we address the situation where only a given finite dataset is available to estimate the quality of the performance our embedded application claimed to have? If the embedded application is claimed to be 95 % accurate, which is the confidence associated with the statement? The chapter provides answers also to these questions.

1.1.7 Intelligent Mechanisms in Embedded Systems

Adaptation mechanisms are related to those automatic processes implemented by the amygdala and thus allowing our brain to make quick decisions without the need to activate conscious controlled processes. The chapter focuses on some examples of adaptation and conscious decision-making that intelligent embedded systems should possess depending on the functional constraints the application is requesting. At the lowest abstraction level, we have those forms of adaptation affecting the voltage/clock frequency of the system, strategies mainly introduced to keep under control the power consumed by the embedded system. Then, we encounter adaptation at the acquisition level, again with the aim of reducing the energy required to carry out data sampling; the issue is particularly relevant in energy-eager sensors. Here, adaptation basically intervenes on the sampling frequency to reduce the energy consumption.

Intelligence plays also a fundamental role in maximizing the energy harvested when the embedded system can scavenge it from the environment as well as in adapting the system clock to have it aligned with those of neighbors' units. Intelligent mechanisms are beneficial to localize the sensor's unit within an environment without requesting a GPS sensor. In order to achieve this goal, other communicating units have to be deployed nearby and collaborate in a coordinated fashion to the localization effort.

Functional reprogrammability is another form of intelligence that allows the embedded application to undergo changes whenever needed. Although this mechanism is mostly carried out at the software level, with code updated remotely as needed, advances in hardware makes available FPGA-based technologies where reprogrammability can be also envisaged at the hardware level.

1.1.8 Learning in Nonstationary and Evolving Environments

A chapter on "learning in nonstationary and evolving environments" is particularly timely and addresses the case, rather frequent in the real world, where the environment changes but our embedded application does not (it was configured by assuming that the environment was time invariant).

The implications of this way of thinking are fruitful. Before releasing the embedded system, we should ask ourselves if the application we have designed is assuming that the environment and the interaction between the device and the external world will change over time or not. Since all physical processes are, at least, subject to aging phenomena and the environment is mostly time variant, unless suitably controlled (and controllable), we should wonder whether during the lifetime of the embedded application a change is expected or not. In case of a positive answer, we should then ask if the change is negligible or will significantly affect the performance of the embedded application. If this is the case, then the application must be revisited to make it able to deal with changes in the environment or intervene to mitigate the effects of this change.

The main methodologies that allow our application to learn in a nonstationary/time variant environment are presented and detailed. This chapter is fundamental if we have to address the big data scenario where the embedded system might be used to extract features and take a first level decision within a hierarchical triggering mechanism (the embedded system quickly detects events and relevant instances and activates an alarm so that more sophisticated and complex agents intervene to accept/reject the hypothesis). Cognitive mechanisms will play a key role since adaptation itself might not be sufficient to grant the performance level the system is expected to have.

1.1.9 Fault Diagnosis Systems

The last section of the book focuses on fault diagnosis systems. In particular, we will investigate the issue of fault tolerance for sensors and how an application can build mechanisms to detect the occurrence of faults. Here, a cognitive approach will be used, since we want to push the difficult and real case where little a priori information is available and changes and fault signatures must be learned along with the fault diagnosis system directly from the data.

It is shown that little can be done at the single sensor level unless strong hypotheses are made. However, the situation is different if the embedded system mounts a rich sensor platform or is inserted in a sensor network. In such case, redundancy in the information content and functional dependencies among sensors can be exploited to classify a change as fault, change in the environment, or inefficiency of the change detection method (model bias).

Chapter 2
From Metrology to Digital Data

2.1 Measure and Measurements

The operation of measuring an unknown quantity x_o can be modeled as taking an instance—or measurement—x_i at time i with an ad hoc sensor S. Although S has been suitably designed and realized, the physical elements composing it are far from ideal and introduce sources of uncertainty in the measurement process. As a consequence, x_i represents only an estimate of x_o. In extreme cases, the value of x_o might not even exist [109] or simply cannot be measured, e.g., think of the Heisenberg's principle of uncertainty stating that it is not possible to exactly measure both the momentum and the position of a particle [112] with arbitrary accuracy.

As a consequence, despite the intuitive formalization of the measurement process, several major aspects need to be investigated and addressed before claiming that a generic measurement x_i is a an accurate and reliable estimate of x_o. For instance, we would rather require subsequent measurements x_i to be somehow centered around x_o, where centering must be intended according to a chosen figure of merit. In other words, we are requesting an accurate sensor that does not introduce some bias error (*accuracy* property). Then, we hope that the sensor is able to provide a long sequence of correct digits of the number associated with the acquired data. Clearly, a weight sensor able to perceive variations of 1 mg is better than a scale providing a resolution of 10 g (*resolution* property). Finally, each measurement represents only an estimate of the true unknown value, the discrepancy between the two—or error—depending on the quality of the sensor and the working conditions under which the measure was taken (*precision* property). Note that we might have an accurate sensor with a high resolution but a poor precision associated with the measurement process, yielding to a poor measurement. Moreover, we might have a precise measurement acquired with a high resolution sensor, again yielding a poor outcome whenever the sensor is not accurate.

There are other properties we should look at when considering a sensor, e.g., repeatability. Repeatability requires that subsequent measurements acquired in the same operational conditions should be indistinguishable within the uncertainty level

C. Alippi, *Intelligence for Embedded Systems*, DOI: 10.1007/978-3-319-05278-6_2,

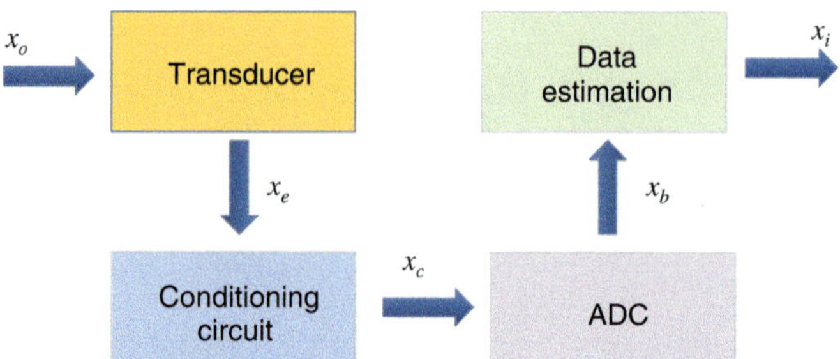

Fig. 2.1 The complete measurement chain of a sensor. The key elements are the transducer, converting an unknown physical entity x_o into the analog electrical entity x_e, the conditioning stage providing an improved analog value x_c, the ADC converting the analog value x_c to a binary codeword x_b, and the final data estimation module leading to the output value associated with the data instance x_i

associated with the sensor. For an in-depth analysis of metrological aspects readers can refer to [180, 182].

In the chapter, we introduce the main actors taking part in the measurement chain which leads, from the physical quantity to be measured x_o, to the final value x_i to be used in the subsequent data processing and decision-making phases. In the following, the measurement framework will be suitably modeled and the properties we expect from the retrieved data formalized.

2.1.1 The Measurement Chain

The main functional elements composing the measurement chain carried out by a sensor are the transduction module, the conditioning circuit, the Analog to Digital Converter (ADC), and the final data estimation module. Figure 2.1 represents a common structure for the measurement chain. The input to the chain is the physical quantity to be measured x_o and the output the digital data x_i.

The functional chain of the figure represents the most common model describing a modern electronic sensor. However, it should be noted that some of the elements composing the chain might be missing in a specific design depending on the cost, the required sophistication level, and where the analog to digital conversion takes place. We will return to these aspects later.

2.1.1.1 The Transducer

A transducer is a device transforming one form of energy into another, here converting a physical quantity x_o into an electric or electric-related quantity x_e (in some cases

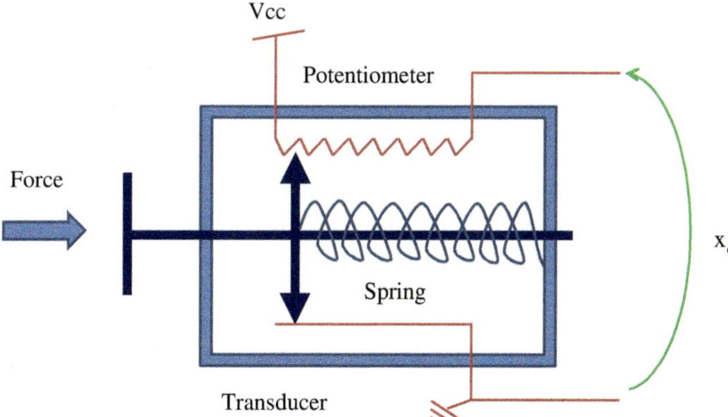

Fig. 2.2 A force transducer composed of a spring and a potentiometer. The force, of intensity x_o, moves a mobile element compressing/releasing the spring. The induced displacement is converted by the potentiometer into voltage x_e thanks to a voltage divider. Vcc represents the reference voltage

the transducer operates with electrical quantities both at the input and output levels). For instance, the temperature of an environment is converted into a voltage (voltage output sensor), the pressure or humidity to a current (current output sensor); the particular target electrical entity depends on the type of the chosen sensor and the way it has been designed. For a detailed analysis of the different typologies of sensors the interested reader can refer to [108]. Clearly, the transduction stage introduces uncertainty on the transduced quantity, which depends on the mechanism used to transform a form of energy into an electrical one.

As an example, and by referring to Fig. 2.2, a sensor of force can be composed, in its transduction principle, of a spring and a potentiometer: the spring converts the force into a displacement and a potentiometer converts the displacement into a voltage variation.

Sensors can be active or passive in their transduction mechanism: an active sensor requires energy to carry out the operation and needs to be powered, whereas a passive sensor does not. Another relevant information is related to the time requested to produce a stable measurement. Such a time depends, for instance, on the dynamics of the transduction mechanisms or the time needed to complete the self-calibration/compensation phase introduced to improve the quality of the sensor outcome.

2.1.1.2 The Conditioning Circuit

The aim of the conditioning circuit [110] is to provide an enhanced electrical quantity x_c of x_e so that the sensitivity of the sensor is amplified, the effect of the noise is mitigated, the interval of definition of the electrical entity is adapted to the requirements of the subsequent ADC. More in detail, the conditioning circuit, which is an analog circuit juxtaposed to the transducer module, at first usually amplifies x_e and then filters its output (e.g., with a low pass filter) to improve the signal-to-noise ratio and the quality of the signal x_c to be passed to the analog to digital conversion stage.

The conditioning circuit might also encompass a module designed to help in compensating parasitic thermal effects, which influence the readout value, as well as introducing corrections to linearize the relationship between the input x_o and x_c. When non-ideal behaviors are compensated by means of a microcontroller, we say that the sensor is enhanced (enhanced sensor). However, it should be pointed out that, in the case of enhanced sensors, the output of the microcontroller is again an analog signal.

In some cases the sensor has an analog output. When this is the case, the output x_i is either x_e or x_c depending on whether the conditioning circuitry is available or not. Analog to digital conversion is carried out later, generally at the microprocessor level, by exploiting the on-chip ADCs. This is a common case in many microprocessors for embedded systems which make available input pins to host analog input signals. Internal on-chip conversion modules are then provided. Clearly, the input signal must be suitably treated and conditioned before it is fed to the microcontroller. For a general presentation of aspects related to embedded system design the reader can refer to [5].

2.1.1.3 The Analog to Digital Converter

The third stage of the functional chain is the conversion module, also known as ADC. The input to the module is the analog electrical signal x_c and the output is a codeword x_b represented in a binary format. There is a large variety of architectures for ADCs [107], all of them having in common the resolution (the number of bits of the codeword) and the sampling rate as target outputs. During the conversion phase, the input x_c must be kept constant, operation carried out by the "sample and hold" mechanism (the analog value is sampled and kept to avoid dangerous fluctuations in the input signal). The conversion introduces an error associated with the quantization level, whose statistical properties may depend on the specific ADC architecture. The source of uncertainty is here variegate and depends, to name a few examples, on the quality of the reference signal (which can change with fluctuations of the powering source), the speed and quality of the conversion step, and the presence of thermal variations that shift the working point of the electronics from a reference ideal one into a different state. The interested reader can refer to [107, 111].

2.1.1.4 The Data Estimation Module

The final module (when present) introduces further corrections on x_b by operating at the digital level. In particular, it generally carries out a further calibration phase aiming at improving the quality of the final data x_i. When a microprocessor is present to address the data estimation module needs, the sensor is defined to be a "smart sensor." The microprocessor can carry out a more sophisticated processing relying on simple but effective algorithms, generally aimed at introducing corrections and structural error compensations. For instance, a thermal sensor can be onboard, in addition to the principal sensor, to compensate the thermal effect on the principal sensor readout. The microprocessor carries out the thermal compensation by reading the temperature value, comparing it with the rated working temperature defined at design time and introducing a correction on the readout value, mostly by considering a polynomial correction function of the discrepancy between the nominal temperature and the current one. The final value x_i shows better properties being closer to x_o. When the dynamics of the signal are known not to change too quickly (compared with the time requested by the ADC to convert a value) or the signal is constant, the microcontroller can instruct the sensor to take a burst of n readings over time. The outcome data sequence $x_{b,j}$ $j = 1, \ldots, n$ can be used to provide an improved final estimate of x_o by averaging

$$x_i = \frac{1}{n} \sum_{j=1}^{n} x_{b,j} \tag{2.1}$$

When the data estimation module is not available, the best estimate of x_o at this level is the value provided by the ADC, i.e., $x_i = x_b$. The designer of the embedded application might decide to carry out this operation later within the application by implementing it in software.

2.1.2 Modeling the Measurement Process

Following the functional description of the sensor given in Sect. 2.1.1 the whole measurement process can now be seen as a black box, suitably described by an input–output model whose simplest, but generally effective form, is

$$x = x_o + \eta \tag{2.2}$$

where $x \in X \subset \mathbb{R}$ is a generic acquired instance, x_o its the ideal, noise-free unknown value, and $\eta = f_\eta(0, \sigma_\eta^2)$ is an independent and identically distributed (i.i.d) random variable with zero mean and finite σ_η^2 variance drawn from probability density function f_η and corrupting the measurement. The additive signal plus noise model (2.2) represents a simple but realistic model describing the measurement process as carried out by the sensor with η accounting for the uncertainty associated with the measurement process. The model implicitly assumes that the noise does not depend on the working point x_o.

Despite the fact that the i.i.d hypothesis is commonly assumed and in fact holds in many circumstances, it might not be satisfied a priori for a specific sensor/application. In fact, we have seen that several sources of uncertainty affect the sensor components and the independency assumption might be violated. It is one of the tasks of the application designer to verify the appropriate model for a sensor as well as determine the existing metrological properties. This is done by first inspecting the sensor datasheet, the operating conditions afterwards, and carrying out suitable acquisitions and metrological analyses whenever requested.

Another common model for the sensor is the multiplicative one where

$$x = x_o + \eta x_o = x_o(1 + \eta). \tag{2.3}$$

In this way, the noise depends on the working point x_o. In absolute terms, the impact of the noise on the signal is $x_o\eta$, but the relative contribution is η and does not depend on x_o. The type of model to be considered depends on the structure of the instrument/sensor available and the way it has been designed and implemented. Working conditions might also have an impact on the selection of the proper model.

In the sequel, we focus on the additive model and introduce other models whenever appropriate. Details related to the validity of the above "signal plus noise" model will be discussed later in the chapter. Despite the particularities of each sensor, we expect some basic properties to hold. The main ones are formalized in the sequel for historical reasons and for their intuitive and common use. However, whenever possible, we should speak about sensor measurement uncertainty. In particular, we need to provide the model adopted for the noise affecting the signal and the pdf function fully associated with the uncertainty. The interested reader can deepen the study of these issues by referring to [180].

2.1.3 Accuracy

Consider the additive signal plus noise model of (2.2). We say that a measure is accurate when the expectation taken w.r.t. the noise satisfies

$$E[x] = x_o. \tag{2.4}$$

In order to have an accurate measurement, the instrument and the measurement process need not introduce any bias contribution. However, this is not always the case: in real-life we all experience problems with sensors providing wrong measurements despite several acquisition attempts, e.g., a room temperature or a badly deployed scale. When this is the case, the simplest model for the sensor becomes

$$x = x_o + k + \eta \tag{2.5}$$

being k the bias value associated with the measure. By taking expectation of (2.5) we have that

$$E[x] = x_o + k \tag{2.6}$$

and, even if we are able to remove the measurement uncertainty, the acquired value is wrong, introducing an unknown offset (bias) value k. When a measurement process is biased we need to subtract the expected value (or its estimate) from the read value. However, since k is unknown, we must rely on a reference value to estimate it. For instance, if we are able to drive the sensor to a controlled state where the expected value is known, say x_o, then, from (2.6) $k = E[x] - x_o$. This phase is called sensor calibration [109, 182].

Accuracy is a main property a measurement system should have since we would like our measurements not to contain any bias error. If we have an accurate measurement system, (2.6) states that, by taking expectation w.r.t. the noise, we remove the impact of noise on the specific value x. During this phase, the value x need not change: in practice, we have to sample at a frequency rate much higher than the dynamics of the signal the sensor is acquiring. This operation is done by the data estimation module if the sensor is smart; otherwise, we have to do it in software with an ad hoc code at the primary microcontroller of the embedded system.

It is always a good practice to take the average of a sequence of n repeated measurements $x_1, x_2 \cdots, x_n$ of the same quantity x to provide a better estimate, $\hat{x} = \frac{1}{n}\sum_{i=1}^{n} x_i$, of x_o compared to that obtainable by using a single instance x_i, leading to $\hat{x} = x_i$. The number of samples n we should consider as well as the convergence properties of the average to the expectation are studied in Sect. 4.2.

Example: Sensor Calibration

We bought a low-cost temperature sensor and are not sure about its accuracy. We wish to quantify the potential bias value so as to zero center subsequent measurements.

For this purpose, we drive our sensor to operate at a known reference value x_o (e.g., set by a laboratory-grade temperature standard) and wait until the dynamics effect associated with the change of state vanishes. In the steady state the sensor shares the same temperature as the environment. We then take n samples, say $n = 40$, from the sensor. An estimate $\hat{k}$ for the bias k is

$$\hat{k} = \frac{1}{n}\sum_{i=1}^{n} x_i - x_o. \tag{2.7}$$

If we iterate the process for different x_o values so as to explore the input domain, we can construct a curve that passes through these points and get a very good calibration curve specialized for the given sensor.

Despite the intuitive example, we comment that calibration is a more complex problem if we look at it closely, in its inner mechanisms. For instance, for an integrated temperature sensor, the read value depends on the value of the power voltage, which is contrasted toward a reference value to identify the change in temperature. Any structural discrepancy between these two values introduces a bias error on the final output. Moreover, the measured voltage is not the voltage powering the sensor since the conditioning and the ADC electronics modify it. In addition, the relationship between the measured voltage and the sensed temperature is nonlinear, depending on the transducing mechanism. Compensations of the above phenomena are known in the literature as offset, gain, and linearization.

2.1.4 Precision

Under the signal plus noise framework and the above assumptions, each taken measurement is seen as a realization of a random variable. Measurements will then be spread around a given value (x_o in the case of accurate sensors, $x_o + k$ in case of an inaccurate one), with the standard deviation defining a scattering level index (other indexes can be defined, e.g., as proposed in [181]). In the sequel, precision is a measure of such scattering and is a function of the standard deviation of the noise σ_η, in the case of both accurate and inaccurate sensors.

Given a confidence level δ, precision defines an interval I for x_o within which all values are indistinguishable due to the presence of uncertainty η. In other words, all values $x \in I$ are equivalent estimates of x_o. The amplitude of the interval depends on the confidence level δ, i.e., $I = I(\delta)$, as it will be immediately clear.

To ease the understanding, let us consider at first η as drawn from a Gaussian distribution $f_\eta(0, \sigma_\eta^2)$ of zero mean and variance σ_η^2. The Gaussian hypothesis holds in many off-the-shelf integrated sensors and can be safely introduced unless differently specified by the sensor data-sheet. Under the Gaussian assumption [181] and by setting a confidence level $\delta = 0.95$, we have that a realization x_i of x_o lies in $I = [x_o - 2\sigma_\eta, x_o + 2\sigma_\eta]$ at least with probability 0.95. With the choice of the confidence interval $I = [x_o - 3\sigma_\eta, x_o + 3\sigma_\eta]$ the confidence level raises to 0.997 (acquired x_i belongs to I with at least probability 0.997). The interval defines the precision (interval) of the measure at a given confidence δ. In this last case, the precision of the sensor (sensor tolerance) is defined as $3\sigma_\eta$, so that $x = x_o \pm 3\sigma_\eta$.

When f_η is unknown, we cannot use the strong results valid for the Gaussian distribution. In this case, we need to define an interval I function of δ within a pdf-free framework. The issue can be solved by invoking the Tchebychev theorem [2] which, given a positive λ value and a confidence δ, grants that inequality

$$\Pr\left(|x_o - x| \leq \lambda\sigma_\eta\right) \geq 1 - \frac{1}{\lambda^2} = \delta$$

holds. By selecting a wished confidence δ, e.g., $\delta = 0.95$, we select the consequent value $\bar{\lambda}$. The precision interval I is now $x = x_o \pm \bar{\lambda}\sigma_\eta$. Clearly, the lack of priors about the distribution is a cost we pay in terms of a larger tolerance interval. This can be clearly seen in Table 2.1 where we compare results provided by a "compact" distribution such as the Gaussian one with those obtainable with a distribution-free approach based on Tchebychev's inequality. By having a priori information about the noise distribution, the precision interval can be easily characterized with a better precision.

Table 2.1 The confidence achievable with precision interval $I = [x_o - \lambda\sigma_\eta, x_o + \lambda\sigma_\eta]$ in the Gaussian and the distribution-free case (Tchebychev inequality)

Distribution	$\lambda = 1$	$\lambda = 2$	$\lambda = 3$	$\lambda = 4$
Gaussian	0.682	0.954	0.997	1
Distribution-free	n.a.	0.750	0.889	0.938

2.1.5 Resolution

Whereas precision is a property associated with a measure, resolution is associated with an instrument/sensor and represents the smallest value that can be perceived and differentiated by others given a confidence level.

If our instrument has a resolution of 1 g, we will not be able to measure values of 1 mg due to the limits of the instrument: the scale will make sense in steps of 1 g (and all values in such interval will be equivalent and indistinguishable). However, having a high resolution neither implies that the measure is accurate nor precise. In fact, the scale can be badly mounted, hence introducing a $k = 100$ g fixed error in the readout (the scale is not accurate). Moreover, if the scale is analog, we might not be able to perceive changes affecting the gram for visualization insufficiency but only something around 10 g (precision error): the size of the pointer might well exceed the gram!

Since our final interest is the accuracy and precision of a sensor, sensor designers mostly provide the precision level (by automatically also considering the resolution impact on the measure in there). That said, the reader must be aware of the confusion present in the market and attention should be paid before selecting a sensor. Moreover, a metrological analysis phase should be carried out if we are not sure about the provided figures.

Example: A Real Sensor

Table 2.2 presents the main features of a temperature sensor for aquatic measurements. The resolution of the instrument is high, but the impact of the noise on the readout value is high as well. The sensor provides values within a [−4 °C, 36 °C] interval with an additive error model influencing the read value up to ±0.3. We immediately derive that $\sigma_\eta = 0.1$ since the sensor is ruled by a Gaussian distribution from data-sheet information and we consider $\lambda = 3$. Otherwise, we should have invoked the Tchebychev's inequality, set a confidence level, e.g., 0.997 (so as to be in line with the Gaussian case) leading to $\lambda = 5.77$. The sensor requires a warm-up time up to 2 s: any value read before the warm-up time has elapsed would produce erroneous data (repeatability is not granted). Engineers should pay attention to this issue.

Table 2.2 A temperature sensor for aquatic measurements

Features	Value
Range	−4–36 °C
Resolution	0.01 °C
Accuracy	±0.3 °C
Response time	≤2 s

2.2 A Deterministic Versus a Stochastic Representation of Data

A common problem we face when designing an embedded application is related to the number of significant digits available within the given codeword. The uncertainty aspect introduced by the binary representation will be studied in detail in Sect. 3.1. Differently, here, we focus on the fact that uncertainty exists and affects somehow the data. We ask ourselves the question: if the output of the data estimation module x_i is represented by means of n bits and hence uncertainty affects the readout, how many bits p are relevant out of the n? The answer to the question requires a deeper analysis and can be addressed by considering two relevant scenarios depending on the nature of the available data, as it will become clear in the sequel.

Consider $x_o = x_o(t) \in X \subset \mathbb{R}$ to be a signal evolving over time and assume that the measurement process is much faster than the dynamics of the signal so that sample $x = x(t)$ can be considered constant during each data acquisition.

2.2.1 A Deterministic Representation: Noise-Free Data

The case covers the situation in which digital data x_i are confined within a deterministic domain, i.e., the feasible values of acquired data are error-free and belong to the closed interval $[a, b]$. If n bits are made available to represent the data and no noise affects them, then each of the 2^n available codewords are worth to be used. By considering a reasonable uniform assignment codeword-information, the distance Δx between two subsequent data instances is

$$\Delta x = \frac{b - a}{2^n - 1}$$

if we also wish to represent both extremes of the interval. In this way the 2^n codewords are respectively assigned as $x_1 = a, x_2 = a + \Delta x, \ldots, x_{2^n} = b$. Clearly, different assignments can be made, also depending on the specific application. Given a value x_o, the maximum representation error is $\frac{\Delta x}{2}$ and the average error is zero. If values are uniformly distributed in the interval $[x_o - \frac{\Delta x}{2}, x_o + \frac{\Delta x}{2}]$, then the variance of the error representation is $\frac{\Delta x^2}{12}$.

Differently, if the data we wish to represent are affected by noise, as it is generally the case, then not all the codewords are meaningful and less than n bits are relevant and should be kept.

2.2.2 A Stochastic Representation: Noise-Affected Data

As we have seen in the measurement chain, data acquired from a sensor are noise-affected. Obviously, we are not interested in spending bits to represent the noise when writing a number. At the same time, precision introduces a constraint on the indistinguishable values we can acquire. In fact, two data are distinguishable and deserve distinct codewords only if their distance is above the precision interval I which, as we have seen, depends on a predefined confidence level δ and acts as the deterministic Δx of the Sect. 2.2.1. The number of independent values can be written as the ratio between the domain interval of the data and the value $I_m = 2\lambda\sigma_\eta$ of the probabilistic indistinguishability interval, σ_η being the uncertainty standard deviation associated with the measurement process. Finally, the number of independent points I_p is

$$I_p = \frac{b-a}{2\lambda\sigma_\eta} + 1$$

if we require the interval extremes to be represented. As before, a straight assignment would be $x_1 = a, x_2 = a + I_m, \ldots, x_{I_p} = b$. The number of significant bits is now

$$p = \lceil log_2\left(I_p\right)\rceil$$

where $\lceil\cdot\rceil$ is the ceiling operator. We have that $p \leq n$ represents the significant bits within the n with the statement holding at least with probability δ. Figure 2.3 shows how values around $x_o = 6$ are affected by noise under the assumption that the noise is normal (zero mean, unitary standard deviation, and $\lambda = 3$). As we get further from x_o, the probability of having a value wrongly assigned to x_o diminishes. Codewords are $x_o = 0, 6, 12, 18$ but the error distribution is shown only in correspondence to codewords $x_o = 6$ and $x_o = 12$. Given the tails of the distribution, we might erroneously assign with probability $1 - \delta$ a wrong codeword to a given value. Such a probability is 0.003 for the example given in Fig. 2.3, e.g., see numbers around 9 which can be assigned both to $x_o = 6$ and $x_o = 12$, though with different probabilities. However, in most reasonable distributions (and in most embedded applications), the introduced error is contained since the mis-assignment probability rapidly diminishes.

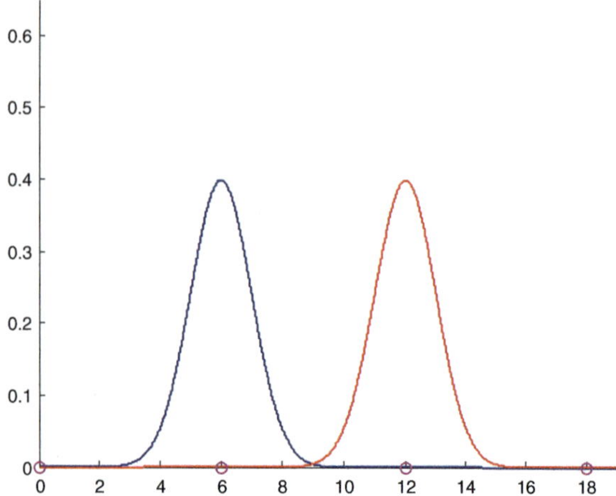

Fig. 2.3 The impact of a normally distributed noise as a function of the distance from $x_o = 6$. The presence of a distribution tail implies that we might wrongly assign the codeword of x_o to a value x which should be assigned to a different codeword instead. As an example, value $x = 9.1$ associated with codeword $x_o = 12$ might have been generated by $x_o = 6$ as well

2.2.3 The Signal-to-Noise Ratio

Consider now the case where the signal is not bounded in deterministic terms and measurements are modeled as instances drawn from a stationary—possibly unknown—pdf. A probabilistic interval can be identified for x_o whose probabilistic extremes are associated with $\lambda_x \sigma_x$, being σ_x the standard deviation of the signal and λ_x the term modulating the width of the interval, chosen to grab confidence level δ. As in previous sections, the number of independent values I_p depends on the interval between two distinct codewords, which are distinguishable according to confidence level δ

$$I_p = \frac{2\lambda_x \sigma_x}{2\lambda \sigma_\eta} + 1$$

By considering the same λs both for the noise and signal we define the Signal-to-Noise ratio (SNR) as

$$SNR = \log \frac{\sigma_x}{\sigma_\eta}$$

where the logarithm base can either be 2 or 10 depending on the subsequent use. Interestingly, $2SNR$ represents the logarithmic ratio of the energy of the signal compared with that of the noise. The SNR is pdf-free and applies to any distribution thanks to the Tchebychev inequality, provided that the same λ value is considered. The number of relevant bits p of the binary codeword finally becomes

$$p = \lceil \log_2 \left(\frac{\sigma_x}{\sigma_\eta} + 1 \right) \rceil \leq \lceil SNR_2 \rceil + 1. \tag{2.8}$$

If $p \geq n$, all bits present in x_i are statistically relevant; otherwise, only p out of n are relevant and $n - p$ are associated with noise. We comment that (2.8) holds for $\frac{\sigma_x}{\sigma_\eta} > 1$, i.e., in all meaningful applications. Alippi and Briozzo [37] show how the SNR can be used to dimension a digital architecture implementing the scalar product between two vectors and then the processing requested by an artificial neuron.

Chapter 3
Uncertainty, Information, and Learning Mechanisms

The real-world is prone to uncertainty. We experience uncertainty in acquiring data, in interacting with the environment through an actuator, in representing information on a finite-precision machine, and in designing an unknown solution to a problem. The chapter formalizes and deals at first with the concept of uncertainty and the way it propagates through a computational flow. Afterwards, the basics of statistical learning are provided. It is shown how different sources of uncertainty, that depend on the chosen model family, the number of available data, and their quality and, ultimately, the complexity of the problem, are introduced when learning from data.

3.1 Uncertainty and Perturbations

3.1.1 From Errors to Perturbations

We have uncertainty whenever we have an approximated entity which, to some extent, estimates the ideal—possibly unknown—one. Such a situation can be formalized by introducing the ideal uncertainty-free entity and the real uncertainty-affected one and evaluating the error, i.e., the discrepancy between the two according to a suitable figure of merit. Since the error is strictly dependent on a specific pointwise instance, e.g., a representation error for a given value, a model error for a specific input, or a sensor error in correspondence of a particular data acquisition, we abstract the pointwise error with the concept of perturbation, a variable defined in a suitable domain with the pointwise error representing a particular realization of it.

In the following, a generic perturbation δA intervenes on the computation by modifying the status assumed by an entity from its nominal configuration A, whose domain and cardinality depends on the specific case, to a perturbed one A_p. The effect induced by the perturbation can be evaluated through a suitable figure of merit $\|A, A_p\|$ measuring the discrepancy between the two states. For instance, if we are looking at the output of a real sensor providing the constant scalar value $a \in \mathbb{R}$,

C. Alippi, *Intelligence for Embedded Systems*, DOI: 10.1007/978-3-319-05278-6_3,

then the discrepancy between the ideal nominal value and the perturbed one can be expressed as the punctual error $\|A, A_p\| = e = |a_p - a|$. Should we read another sensor instance, the pointwise error would assume a different value. In this case, the mechanism inducing uncertainty can be modeled with the signal plus noise model $a_p = a + \delta_a$ and $\|A, A_p\| = |a_p - a| = |\delta_a| = |e|$. It is evident from this example that δ_a can be described in many cases as a random variable with its probability density function fully characterizing the way uncertainty disrupt the information.

3.1.2 Perturbations

In Sect. 3.1.1 we have intuitively introduced the concept of perturbation as a random variable. More formally, the perturbation δA can be defined as a result of the perturbation operator applied to a structured variable A.

Given a generic variable $\psi \in \Psi \subset \mathbb{R}^d$, perturbation $\delta\psi$ moves ψ into a perturbed state ψ_p according to some perturbation model. Not rarely, we can model $\delta\psi$ as a multivariate random variable drawn from a perturbation probability density function $f_\psi(M, C_{\delta_\psi})$ characterized by mean M and covariance matrix C_{δ_ψ}. Ψ can either be discrete or continuous, the latter being the most common situation in signal/image processing.

Definition: Continuous Perturbations

We say that a perturbation $\delta\psi$ is continuous if $\Pr(\delta\psi = \delta\bar{\psi}) = 0, \forall\psi \in \Psi$. The definition tells us that the probability to sample a continuous perturbation space and get exactly a given perturbation is an event whose probability is null.

Definition: Acute Perturbations

We say that the square matrix A_p obtained by perturbing matrix A is acute (and the associate perturbation δA is said to be acute) if and only if

$$\lim_{A_p \to A} \operatorname{rank}(A_p) = \operatorname{rank}(A).$$

In other words, an acute perturbation does not change the rank of a matrix [51]. If perturbation δA is induced by perturbation $\delta\psi$, i.e., $\delta A = \delta A(\delta\psi)$, we also say that $\delta\psi$ is acute.

We will use the definitions above along the book and, in particular, in Chap. 5.

3.2 Perturbations at the Data Representation Level

Numerical data acquired by sensors and digitalized through an ADC are represented as a sequence of bits coded according to a given transformation which depends on the numerical information we need to represent. In the following sections, we will introduce the main transformations used in numerical representations as well as the types and characterization of uncertainty introduced when representing data in a digital format.

3.2.1 Natural Numbers $\mathbb{N}$*: Binary Natural*

Assume we are willing to spend n bits to represent a finite value $a \in \mathbb{N}$. It immediately comes out that we can represent only numbers belonging to a subset $\mathbb{N}(n) \subset \mathbb{N}$ given the finiteness of n. Since n bits provide 2^n independent codewords, the subset $\mathbb{N}(n)$ contains 2^n instances, e.g., the first 2^n symbols $\mathbb{N}(n) = 0, 1, \ldots, 2^n - 1$. For instance, if we have $n = 8$ we can represent the first 256 natural numbers, starting from 0 (or any other 256 numbers depending on the number of information-codeword association). The representation of values in $\mathbb{N}(n)$ follows the binary natural code representation [206], which can be easily derived once we comment that the numeric representation we are looking at is positional and weighted.

In this section we assume that the information associated with the natural number is not affected by uncertainty (the numeric instance is noise-free) and that the unique source of uncertainty is introduced by finite precision operators, such as truncation or rounding, used in order to reduce the number of bits associated with the information from n to its most significant $q \leq n$ bits.

3.2.1.1 Projection to a Subspace

Define the space of a representation as the space spanned by the vector containing as components the bits/digits considered to represent a value. If n are the bits, then $\mathbb{N}(n) = \{0, 1, 2, \ldots, 2^n - 1\}$ is the set of points in the space, with each point referenced by the generic vector in the form $[a_{n-1}, \ldots, a_1, a_0]$.

As such, an interesting projection to a lower dimensional space can be achieved by simply setting to zero the least significant $n - q$ bits of the n bits codeword associated with a (the least significant q bits are set to zero leading to value $a(q)$). The projection introduces an absolute error whose value is

$$e(q) = a - a(q) < 2^q.$$

The pointwise error is a function of the particular number instance and its stochastic characterization depends on the particular nature of the envisaged application, i.e., on the probability density function of the process generating value a. However, it is common to assume a uniform distribution for a. The projection operator introduces an absolute error that can be modeled as a uniformly distributed random variable defined in interval $[0, 2^q)$. The expected value of the error is $\frac{2^q-1}{2}$ and its variance is bounded by $\frac{2^{2(q-1)}}{3}$ [208].

3.2.1.2 Truncation

Truncation operates as a chopping operator that removes the least significant q bits from a n bits codeword. However, if truncation would simply mean chopping bits from a number, then it would not make any sense. For instance, if we consider the decimal value 123, truncation of the least significant digit would generate the number 12, per se a number not even related to the original one in terms of the absolute information content. The error would not even make sense in relative terms: for the case above, the relative error would be $\frac{123-12}{123}$. However, truncation is a key operator in embedded systems, for the reasons we will now explain.

The notation associated with the binary natural codeword is positional and weighted: value 1 in a_0 has a different meaning of a 1 in a_{n-1} (positional notation). In correspondence to bit a_i, we have a weight quantifying the information contribution carried by the bit, which is 2^i (weighted notation). What does make sense is to apply in turn the two steps

- Projection of the codeword in the subspace of dimension $n - q$;
- Apply the truncation operator so as to remove the q rightmost bits.

The final result of the transformation is that the number is now defined in an $n-q$ dimensional space. We save q bits in representing the information at the cost of an introduced source of uncertainty in the data representation (and a loss in information if the original number was uncertainty-free). Consider, for instance, decimal numbers 1234 and 2545, defined in an $n = 4$ dimensional space. We wish to reduce the space to $n - q = 2$ digits. By applying the projection to a subspace transformation, we get numbers 1200 and 2500 and, after truncation, the numbers become 12 and 25. In other words, we are keeping the most relevant part of the information content by operating into a two-dimensional subspace which somehow keeps the distance between the numbers at the net of the truncated information. The number can then be compared with other numbers defined in the same subspace.

The relative distance between the two codewords is mostly kept, although an error is introduced. In fact, by inspecting the numbers after the transformation, it is clear that they can be intended as generated with q right shifts, with the consequence that each instance of the reduced space should be weighted 10^2 to move back to the original one. The binary number $[a_{n-1}, a_{n-2} \dots, a_q, \dots, a_1, a_0]$ becomes $[a_{n-1}, a_{n-2} \dots, a_q]$ after the transformation. Each instance of the reduced space can be brought back to the original space dimension by multiplying it by value 2^q. The introduced absolute error in the original space is $e(q) = a - 2^q a(q) < 2^q$, hence inducing an error uniformly distributed in the interval $[0, 2^q)$.

3.2.1.3 Rounding

Rounding of a positive number truncates the q least significant bits and adds 1 to the unchopped part if and only if the most significant bit of the truncated segment is 1. Otherwise, the rounded value is the one defined over $n - q$ bits. In a binary natural representation, rounding provides a biased uniform error following the comments made for the projection to a subspace and the truncation operator. The advantage of rounding is that the variance $\frac{2^{2(q-2)}}{3}$ of $e(q)$ is half the truncation one.

3.2.2 *Integer Numbers* $\mathbb{Z}$*: 2's Complement*

3.2.3 *2cp Notation*

We are now interested in representing a value $a \in \mathbb{Z}(n) \subset \mathbb{Z}$ over n bits. A straight representation for the generic number a would be the sign and modulus notation. Such a notation is based on the fact that, although formally inaccurate, $Z = -\mathbb{N} \cup \mathbb{N}$. A generic number can then be represented with its sign (requesting a bit) and its modulus (which, being a natural number, can be represented with the binary natural representation). The sign and modulus representation is redundant, in the sense that it uses two codewords to represent the zero (-0 and $+0$) and requires differentiated hardware architectures to carry out additions and subtractions. A different approach, which is used in most embedded systems, is to use a two's complement notation (2cp) that solves both problems.

Given n bits, we have a total of 2^n available codewords, and we decide to assign half of them to represent negative numbers, and the remaining half to code positive numbers (zero included). That said, subset $\mathbb{Z}(n)$ becomes

$$\mathbb{Z}(n) = -2^{n-1}, \ldots, 0, \ldots, 2^{n-1} - 1.$$

The 2cp representation for number $a \in \mathbb{Z}(n)$ is defined as

$$a_{2cp} = \begin{cases} a_{b,n} & \text{for } a \geq 0 \\ (2^n - |a|)_{b,n} & \text{for } a < 0 \end{cases}$$

where subscript b, n stands for a binary natural representation on n bits. The transformation has remarkable properties that make the 2cp notation the one most used in embedded systems. Other expressions that can be derived from the above transformation are more immediate to generate 2cp codewords. One of these is particularly interesting since it exploits the concept of opposite $-a$ of number a. Having a generic number a_n obtained from a with a 2cp transformation over n bits, its opposite $-a_n$ is $-a_n = \bar{a}_n + 1$, where $\bar{a}$ is the bit-wise complement operator applied to the codeword a_n (1s and 0s are toggled in a_n). The immediate consequence is that the subtraction operation can be reduced to the addition one. In fact, given two numbers a_n, b_n in 2cp and defined on n bits, the subtraction $a_n - b_n$ becomes $a_n - b_n = a_n + (-b_n) = a_n + \bar{b}_n + 1$. Both addition and subtraction operators reduce to the algebraic sum, whose simple algorithm is that used for the addition operator.

In order to characterize the nature of the finite precision representation error, let us consider at first the truncation operator, chopping q bits from the n of the original representation. The limits of the truncation operator for $\mathbb{Z}$ are those presented for $\mathbb{N}$. Truncation should be intended as an operator transforming the n-dimensional space of the data into the reduced one of dimension $n - q$. Under this framework, the truncation error associated with truncated value $a(n - q)$ is always positive, assuming values $0 \leq a - 2^q a(n - q) < 2^q - 1$. The error introduced by the truncation operator is uniformly distributed in the interval $[0, 2^q - 1)$ and introduces a bias value. On the contrary, rounding introduces an unbiased error, a very welcome property in any computation: clearly, we would appreciate the outcome of a computation to be accurate, or, in the worst case, characterized by a small bias. Thus, rounding outperforms truncation in the 2cp representation. This makes it a very interesting operator for embedded systems despite the required extra computational cost. If rounding is applied, it can be shown that the representation error is uniformly distributed in interval $[-2^{q-1}, 2^{q-1})$.

3.2.4 Rational $\mathbb{Q}$ and Real $\mathbb{R}$ Numbers

As pointed out in the previous sections, the finiteness of the machine limits the number of codewords to 2^n if n is the number of available bits. As a consequence, we can only approximate a generic number a, either belonging to $\mathbb{Q}$ or $\mathbb{R}$, with the number $a(n)$ which, for its finite nature, belongs to $\mathbb{Q}$.

3.2.4.1 Fixed Point Representation

Any rational number $a \in \mathbb{Q}$ can be seen as composed of an integer part and a fractional one. A natural approximation $a(n)$ of a is a number where $l = n - k - 1$ bits are assigned to the integer part, one to the sign bit, and k to the fractional one. This notation is called fixed point since the "dot" separating the integer from the fractional part is conventionally fixed in the notation, being k bits leftwards from the least significant bit (note that the point is only virtual and such information is not stored). This said, we also note that the number $a(n)2^k$ is an integer number and, as such, it can be represented with a 2cp notation. De facto, there is no difference between a generic fixed point number and an integer one!

Example: Fixed Point Representation

Consider, as an example, decimal value $a = 1.56$ and say that we are willing to spend $n = 5$ bits to represent it in 2cp. We decide to use 2 bits for the fractional part ($k = 2$). The number can then be represented with the fixed point binary sequence [00110], e.g., codeword [001.10] associated with the approximated decimal number $a(n) = 1.5$. If we multiply the binary codeword by factor 2^k, the fractional point disappears and we obtain codeword [00110], which is associated to the binary value $a(n)2^2$. The introduced absolute error is $|e(q)| = |a - a(n)| = 0.06 < 2^{-2}$.

Let us consider now number a coded in 2cp over n bits with l bits associated with the non-fractional part (without including the sign bit). We wish to reduce the n bits to $n - q$ bits at first through a q least significant bits truncation (truncation might also affect the integer information, and not only the fractional one).

Multiply a by 2^{-l} so that $2^{-l}a$ becomes a totally fractional number. We have seen that if we keep k bits for the fractional part, then the introduced error is lower than 2^{-k}. Given the fact that we wish to keep $n - q$ bits and a bit is used for the sign, we have that the truncation error $e(q)$ is always positive (also for negative numbers) and satisfies the inequalities

$$0 \leq e(q) < 2^l(2^{-(n-q-1)}).$$

As an interesting example, let us consider the decimal number 0.45 represented on $n = 5$ bits, $l = 0$. The 2cp representation becomes [00111]. We wish to represent the number on a smaller space by choosing $q = 2$. The obtained number after truncation is [001], e.g., decimal number 0.25. Since $l = 0$, we have that the representation error must satisfy $0 \leq e(q) = 0.2 < 2^{-2} = 0.25$. If rounding is applied, then it can be shown that the error $e(q)$ satisfies

$$-2^l(2^{-(n-q)}) \leq e(q) < 2^l(2^{-(n-q)}).$$

The error owed to rounding is independent of the binary representation and its mean is zero. Moreover, rounding introduces a lower variance compared to truncation.

Let us consider, as a second example, decimal number 6.9, to be represented in 2cp fixed point notation on $n = 7$ bits. The 2cp representation becomes [0110111]. We wish to represent the number on a smaller space by choosing $q = 2$ and rounding. The obtained number after rounding is [01110], e.g., decimal number 7. Since $l = 3$, $n = 7$, and $q = 2$, we have that the representation error $e(q) = 6.9 - 7$ has to be in magnitude smaller than 2^{-2}, as it is.

As a last example, consider decimal number -6.666 to be represented in a 2cp fixed point notation on $n = 7$ bits. The 2cp representation becomes [1001011]. We wish to represent the number on a smaller space by choosing $q = 1$ and rounding as space reduction technique. The codeword of the positive number is [0110101]; after rounding with $q = 1$ we get codeword [011011], to which the rounded negative codeword [100101] is associated. Since $l = 3$, $n = 7$, and $q = 1$, we have that the representation error $e(q) = -6.666 - (-6.75) = -0.084$ has to be of magnitude smaller than 2^{-3}, as it is.

Summarizing, in a 2cp notation, the error introduced by quantization is uniformly distributed [208] in the interval

$$[0, 2^{l-n+q+1})$$

for the truncation operator and uniformly distributed in the interval

$$[-2^{l-n+q}, 2^{l-n+q})$$

for the rounding operator. The above distributions should be used to test the effects of noise on the embedded computation as requested in Chap. 7 or to evaluate the intrinsicrobustness of the computational flow as detailed in Chap. 5.

3.3 Propagation of Uncertainty

We analyze in this section the way perturbations affecting sensor data propagate within a computational flow $y = f(x), x \in X \subset \mathbb{R}^d, y \in Y \subset \mathbb{R}$. The sensitivity analysis provides closed-form expressions for the linear function case and approximated results for the nonlinear one, provided that the perturbations affecting the inputs are small in magnitude compared to the inputs (small perturbation hypothesis). The analysis of *Perturbations in the large* i.e., perturbations of arbitrary magnitude, for the nonlinear case, cannot be obtained in a closed form unless $y = f(x)$ assumes a particular structure and has properties that make the mathematics amenable. The extended analysis dealing with the perturbation in the large framework will be presented in Chap. 7.

3.3.1 Linear Functions

Let us consider linear function $y = f(x) = \theta^T x$, where $\theta \in \Theta \subset \mathbb{R}^d$ and x are d-dimensional column vectors representing the parameters and the inputs of the linear function, respectively. In the sequel, the parameter vector θ is assumed to be constant and given, otherwise differently specified.

3.3.1.1 The Additive Perturbation Model

A perturbation δx affecting the inputs, say according to an additive signal plus noise perturbation model $x_p = x + \delta x$, generates the perturbed value $y_p = \theta^T x_p$. The pointwise error $\delta y = y_p - y$ can be rewritten, thanks to linearity, as

$$\delta y = \theta^T \delta x. \tag{3.1}$$

Note that the linear function, characterized by its parameter vector θ, is not structurally affected by perturbations, which only influence the function inputs. The (3.1) tells us that the propagated error at the function output is linear with the perturbation vector. The perturbation δx can be modeled as a random variable subject to pdf $f_{\delta x}(0, C_{\delta x})$, where $C_{\delta x}$ is the covariance matrix of the perturbation.

Characterization of the perturbation error δy, which also becomes a random variable, can be done by providing the mean and the standard deviation of the propagated error at the function output and then, where possible, its pdf. We have that

$$E_{\delta x}[\delta y] = E_{\delta x}[\theta^T \delta x] = \theta^T E_{\delta x}[\delta x] = 0$$

and

$$\mathrm{Var}(\delta y) = E_{\delta x}[\theta^T \delta x \delta x^T \theta] = \theta^T E_{\delta x}[\delta x \delta x^T]\theta = \theta^T C_{\delta x}\theta = \mathrm{trace}\left(\theta^T \theta C_{\delta x}\right).$$

Under the independence assumption for the perturbation affecting the inputs, $C_{\delta x}$ happens to be a diagonal matrix with the i-th entry characterized by variance $\sigma^2_{\delta x,i}$. Then, defining θ_i to be the i-th component of vector θ,

$$\mathrm{Var}(\delta y) = \sum_{i=1}^{d} \theta_i^2 \sigma^2_{\delta x,i}.$$

In the particular case where all perturbations have the same variance $\sigma^2_{\delta x}$, e.g., perturbations are uniformly defined within the same bounded interval, the above expression becomes

$$\mathrm{Var}(\delta y) = \sigma^2_{\delta x}\theta^T\theta. \tag{3.2}$$

The pdf of the propagated error cannot be evaluated a priori in a closed form unless we assume that the dimension d is large enough. In such a case, we can invoke the Central Limit Theorem (CLT) under the Lyapunov assumptions [35] and δy can be modeled as a random variable drawn from a Gaussian distribution.

CLT Under the Lyapunov Condition

Let $Y_i, i = 1 \ldots d$ a set of independent random variables characterized by finite expected value $E[Y_i]$ and variance $\mathrm{Var}(Y_i)$. Denote $s_d^2 = \sum_{i=1}^{d} \mathrm{Var}(Y_i)$ and $Y = \sum_i Y_i$. If there exists number $l > 0$ such that

$$\lim_{d\to\infty}\left(\frac{1}{s_d^{2+l}} \sum_{i=1}^{d} E\left[|Y_i - E[Y_i]|^{2+l}\right]\right) = 0,$$

then $Z = \frac{(Y - E[Y])}{\sqrt{Var(Y)}}$ converges to the standard normal distribution.

From the intuitive point of view, the CLT tells us that the sum of many, not-too-large, and not-too-correlated random terms, average out. The Lyapunov condition is one way for quantifying the not-too-large term request by inspecting the behavior on some $2 + l$ moments. In most cases, one tests the satisfaction of the condition for $l = 1$.

From the theorem, with the choice $Y_i = \theta_i \delta x_i$, δy can be approximated as a random variable drawn from Gaussian distribution $\delta y = \mathcal{N}(0, \sum_{i=1}^{d} \theta_i^2 \sigma_{\delta x,i}^2)$ provided that the Lyapunov condition holds. If all $\sigma_{\delta x,i}^2$ terms are identical to $\sigma_{\delta x}^2$, then $\delta y = \mathcal{N}(0, \sigma_{\delta x}^2 \theta^T \theta)$.

It is easy to show that the Lyapunov condition holds if each component of random variable δx is uniformly distributed within a given interval, as it happens in many application cases (think of the error distribution introduced by the rounding and truncation operators operating on binary 2cp codewords). Let us show it with an example that models the situation where all inputs of our embedded system are represented on the same number of bits and a 2cp notation is adopted. Rounding is the considered chopping operator for which we know that the induced error distribution is uniform and centered.

Example: The CLT Under the Lyapunov Condition

Let δx be an i.i.d random variable with each component uniformly defined in interval $[-1, 1]$ (if we consider the case of an embedded system, the error introduced by rounding is defined in such an interval). Assume that $\theta_m^2 \neq 0 \leq \theta_i^2 \leq \theta_M^2$, i.e., the generic parameter is bounded by the same minimum and maximum values. Let $\delta y = \theta^T \delta x$ and define $Y_i = \theta_i \delta x_i$. We have that $E[Y_i] = 0$ and variance $\mathrm{Var}[Y_i] = \frac{\theta_i^2}{3}$.

Denote

$$s_d^2 = \sum_{i=1}^{d} \mathrm{Var}[Y_i] = \frac{1}{3} \sum_{i=1}^{d} \theta_i^2.$$

Let us compute $E\left[|Y_i|^{2+l}\right]$ for $l = 2$

$$E\left[|Y_i|^4\right] = \theta_i^4 \int_0^1 \delta x_i^4 d\delta x_i = \frac{\theta_i^4}{5}$$

Since

$$\frac{1}{(s_d^2)^2} \sum_{i=1}^{d} E\left[|Y_i|^4\right] = \frac{1}{(s_d^2)^2} \sum_{i=1}^{d} \frac{\theta_i^4}{5}$$

and $\sum_{i=1}^{d} \frac{\theta_i^4}{5} \leq \frac{d}{5} \theta_M^4$ and $(s_d^2)^2 = \left(\frac{1}{3} \sum_{i=1}^{d} \theta_i^2\right)^2 \geq \frac{d^2 \theta_m^4}{9}$ we can bound

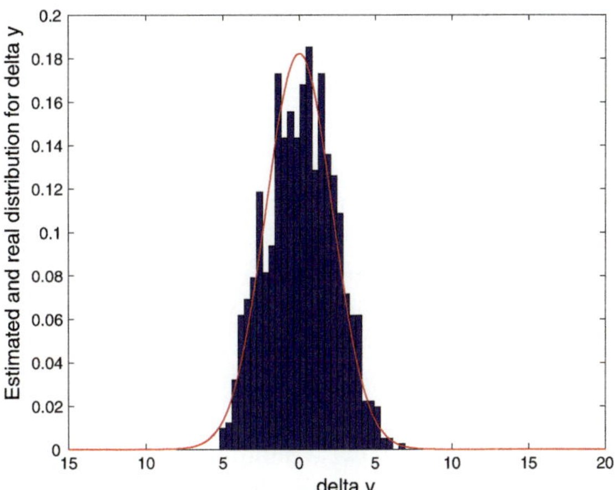

Fig. 3.1 The empirical distribution of δy compared with the set by the CLT. The dimension of the parameter space is $d = 5$

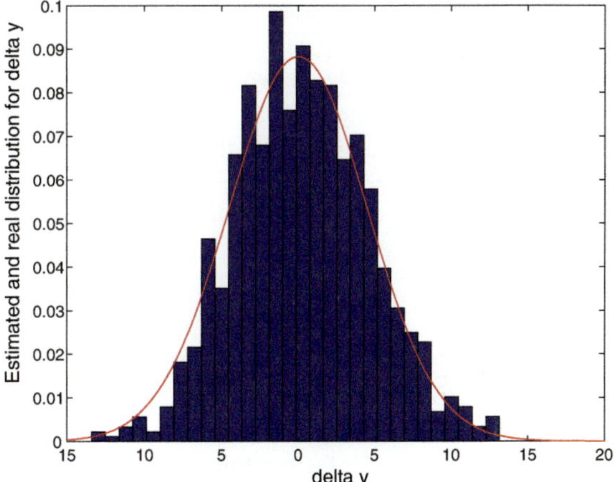

Fig. 3.2 The empirical distribution of δy compared with the set by the CLT. The dimension of the parameter space is $d = 15$

$$\left(\frac{1}{(s_d^2)^2} \sum_{i=1}^{d} E\left[|Y_i|^4 \right] \right) \leq \frac{9\theta_M^4}{5d\theta_m^4}$$

which scales as $O(\frac{1}{d})$ and, when $d \to \infty$, goes to zero, hence satisfying the Lyapunov condition. This grants that $\delta y = N(0, \sigma_{\delta x}^2 \theta^T \theta)$.

In the example above, a sufficiently large d, e.g., $d > 10$, proves to be a good approximation in many cases. Figures 3.1 and 3.2 present an example showing the quality of the approximation for $d = 5$ and $d = 15$, respectively. The application is configured so that $\theta_{d=5} = [2.47, -2.55, 0.52, 1.10, -0.50]$ and $\theta_{d=15} = [-2.81, 1.23, -2.65, -2.66, -1.99, -2.32, -1.50, 0.13, -1.48, -0.30, -1.67, -2.55, -2.89, 0.45, 2.47]$. 1000 δx vectors have been extracted from the $[-1, 1]^d$ hypercube according to the uniform distribution. The histogram for δy is plotted and contrasted with the Gaussian curve granted by the CLT ($\delta y = \mathcal{N}(0, \frac{1}{3}\theta^T\theta)$). We can observe that, also with a low d, the empirical distribution approximates well the Gaussian one.

3.3.1.2 The Multiplicative Perturbation Model

Within a multiplicative model δx affects inputs to yield perturbed value $x_p = x(1 + \delta x)$. The pointwise error $\delta y = y_p - y$ can be rewritten as

$$\delta y = \theta^T (x \circ \delta x)$$

where $\circ$ is the elementwise multiplication operator (multiplication is carried out at the component by component level). As done before, we are interested in characterizing the first two moments of the error distribution and assume that both inputs and perturbations, that are supposed to be independent, are drawn according to distributions $f_x(0, C_x)$ and $f_{\delta x}(0, C_{\delta x})$, respectively. We only require the covariance matrices C_x and $C_{\delta x}$ to be known (or that an estimate can be provided), but not the pdf. Inputs are zero centered only to ease the derivation (a zero mean subtraction can be introduced before carrying out the analysis). Expectation, now taken w.r.t. inputs and perturbations, leads to

$$E_{x,\delta x}[\delta y] = E_{x,\delta x}[\theta^T x \circ \delta x] = \theta^T E_x[x] \circ E_{\delta x}[\delta x] = 0$$

and variance

$$\mathrm{Var}(\delta y) = E_{x,\delta x}[\theta^T x x^T \circ \delta x \delta x^T \theta] = \theta^T C_x \circ C_{\delta x} \theta \tag{3.3}$$

$C_{\delta x}$ is diagonal under the assumption that perturbations affecting inputs are independent. If that is the case, the (3.3) becomes

$$\mathrm{Var}(\delta y) = \sum_{i=1}^{d} \theta_i^2 \sigma_{\delta x,i}^2 \sigma_{x,i}^2.$$

In the particular case that all the input variances are identical to σ_x^2 and perturbation variances to $\sigma_{\delta x}^2$, the variance at the output level simplifies as

$$\mathrm{Var}(\delta y) = \sigma_{\delta x}^2 \sigma_x^2 \theta^T \theta. \tag{3.4}$$

If we compare the variance of (3.4) with that given in (3.2), we see that the former, generated according to the multiplicative model, is equal to the latter (additive model) amplified by term σ_x^2. As with the additive model case, the error distribution can be approximated with a Gaussian one provided the Lyapunov condition is met.

3.3.2 Nonlinear Functions

Let now function $y = f(x)$ be at least twice differentiable w.r.t. x. Again, x is a column vector that, once affected by perturbation δx, assumes value x_p. By adopting the perturbation model, the effect of δx at the output δy is

$$\delta y = f(x_p) - f(x)$$

δy can be hardly described in a closed form unless strong hypotheses about the nature of function $f(\cdot)$ or the perturbation δx are assumed. Perturbation propagation analysis within a nonlinear function is carried out in the literature by assuming the small perturbation hypothesis, e.g., as done with the sensitivity analysis that studies the effects of perturbation affecting inputs on the function output, e.g., [129].

Although the small perturbation hypothesis might hold in several cases, it represents, in general, a strong assumption that needs to be weakened, e.g., as proposed in Chap. 5. However, the small perturbation assumption makes the mathematics amenable and we can expand $f(x_p) = f(x + \delta x)$ according to Taylor around x and stop the expansion at the quadratic term

$$f(x + \delta x) = f(x) + J(x)^T \delta x + \frac{1}{2} \delta x^T H(x) \delta x + \mathrm{o}(\delta x^T \delta x)$$

where $J(x) = \frac{\partial f(x)}{\partial x}$ is the gradient vector and $H(x) = \frac{\partial^2 f(x)}{\partial x^2}$ the Hessian matrix.

By discarding terms of order higher than two the perturbation propagated at the output takes the form

$$\delta y = J(x)^T \delta x + \frac{1}{2} \delta x^T H(x) \delta x. \tag{3.5}$$

Not much more can be said within a deterministic framework unless we introduce strong assumptions on $f(x)$ or δx. However, by moving to a stochastic framework, which considers x and δx mutually independent and identically distributed i.i.d random variables drawn from distributions $f_x(0, C_x)$ and $f_{\delta x}(0, C_{\delta x})$, respectively, the first two moments of the distribution of δy can be computed.

In fact, under the above assumptions and by taking expectation w.r.t. x and δx, the expected value of the perturbed output (3.5) becomes

$$E[\delta y] = \frac{1}{2} E[\delta x^T H(x) \delta x] = \frac{1}{2} \mathrm{trace}\left(E[H(x) \delta x \delta x^T] \right) = \frac{1}{2} \mathrm{trace}\,(E[H(x)] C_{\delta x}).$$

If the quasi-Newton approximation for the Hessian $H(x) = \frac{\partial f(x)}{\partial x} \frac{\partial f(x)^T}{\partial x}$ holds, then $H(x)$ is a semidefinite positive quadratic form and

$$E[\delta y] = \frac{1}{2}\, \mathrm{trace}\,(C_x C_{\delta x}). \tag{3.6}$$

From (3.6), each perturbation introduces an increase in $E[\delta y]$ if we consider the quadratic form expansion (a first-order approximation, obtained by solely maintaining the linear term, provides a null value). In order to compute $\mathrm{Var}(\delta y)$, we consider only the first term of the expansion (the quadratic term does not allow us to advance the mathematics), which means that we only keep the linear approximation for function $f(x)$. Under the above assumptions and by taking expectation w.r.t. x and δx, the variance of the perturbed output becomes

$$\mathrm{Var}(\delta y) = E\left[J(x)^T \delta x \delta x^T J(x)\right] = \mathrm{trace}\left(E\left[J(x)J(x)^T\right] C_{\delta x}\right).$$

Obviously, if $f(x) = \theta^T x$ the derivation reduces to that of the linear function case.

3.4 Learning from Data and Uncertainty at the Model Level

This section studies the case where parameterized models are built from a series of noisy data. The use of a limited number of data to estimate the model, i.e., to determine an estimate of the optimal parameter configuration, introduces an extra source of uncertainty on the estimated parameters in addition to the noise (in previous sections, the parameters were given). In fact, given a different data set with the same cardinality, we will obtain a different parameter configuration with probability one, also in the linear model case. What happens when we select a non-optimal ("wrong") model to describe the data? Which is the relationship between the optimal parameter configuration, constrained by the selected model family, and the current one configured on a limited data set? Since the estimated parameter vector is a realization of a random variable centered on the optimal one, the model we obtain from the available data can be seen as a perturbed model induced by perturbations affecting the parameter vector. Which are then the effects of this perturbation on the performance of the model? This section aims at addressing the above aspects.

3.4.1 Basics of Learning: Inherent, Approximation, and Estimation Risks

Let $Z_N = \{(x_1, y_1), ..., (x_N, y_N)\}$ be the set composed of N (input-output) couples. The goal of machine learning is to build the simplest approximating model able to explain past Z_N data and future instances that will be provided by the data generating process.

Consider then the situation where the process generating the data (system model) is ruled by

$$y = g(x) + \eta, \tag{3.7}$$

where η is a noise term modeling the existing uncertainty affecting the unknown nonlinear function $g(x)$, if any. Once the generic data instance x_i is available, (3.7) provides value $y_i = g(x_i) + \eta_i$, η_i being a realization of the random variable η. In practical cases, the system for which we aim to create a model, by receiving input x_i, provides value y_i. We comment that both inputs and outputs are quantities measurable through sensors. The ultimate goal of learning is to build an approximation of $g(x)$ based on the information present in dataset Z_N through model family

$$f(\theta, x) \tag{3.8}$$

parameterized in the parameter vector $\theta \in \Theta \subset \mathbb{R}^p$. Selection of a suitable family of models $f(\theta, x)$ can be driven by some a priori available information about the system model. If data are likely to be generated by a linear model—or a linear model suffices—then this type of model should be considered. In this case, learning relies on vast results provided by the system identification theory, e.g., see [130]. The outcome of the the learning procedure is the parameter configuration $\hat{\theta}$ and, hence, model $f(\hat{\theta}, x)$, whose quality/accuracy must be assessed.

If the accuracy performance is not met, and margin for improvement exists, we have to select a new model family and reiterate the learning process. For instance, if the difference—residual— between the reconstructed value $f(\hat{\theta}, x)$ and the measured $y(x)$ one on a new data set in not a white noise (test procedure), then there is information that model $f(\hat{\theta}, x)$ was not able to capture. A new richer model family should be chosen and learning restarted. In this direction, feedforward neural networks have been shown to be universal function approximators [131], i.e., can approximate any nonlinear function, and are ideal candidates to solve the above learning problem [39]. However, the complexity of the neural model family must be balanced with the information content provided by the data, otherwise we might experience poor approximating accuracy. Such performance loss is either associated with overfitting (the degrees of freedom exposed by the family model are overdimensioned compared to the effective needs, so that noise affecting the data instances is learned as well) or underfitting (the model is underdimensioned w.r.t. the available data and the model cannot extract all the information present in the data).

In the sequel, we present the classic learning from data mechanism based on the statistical formulation set by Vapnik [132–134].

Define as *Structural risk* the function

$$\bar{V}(\theta) = \int L\left(y, f(\theta, x)\right) p_{x,y} dxy \tag{3.9}$$

where $L\left(y, f(\theta, x)\right)$ is a discrepancy loss function evaluating the closeness between $g(x)$ and $f(\theta, x)$ and $p_{x,y}$ is the probability density function associated with the i.i.d. (x, y) random variable vector. The structural risk (3.9) assesses the accuracy of a given model according to the loss function $L\left(y, f(\theta, x)\right)$.

The optimal parameter θ^{o} yielding the optimal model $f(\theta^{\mathrm{o}}, x)$ constrained by the particular choice of the model family $f(\theta, x)$, is

$$\theta^o = \arg\min_{\theta \in \Theta} \bar{V}(\theta).$$

However, we do not have access to $p_{x,y}$ and only the data set Z_N is available. Such an information allows us to construct the empirical distribution

$$\hat{p}_{x,y} = \frac{1}{N} \sum_{i=1}^{N} D_\delta(x - x_i, y - y_i) \tag{3.10}$$

where $D_\delta(x - x_i, y - y_i)$ is the Dirac function. The use of the estimate $\hat{p}_{x,y}$ of (3.10) in (3.9) leads to the *Empirical Risk*

$$V_N(\theta) = \frac{1}{N} \sum_{i=1}^{N} L(y_i, f(\theta, x_i)). \tag{3.11}$$

Finally, minimization of the empirical risk provides the estimate $\hat{\theta}$

$$\hat{\theta} = \arg\min_{\theta \in \Theta} V_N(\theta) \tag{3.12}$$

and, in turn, the model $f(\hat{\theta}, x)$ approximating $g(x)$ whose accuracy performance is $\bar{V}(\hat{\theta})$. Minimization of the empirical risk defined in (3.12) is also called the *learning process* and the minimization procedure *learning algorithm*.

Conditions granting $\hat{\theta}$ to converge to θ^o as well as observations regarding the speed of convergence will be given later in the chapter. Here, we introduce at first the concepts of *inherent risk, approximation risk*, and *estimation risk* relevant to subsequent analyses.

Define $V_I = \bar{V}(\theta^o)|_{g(x)=f(\theta^o,x)}$ to be the inherent risk, i.e., the a priori non-null intrinsic risk we have when unknown function $g(x)$ belongs to the chosen model family, i.e., $g(x) = f(\theta^o, x)$. Rewrite the structural risk $\bar{V}(\hat{\theta})$ associated with model $f(\hat{\theta}, x)$, i.e., the performance of the obtained model, as

$$\bar{V}(\hat{\theta}) = \left(\bar{V}(\hat{\theta}) - \bar{V}(\theta^o)\right) + \left(\bar{V}(\theta^o) - V_I\right) + V_I. \tag{3.13}$$

The risk associated with the model is composed of three terms

- The inherent risk V_I. The risk depends only on the structure of the learning problem and, for this reason, can be improved only by improving the problem itself, i.e., by acting on the process generating the data, (e.g., by designing a more precise sensor architecture). Nothing else can be done. This is the minimum risk we can have and we reach it—implying optimal accuracy performance in function approximation—when the other sources of uncertainty leading to the two other risks are null;
- The approximation risk $\bar{V}(\theta^o) - V_I$. The risk depends on how close the model family (also named hypothesis space) is to the process generating the data. To

improve it we need to select model families that are more and more expressive, i.e., either contain or are very close to $g(x)$ according to the figure of merit $L(\cdot, \cdot)$. Given an unknown $g(x)$ function, we need to select families of approximating functions that are universal function approximator, e.g., feedforward neural networks.

- The estimation risk $\bar{V}(\hat{\theta}) - \bar{V}(\theta^o)$. The risk depends on the ability of the learning algorithm to select a parameter vector $\hat{\theta}$ close to θ^o. If we have an effective learning process, we hope to be able to get a $\hat{\theta}$ close to θ^o so that the contribution to the model risk is negligible.

The theory allows us to understand the intrinsic limits of learning. A learning problem is affected by three sources of error; of these, the inherent one is determined by the nature of the problem and, for this reason, cannot be improved by learning. The remaining error sources, i.e., those introduced by the approximation and the estimation processes, are the true target of any learning procedure.

Asymptotically with the number of available data N, the approximation and the estimation errors can both be controlled if the learning method has some basic consistency features (which most practical methods have). But when the available data set is small, the dominating component of the learning error is determined if the method is consistent by the approximation error, i.e., by how well the model family $f(\theta, x)$ is close to the process generating the data $g(x)$. In other words, the model risk is mainly determined by the choice of the approximating function $f(\theta, x)$, rather than by the training procedure. As a consequence, in the absence of a priori information, we have no basis to prefer a consistent learning method to another one. Further details applied to the particular case where the learning problem is of classification type can be found in [135].

Example: Inherent, Approximation, and Estimation Risks

Consider a quadratic loss function $L(y, f(\theta, x)) = (y - f(\theta, x))^2$ and a process generating the data ruled by $g(x) = x, x \in [0, 1]$ affected by a Gaussian noise so that $\eta = \mathcal{N}(0, \sigma_\eta^2)$

$$y = x + \eta.$$

Consider the function family $f(\theta, x) = k, \theta = [k]$. The structural risk becomes

$$\bar{V}(\theta) = \int (x + \eta - k)^2 \frac{1}{\sqrt{2\pi}} e^{\frac{-\eta^2}{2}} dx d\eta \tag{3.14}$$

that, after some calculus, leads to

$$\bar{V}(\theta) = \frac{1}{3} + \sigma_\eta^2 + k^2 - k. \tag{3.15}$$

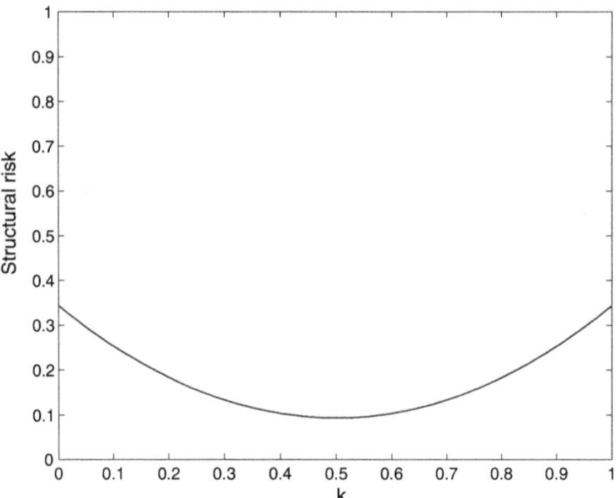

Fig. 3.3 The structural risk as function of k for the example. The structural risk presents a unique minimum θ^{o} in correspondence with $k = \frac{1}{2}$

Figure 3.3 presents the structural risk as function of k for the case $\sigma_\eta^2 = 0.01$. We see that the curve is characterized by a unique minimum θ^{o}.

The optimal point

$$\theta^{\mathrm{o}} = \arg\min_{\theta\in\Theta} \bar{V}(\theta)$$

can be obtained by imposing the stationary relationship $\frac{\partial \bar{V}(\theta)}{\partial\theta} = 0$ and leads to $\theta^{\mathrm{o}} = [\frac{1}{2}]$. Assume that the learning procedure has provided value $\hat{\theta} = [\frac{1}{4}]$. The learning situation is that of Fig. 3.4, where we have some points generated by the system model $y = x + \eta$, the optimal model minimizing the structural risk $y = \frac{1}{2}$, and the available one $y = \frac{1}{4}$.

It is now easy to derive from (3.13)

- The inherent risk $V_I = \bar{V}(\theta^{\mathrm{o}})|_{g(x)=\frac{1}{2}} = \sigma_\eta^2$;
- The approximation risk $\bar{V}(\theta^{\mathrm{o}}) - V_I = \frac{1}{12}$;
- The estimation risk $\bar{V}(\hat{\theta}) - \bar{V}(\theta^{\mathrm{o}}) = \frac{3}{48}$.

By adding the three risks we obtain the accuracy performance $\bar{V}(\hat{\theta}) = \sigma_\eta^2 + \frac{7}{48}$ of the available model, in line with what is expected by using (3.15) evaluated for $\hat{\theta} = [\frac{1}{4}]$.

3.4.2 The Bias-Variance Tradeoff

Following the previous section, consider now the case where we aim at determining the *expected prediction error* at a given—fixed—input value x for a quadratic loss

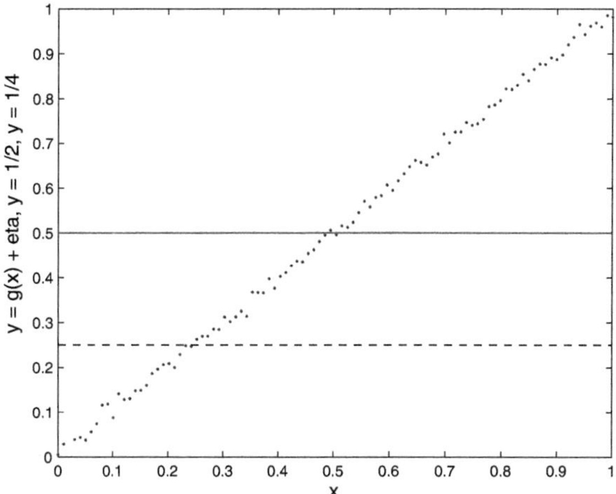

Fig. 3.4 The key elements of the learning process. The process generating the data $y = x + \eta$, the optimal model minimizing the structural risk $y = f(\theta^o, x) = \frac{1}{2}$, and the one provided by the learning procedure $y = f(\hat{\theta}, x) = \frac{1}{4}$

function Squared Error (SE), namely the error we should expect in test on an unknown instance x. The trained model $f(\hat{\theta}, x)$ has been derived by following the minimization of the empirical procedure (3.12). In the following, expectation is taken with respect to the noise on a *given* sample x, namely

$$SE_{PE}(x) = E\left[y(x) - f(\hat{\theta}, x)\right]^2. \tag{3.16}$$

We present a main result known in the literature as the bias-variance tradeoff. The SE can be seen as decomposed in the intrinsic error level, that introduced by the model or approximation error, and that introduced by the estimation procedure. Let us elaborate (3.16) as

$$\begin{aligned} SE_{PE} &= E\left[y(x) - f(\hat{\theta}, x)\right]^2 \\ &= E\left[y(x) - g(x) + g(x) - f(\hat{\theta}, x)\right]^2 \\ &= E\left[(y(x) - g(x))^2\right] + E[(g(x) - f(\hat{\theta}, x))^2] \\ &\quad + 2E\left[(y(x) - g(x))\,(g(x) - f(\hat{\theta}, x))\right] \\ &= E\left[\eta^2\right] + E\left[\left(g(x) - f(\hat{\theta}, x)\right)^2\right] \end{aligned}$$

since $E\left[(y(x)-g(x))\,(g(x)-f(\hat{\theta},x))\right]=0$. In fact, we can rewrite the term as

$$E\left[y(x)g(x)\right]+E\left[y(x)f(\hat{\theta},x)\right]-E\left[g(x)g(x)\right]+E\left[g(x)f(\hat{\theta},x)\right]$$

and

$$\begin{aligned}
E\left[y(x)g(x)\right]&=g^2(x)\\
E\left[y(x)f(\hat{\theta},x)\right]&=E\left[(g(x)+\eta)\,f(\hat{\theta},x)\right]=E\left[g(x)f(\hat{\theta},x)\right]\\
E\left[g(x)g(x)\right]&=g^2(x).
\end{aligned}$$

Thus, the SE can be decomposed in the variance of the noise and the SE between the true function and the estimate

$$E\left[\mathrm{SE}\right]=\sigma_\eta^2+E\left[\left(g(x)-f(\hat{\theta},x)\right)^2\right]. \tag{3.17}$$

The second term of Eq. (3.17) can be further refined by using the same trick used above, which requires adding and subtracting $E\left[f(\hat{\theta},x)\right]$

$$\begin{aligned}
&E\left[\left(g(x)-E\left[f(\hat{\theta},x)\right]+E\left[f(\hat{\theta},x)\right]+f(\hat{\theta},x)\right)^2\right]\\
&\quad=E\left[\left(g(x)-E\left[f(\hat{\theta},x)\right]\right)^2\right]+E\left[\left(E\left[f(\hat{\theta},x)\right]-f(\hat{\theta},x)\right)^2\right]\\
&\qquad+2E\left[\left(g(x)-E\left[f(\hat{\theta},x)\right]\right)\left(E\left[f(\hat{\theta},x)\right]-f(\hat{\theta},x)\right)\right]
\end{aligned}$$

The double product cancels since

$$\begin{aligned}
E\left[g(x)E\left[f(\hat{\theta},x)\right]\right]&=g(x)E\left[f(\hat{\theta},x)\right]\\
E\left[g(x)f(\hat{\theta},x)\right]&=g(x)E\left[f(\hat{\theta},x)\right]\\
E\left[E\left[f(\hat{\theta},x)\right]^2\right]&=E\left[f(\hat{\theta},x)\right]^2\\
E\left[f(\hat{\theta},x)E\left[f(\hat{\theta},x)\right]\right]&=E\left[f(\hat{\theta},x)\right]^2
\end{aligned}$$

Equation (3.17) can be finally rewritten as

$$\mathrm{SE}_{\mathrm{PE}}=\sigma_\eta^2+E\left[\left(g(x)-E\left[f(\hat{\theta},x)\right]\right)^2\right]+E\left[\left(E\left[f(\hat{\theta},x)\right]-f(\hat{\theta},x)\right)^2\right]. \tag{3.18}$$

Equation (3.18) states that the accuracy performance of the approximating function $f(\hat{\theta}, x)$ can be described by means of three terms. The first one is the variance of the intrinsic noise and cannot be canceled, independently of how good our approximation is. The second term is the square of the bias and represents the quadratic error we have in approximating the true function $g(x)$ when our model generation procedure is able to provide the best model of the model family $f(\theta^{o}, x)$ (recall again that the best model is the one that minimizes the distance between the true function and the optimal approximation built in a noise-free environment having an infinite number of training points). De facto, the bias represents a discrepancy between the two functions according to the SE_{PE} figure of merit. The last term is known as variance and it represents the variance introduced by having considered the approximating model $f(\hat{\theta}, x)$ instead of the optimal one within the family $f(\theta^{o}, x)$. Of course, if our model generating process is capable of providing a more accurate model, the variance term reduces.

Comments

Recall that the SE is the integral of the quadratic pointwise discrepancy only in the case that the distribution of the inputs is uniform. When this is not the case, the quadratic discrepancy is weighted by the pdf f_x to differentiate the interest of the error toward more likely inputs. For instance, if we consider interval $X = [0, 1], g(x) = x^2$ and the approximating function $f(\hat{\theta}, x) = x$ then the quadratic discrepancy is SE $= 1/30$. Differently, if we induce the probability density function $f_x = 2$ if $x \leq 0.25$, $f_x = 2/3$ if $x > 0.25$, then the SE $\simeq 0.027$. Results related to convergence of the Mean Squared Error (MSE) to the SE are given in Chap. 4.

From Eq. (3.18) we observe that, in order to minimize the SE_{PE}, we need to minimize both the bias and the variance terms over the input space. However, this is not trivial. For instance, if the model family is rich in terms of degrees of freedom (say overdimensioned w.r.t. the problem) then $f(\hat{\theta}, x)$ would perfectly interpolate the training data. This will make the bias term vanish entirely at the cost of a high variance. More in general, finding an optimal bias-variance tradeoff is a difficult task but acceptable solutions can be found, e.g., by relying also on early stopping techniques, by cross-validation methods, or by introducing regularisers in the learning process [39].

3.4.3 Nonlinear Regression

The regression problem is a particular case of learning and aims at determining the best static model approximating an unknown static function. The framework assumes that there exists a time invariant model generating the couples (x_i, y_i) populating the data set Z_N. The learning framework is that of the empirical/structural risks presented in previous sections, here presented in the asymptotic formulation to give the reader a different prospective on how the learning framework can also be formalized.

Define the structural risk in the form

$$\bar{V}_N(\theta) = \frac{1}{N}\sum_{i=1}^{N} E\left[L(\varepsilon_i(\theta))\right]$$

and the empirical risk

$$V_N(\theta) = \frac{1}{N}\sum_{i=1}^{N} L(\varepsilon_i(\theta))$$

where $\varepsilon_i(\theta) = y_i - f(\theta, x_i)$ is the prediction error at sample (x_i, y_i). The optimal parameter point is defined as

$$\theta^{\mathrm{o}} = \arg\min_{\theta\in\Theta}\left[\lim_{N\to+\infty} \bar{V}_N(\theta)\right]$$

and estimated by

$$\hat{\theta} = \arg\min_{\theta\in\Theta} V_N(\theta).$$

The following analysis holds around the optimal point θ^{o} which minimizes the structural risk (note that, under the regularity hypothesis, $\bar{V}_N \to \bar{V}$ when $N \to \infty$). If local minima exist and we find ourselves in one of these, then we require that there exists a neighborhood for which the optimal point is unique. The analysis confines us within this neighborhood.

As with [137], we require that the approximating function $f(\theta, x)$ is Lipschitz and that the partial derivatives up to the third order w.r.t θ and ε are bound by a constant (regularity conditions). Under these hypotheses, $\hat{\theta}$ converges in probability to θ^{o} when $N \to \infty$ and the distribution of the parameter vector follows a multivariate Gaussian distribution

$$\lim_{N\to\infty} \sqrt{N}\Sigma_N^{-\frac{1}{2}}(\hat{\theta} - \theta^{\mathrm{o}}) \sim \mathcal{N}(0, I_p) \tag{3.19}$$

where, the Hessian of $\bar{V}_N$ is defined as $\bar{V}_N''$

$$\Sigma_N = \left[\bar{V}_N''(\theta^{\mathrm{o}})\right]^{-1} U_N \left[\bar{V}_N''(\theta^{\mathrm{o}})\right]^{-1},$$

and the squared matrix U_N of order d is

$$U_N = NE\left[\left(\frac{\partial V_N(\theta)}{\partial\theta}\right)\left(\frac{\partial V_N(\theta)}{\partial\theta}\right)^T\right].$$

I_p is the identity matrix of order p.

Comments

The framework designed for the nonlinear case is general but nonlinearity does not allow us to obtain closed-form results unless particular assumptions are made. In nonlinear regression, we identify a suitable nonlinear model family $f(\theta, x)$ characterized by enough expressive power to keep as small as possible the approximation risk. This can be achieved by resorting to universal function approximators, e.g., feedforward neural networks or radial basis functions [39, 131] to be used as $f(\theta, x)$. Then, we have to control the estimation error for the chosen neural network, an operation that can be carried out by selecting an effective learning algorithm, such as a second-order Levenberg–Marquardt, a DFP, or a BFGS one, to be applied to the empirical risk (see [125] for a comprehensive treatment). We might even need to run the algorithm several times to mitigate the presence of local minima in the V_N function.

Once training is perfected, the validity of the above must be intended in the neighborhood of the found local minimum that, under the regularity assumptions, is unique. However, it is thought that such univocity for the minimum does not exist for overdimensioned neural networks where the minimum provided by the learning procedure might be a saddle point. It should also be emphasized that if we run again the learning algorithm we will end up in a different minimum with probability one. This is a consequence of the complexity of the parameter space and the random selection for the initial weights. The final minimum also depends on the particular choice of Z_N. Although the derivation might seem to have a small impact in real applications, it is relevant when the approximating function is linear, either static or dynamic, as presented in the next two sections.

3.4.4 Linear Regression

The linear regression case is a particularly relevant case of nonlinear regression, where the system generating the data is linear

$$y = g(x) + \eta = \theta^{o^T} x + \eta \tag{3.20}$$

with θ^o being an unknown parameter vector to be estimated. $\eta \sim \mathcal{N}(0, \sigma_\eta^2)$ is a white noise of σ_η^2 variance. The Gaussian request for the noise is amply satisfied whenever data come from sensors. In other words, we assume that scattered data coming from the sensor (or otherwise available) can be optimally described by a linear model. However, only the data set Z_N is available and we wish to provide, starting from the finite data set, an estimate $\hat{\theta}$ of θ^o.

We choose the linear model family $f(\theta, x) = \theta^T x$ and the loss function to be an SE. All requested hypotheses leading to (3.19) are satisfied and a unique minimum θ^o is achieved if inputs are linearly independent (otherwise we simply have to remove the dependent inputs).

$\bar{V}_N$ and V_N are chosen as

$$\bar{V}_N(\theta) = \frac{1}{2N}\sum_{i=1}^{N} E\left[\varepsilon_i(\theta)^2\right]$$

$$V_N(\theta) = \frac{1}{2N}\sum_{i=1}^{N} \varepsilon_i(\theta)^2.$$

Since

$$\left.\frac{\partial V_N(\theta)}{\partial\theta}\right|_{Z_N} = -\frac{1}{N}\sum_{i=1}^{N}\varepsilon_i(\theta)x_i,$$

by exploiting the independence between x and ε and recalling that $E[\varepsilon_i\varepsilon_j] = 0$ since η is an i.i.d. random variable

$$\begin{aligned} U_N(\theta) &= \frac{1}{N}E\left[\sum_{i=1}^{N}\varepsilon_i(\theta)x_i\sum_{j=1}^{N}\varepsilon_j(\theta)x_j^T\right] = \frac{1}{N}\sum_{i=1}^{N}\sum_{j=1}^{N}E\left[\varepsilon_i(\theta)\varepsilon_j(\theta)x_ix_j^T\right] \\ &= \frac{1}{N}\sum_{i=1}^{N}\sum_{j=1}^{N}E\left[\varepsilon_i(\theta)\varepsilon_j(\theta)\right]E\left[x_ix_j^T\right] = \frac{1}{N}\sum_{i=1}^{N}E\left[\varepsilon_i(\theta)^2\right]x_ix_i^T \end{aligned}$$

while

$$\bar{V}_N'' = \frac{\partial^2\bar{V}_N(\theta)}{\partial\theta^2} = \frac{1}{N}\sum_{i=1}^{N}x_ix_i^T.$$

Since $E\left[\varepsilon_i(\theta)^2\right] = \sigma_\eta^2$

$$U_N(\theta) = \sigma_\eta^2\frac{1}{N}\sum_{i=1}^{N}x_ix_i^T = \sigma_\eta^2\bar{V}_N''(\theta).$$

The expression of Σ_N simplifies as

$$\begin{aligned} \Sigma_N &= \left[\bar{V}_N''(\theta^o)\right]^{-1}U_N(\theta^o)\left[\bar{V}_N''(\theta^o)\right]^{-1} \\ &= \left[\bar{V}_N''(\theta^o)\right]^{-1}\sigma_\eta^2\bar{V}_N''(\theta^o)\left[\bar{V}_N''(\theta^o)\right]^{-1} \\ &= \sigma_\eta^2\left[\bar{V}_N''(\theta^o)\right]^{-1} = \sigma_\eta^2\left[\bar{V}_N''\right]^{-1} \end{aligned}$$

Also, in the linear case the distribution of the parameters is Gaussian following (3.19), and reduces to

$$\lim_{N\to\infty} (\hat{\theta} - \theta^o) \sim \mathcal{N}(0, \frac{\sigma_\eta^2}{N}\left[\bar{V}_N''\right]^{-1}).$$

The covariance depends on the variance of the noise and the inverse of the input Hessian $\bar{V}_N'' = \frac{1}{N}\sum_{i=1}^N x_i x_i^T$.

Comments

We comment that linearity must be intended w.r.t. parameters θ. As such, any model family $f(\theta, x) = \theta^T\phi(x)$, with ϕ a generic function, can be used, assumptions are automatically valid, and the results hold. This comment has some relevance to machine learning, where the ϕ function can be intended as a transformation of the inputs (feature extraction).

3.4.5 Linear Time-Invariant Predictive Models

One of the most common applications we can consider given a datastream is the design of dynamic models able to approximate it over time. This can be carried out by different methods, e.g., by limiting the analysis to the most common linear techniques, by considering space–state descriptions or predictive input–output models. In the following, we will focus on predictive models since they will be used in Chap. 10. Assume that a physical description for the system model is unavailable and that the unknown dynamic system generating the data is time invariant.

Following the notation set up by Ljung [130], given a parameter vector $\theta \in \Theta$, the input and the output sequences $u^t = (u(1), \ldots, u(t)) \in \mathbb{R}^{t\times m}$, $u(\cdot) \in \mathbb{R}^m$ and $y^{t-1} = (y(1), \ldots, y(t-1)) \in \mathbb{R}^{t-1}$, respectively, the predictive model in a one-step prediction form for output $y(t)$, at time t is

$$\hat{y}(t, \theta) = f(\theta, u^t, y^{t-1}).$$

The prediction error for model $f(\theta, u^t, y^{t-1})$, at time t is

$$\varepsilon(\theta, u^t, y^{t-1}) = \varepsilon(t, \theta) = y(t) - \hat{y}(t, \theta).$$

The structural risk $\bar{V}_N(\theta)$ can be defined as

$$\bar{V}_N(\theta) = \frac{1}{N}\sum_{t=1}^{N} E\left[L(\theta, \varepsilon(t, \theta))\right]$$

where $L(\cdot, \cdot) \in \mathbb{R}$ is a suitable loss function and expectation is taken w.r.t. the distribution of u and y.

The optimal parameter configuration is the one minimizing the structural risk

$$\theta^{o} = \arg\min_{\theta \in \Theta} \left[\lim_{N \to +\infty} \bar{V}_N(\theta) \right].$$

Following the procedure presented in Sect. 3.4.1 and given the input–output training sequence $Z_N = \{(u(t), y(t))\}_{t=1}^{N}$, the empirical risk becomes

$$V_N(\theta, u^t, y^{t-1}) = \frac{1}{N} \sum_{t=1}^{N} L(\theta, \varepsilon(t, \theta)).$$

Minimization of the empirical risk leads to the parameter configuration $\hat{\theta}_N$

$$\hat{\theta}_N = \arg\min_{\theta \in \Theta} V_N(\theta, u^t, y^{t-1}).$$

In the following, we assume that the obtained model has been suitably chosen and does not degenerate in the identification phase (otherwise, it must be scaled to a smaller model).

By relying on the theoretical framework developed in [130, 136, 137], under the mild hypotheses that recent past data suffice to generate accurate approximations of $u(t)$ and $y(t)$, that $f(\cdot)$ is three time differentiable w.r.t. θ and satisfies Lipschitz conditions, and that the structural risk is a convex function, minimization of $W_N(\theta)$ provides a unique point θ^{o}, it is true that:

$$\lim_{N \to \infty} V_N - \bar{V}_N \to 0 \quad \text{w.p. } 1$$

thus

$$\lim_{N \to \infty} \hat{\theta}_N \to \theta^{o} \quad \text{w.p. } 1$$

and

$$\lim_{N \to \infty} \sqrt{N} \Sigma_N^{-\frac{1}{2}} (\hat{\theta}_N - \theta^{o}) \sim \mathcal{N}(0, I_p)$$

where Σ_N is the covariance matrix

$$\Sigma_N = \left[W_N''(\theta^{o}) \right]^{-1} U_N \left[W_N''(\theta^{o}) \right]^{-1},$$
$$U_N = NE\left[V_N'(\theta^{o}) V_N'(\theta^{o})^T \right]$$

$W_N'' = \frac{\partial^2 W_N}{\partial \theta^2}$ is the Hessian matrix of W_N and $V_N' = \frac{\partial V_N}{\partial \theta}$ is the gradient of V_N. The theorem assures that, given a sufficiently large N, $\hat{\theta}_N$ follows a multivariate Gaussian distribution with mean vector θ^{o} and covariance matrix $\frac{\Sigma_N}{N}$.

The theorem contemplates the situation where there exists model bias between the optimal model and the process generating the data. In fact, under the aforementioned hypotheses, a unique optimal point $\theta^o \in \Theta$ exists even in case of model bias.

The above equations grant that the tolerated perturbation space for parameters is ruled by a Gaussian distribution. Having an application we can compute an estimate of matrix Σ_N, hence knowing the uncertainty we should expect on parameters. Having the uncertainty distribution on $\hat{\theta}$ we can now estimate the expected uncertainty in accuracy, e.g., $\bar{V}(\hat{\theta})$ by using randomization techniques as explained in Chap. 4.

3.4.6 Uncertainty at the Application Level

As a final note we comment that, in addition to the different sources of uncertainty we might experience, and that were summarized in the previous sections, we also have uncertainty at the application level. This source of uncertainty derives from the fact that, often, we do not know the solution to our application and we design it by exploiting some a priori information, whenever available. The outcome is a numerical algorithm describing the application.

However, we might have considered another solution, better, worse, or equivalent to the one we have. Clearly, we will keep the best performing solution also satisfying some extra application constraints such as computational complexity, power/energy consumption, or memory requirements, to name the few. The application solution might be complex, since it might come from a partitioning approach where the application is partitioned in parts, each of which has to be solved with the most appropriate signal/image processing or computational intelligence tool.

However, by looking at the problem from a high-level perspective the key elements are the unknown ideal algorithm, optimal in its own sense $g(x)$ and the best solution we found $f(x)$, x representing the input vector feeding my application. The uncertainty at the application level can be evaluated by introducing a discrepancy $L(\cdot, \cdot)$ between the two functions and computing the functional

$$\bar{V}(f, g) = \int L\left(g(x), f(x)\right) p_x dx$$

where p_x is the pdf induced on the input space. Function $g(x)$ is unknown but, as an Oracle, it provides the noisy value $y = g(x, \eta)$ once queried with input x. η is an unknown noise affecting the y generation process.

Clearly, the application bias increases if function $f(x)$ has not been suitably selected and it badly approximates $g(x)$. All aspects related to approximation and inherent risks can be extended as well to this framework if we assume that function $f(\cdot)$ belongs to a functional space F, possibly, but not necessarily, parameterized.

The problem requires now to estimate $\bar{V}(f, g)$ with $\hat{V}(f, g)$ having only data set Z_N but being able to invoke the Oracle all the needed time. Chapter 7 will identify the optimal N so that $\hat{V}(f, g)$ will be a good estimate of $\bar{V}(f, g)$. Let us now assume that we have found a set of solutions $F_f = \{f_1(x), f_2(x), \ldots, f_n(x)\}$. Since it is hard to guarantee that solutions are i.i.d we can only select the best solution $\bar{f}(x)$ as the one minimizing the discrepancy function over set F_f

$$\bar{f}(x) = \arg \min_{f_i(x) \in F_f} \hat{V}(f, g).$$

Not much more than this can be done unless a priori information is made available by someone.

Chapter 4
Randomized Algorithms

There exists a very large class of problems that are computationally prohibitive when formalized in deterministic terms, but may become manageable when a probabilistic formulation can be derived and considered instead. For those problems, we are no more requesting to find *the* problem solution but *a* solution that, according to some probabilistic figure of merit, solves the problem.

Examples are the evaluation of the performance of a system when its computation is affected by perturbations (robustness analysis), verification of the satisfaction of the performance level of an embedded system or an algorithm (performance verification problem), identification of extrema of functions (function optimization problem), and design and analysis of robust controllers, just to name the few applications. The cost we have to pay to abandon determinism is that derived results will hold in probability.

Since the focus here is on embedded computation, we will see that there are some particular cases that might arise during the operational life of the embedded system violating the application constraints. However, such situations might be acceptable provided that the constraints are violated for a short time and constraints violation is a rare event. These aspects will be addressed in Chaps. 5 and 7.

Here, we request to be able to address a very large class of numerical-based problems and applications and find in the space of Lebesgue measurable functions the appropriate mathematical framework.

Definition: Lebesgue measurability

We say that a generic function $u(\psi)$, $\psi \in \Psi \subseteq \mathbb{R}^l$ is Lebesgue measurable with respect to Ψ when its generic step-function approximation S_N obtained by partitioning Ψ in N arbitrary domains grants that

$$\lim_{N\to\infty} S_N = u(\psi)$$

holds on set $\Psi - \Omega$, $\Omega \subseteq \mathbb{R}^l$ being a null measure set [20].

C. Alippi, *Intelligence for Embedded Systems*, DOI: 10.1007/978-3-319-05278-6_4,

We point out that no functions generated by a finite-step, finite-time algorithm, such as *any* engineering-related mathematical computations, can be Lebesgue nonmeasurable. Indeed (see, e.g., [21]), the only way to produce nonmeasurable functions is to invoke the Axiom of Choice over an uncountable family of sets. This procedure is purely theoretical, and the objects obtained in this fashion are necessarily nonconstructible since the construction procedure would involve an uncountable number of arbitrary choices.

Under the Lebesgue measurability hypothesis and by defining a probability density function f_ψ with support Ψ, it comes out that we can transform computationally hard problems into manageable problems by sampling from Ψ and resorting to probability. Randomization comes as the main ingredient of the recipe and grants that obtained results, valid in probability, are characterized by an arbitrary accuracy and confidence levels function of the number of drawn samples. The loss in determinism is largely paid back by the possibility to solve our problem with a polynomial time algorithm.

In fact, all useful algorithms to be executed on embedded systems can be described as Lebesgue measurable functions and many interesting problems can be cast in the same formalism. However, by setting a general framework for a problem solution we can neither expect to find results in a closed form for all Lebesgue measurable applications nor pretend to solve deterministically the computationally hard problem associated with the application solution. To tackle such an issue we reformulate the deterministic problem in a probabilistic one which can be solved by Monte Carlo sampling under the control of the probabilistic theory of learning.

The chapter introduces the randomization mechanism for problem solution whose algorithmic description is known as *Randomized algorithm*.

The structure of the chapter is as follows: At first, we briefly introduce the complexity aspect associated with algorithms and problems. Since solutions will mostly be unmanageable given a generic problem described as a Lebesgue measurable function, we will resort to randomization to solve it. Monte Carlo is then presented followed by such fundamental results that are asymptotic in the number of samples n that grants convergence of some estimates to their expected values (laws of large numbers). Since asymptotic results are of scarce use in real applications (we cannot obtain a solution for a problem by taking an infinite number of sample), we need to search for bounds that grant some results to hold for a finite n. This can be achieved with randomized algorithms that integrate Monte Carlo with results coming from the theory of learning.

4.1 Computational Complexity

The computational complexity theory studies the intrinsic difficulty associated with the solution of a computable problem. Since for a computable problem an algorithm exists, i.e., the problem solution can be obtained in a finite time with a finite number of steps, it is our interest to identify "the best" algorithm solving the problem, with optimality intended according to a given figure of merit. The complexity of an algorithm is generally evaluated as the time execution and memory resource required by an abstract machine to execute it. If time execution and memory resources are the figures of merit considered to assess the performance of the algorithm, say for solving the sorting problem, we might be interested in

- evaluating the complexity of the sorting algorithm;
- asking whether it is possible to identify a better solution for it or not.

If we focus on memory and execution time we can ask several questions whose answers are, a priori, not trivial. Which algorithm is using less memory among the ones we have? Which one is best performing on the average (i.e., the expected execution time w.r.t. random data in the sequence)? And when the sequence is ordered in the opposite way (worst case), which is the time complexity of our algorithm? Answering to these questions—and many others scholars or practitioners might raise—is fundamental if we wish to execute the algorithm on a real machine characterized by finite resources.

We comment that the questions posed above represent specific problems we wish to solve either deterministically or in probability and are of paramount relevance. In fact, even if a problem is computationally solvable in principle, it may not be addressable in practice whenever the algorithm requires an unfeasible amount of execution time or storage. The problem is general: any computer or embedded system introduces at some point hardware constraints which might make the practical execution of a given algorithm unfeasible.

4.1.1 Analysis of Algorithms

Computational complexity, in its analysis of an algorithm realm, approaches an algorithm to be investigated by observing how some extensive variables scale, e.g., the cardinality of a data set n or the dimension of the input space n.

Do we need to store the whole data set? If the answer is positive then we need n cells for storage and the Big Data paradigm is likely to become a problem. Do we need extra data structures to execute the algorithm? Then the needed storage space is the sum of all requested memory resources.

The algorithm time complexity can be decomposed in the time requested to address the basic sequences of operations (or instructions) and similarly to the memory complexity case becomes function of extensive variables.

Consider, for instance, the algorithm A given in Algorithm 1 evaluating the scalar product of two n-dimensional integer vectors $x, y \in \mathbb{N}^n$.

The complexity of algorithm A can be computed by evaluating the memory requirements $M(A)$ and the abstract computation execution time $C(A)$. For simplicity, we do not consider the complexity associated with memory assignment to the vectors and data acquisition since we wish to focus the attention on the algorithm itself. The memory requirement is simply the sum of requested variables (e.g., in memory cells, words, or bytes)

Algorithm 1: Algorithm A: a simple algorithm computing the scalar product between two vectors

```
scalar_product = 0;
i = 0;
assign memory to vectors x and y and populate the content;
while i < n do
    scalar_product = scalar_product + x[i]y[i];
    i = i+1;
end
```

$$M(A) = 2n + 2$$

while the computational complexity is

$$C(A) = (2n + 2)T_a + (n + 1)T_c + n(2T_+ + T_*)$$

where T_a is the time requested for an assignment, T_c that associated with the evaluation of a condition, T_+ and T_* represent the times requested to execute an addition and a multiplication, respectively.

We comment that all time components T assume constant values on a given processor (to ease the understanding we assume a sequential execution on a single core processor having independent instructions, e.g., for assignment, addition and multiplication); the faster the processor the shorter the execution time. It is clear that the average, the worst case or a generic case analysis coincide since complexity is not dependent here on the specific data instances but solely on the cardinality of the sequence.

The complexity of an algorithm is defined by means of the asymptotic character of the complexity figures of merit when the extensive variable n goes to infinity. The consequence is that the algorithm complexity is assessed by investigating how it scales with the problem complexity. Here, dependencies introduced by a specific machine assume constant values and can be neglected.

By referring to Algorithm 1, $M(A)$ scales as $2n$, that is to say its order is $O(n)$, while $C(A) = n(2T_+ + T_c + T_* + 2T_a)$ yields to an $O(n)$ order. It comes out that both functions C and M are linear with n. When Big Data are available we know that both the execution time and the memory complexity of the algorithm will scale linearly with n.

The "big Oh" notation characterizes the complexity of a given algorithm by hiding smaller terms contributions. The advantage in its use is that it makes the evaluation of complexity independent of the specific hardware platform or the computational model used. Other approaches evaluate the complexity of an algorithm by providing lower and upper bounds for it [209].

Let us evaluate the scalar product in a different way, within a sequential approach. The complexity of Algorithm 2 according to the two figures of merit is $M(B) = O(1)$, $C(B) = O(n)$ since the memory occupation does not scale with n and the loop is iterated $n + 1$ times.

Algorithm 2: Algorithm B: a sequential scalar product computation

```
scalar_product = 0;
i = 0;
assign memory to scalars x and y;
while i < n do
    input x and y;
    scalar_product = scalar_product + xy;
    i = i+1;
end
```

The comparison between the two algorithms is then carried out at the big Oh notation level by ordering their complexity according to the rank

$$\cdots O(k^{-n}) \prec O(n^{-k}) \prec O(n^{-1}) \prec O(1) \prec \cdots$$

$$\prec \cdots O(\log n) \prec O(n) \prec O(n^k) \prec O(k^n) \prec \cdots$$

where k is a strictly positive real value. Clearly, algorithms A and B have the same computational complexity in terms of execution time but algorithm B does not require storing the two vectors and, hence, is to be preferred to A if memory is a problem and n increases. For a small number of data the opposite might hold since the constant terms associated with arithmetic operations, the memory assignment and the input readout operations might introduce a strong influence on the final time execution. However, these situations are of no interest to computational complexity.

Complexity can be evaluated also by inspecting the behavior of the algorithm in the worst or the average case; the worst case is generally considered to compare two algorithms when their average complexity is identical.

4.1.2 *P, NP-Complete, and NP-Hard Problems*

We say that a problem A belongs to class P if its computational time complexity is polynomial $O(n^k)$ with constant k. In other words, the algorithm solves the problem in a polynomial execution time. Some authors, e.g., [23] claim that such a property characterizes problems that can be considered "efficiently solvable" or "tractable." This statement is only qualitatively true but sheds some light on the intrinsic complexity behind algorithms.

The problem of evaluating a scalar product belongs to class P. Many other algorithms belong to the P class: from ordering a vector of finite dimension, to verifying the presence of a given pattern within an image and carrying out a digital filtering of a signal.

Consider, as an example, the problem of sorting a numerical vector of cardinality n. *Bubblesort* shows complexity $O(n^2)$, *merge sort* $O(n \log n)$ both for the worst and

Fig. 4.1 The classes of problems *P*, *NP*, *NP-complete*, and *NP-hard* and their inclusions provided that P $\neq$ NP. The complexity increases when we leave the P problems and move toward the NP-hard ones

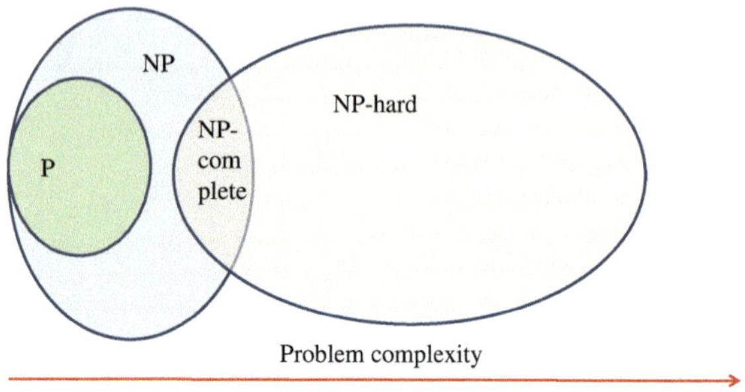

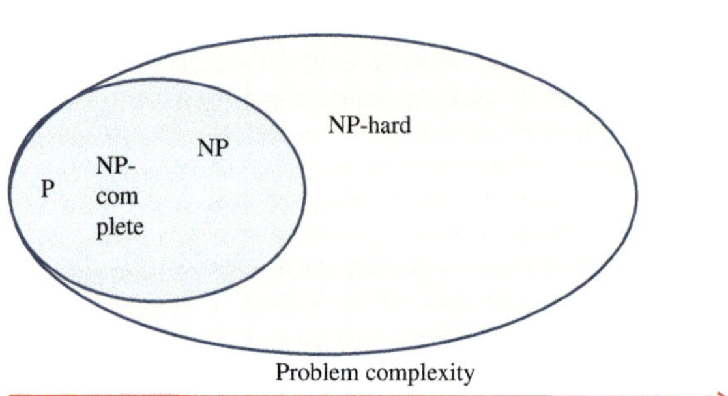

Fig. 4.2 The classes of problems *P*, *NP*, *NP-complete*, and *NP-hard* and their inclusionsprovided that P = NP. The complexity increases when we move toward NP-hard problems

the average cases, *quicksort* $O(n \log n)$ for the average case, and $O(n^2)$ for the worst case [24]; different algorithms have different complexities.

With reference to Fig. 4.1, the class of *nondeterministic polynomial time* problems *NP* is larger than the *P* one. *NP* contains the class of decision problems for which, given a candidate solution, we can verify in polynomial time if the solution solves the problem or not (in other words, we sample a candidate solution from the solution space and verify in polynomial time whether the selected solution is effective or not). *NP* contains many important problems with the hardest called *NP*-complete. For *NP*-complete problems no polynomial-time algorithms are known to solve them. A different way to characterize a *NP*-complete problem is the following: a decision problem is said to be *NP*-complete if it is *NP* and any other *NP* problem can be reduced to it so that its complexity is bounded by a polynomial in the complexity of the original problem.

A problem *H* is said to be *NP*-hard if and only if there exists a *NP*-complete problem *L* that is reducible to *H* in polynomial time. In other words, problem *L* can be solved in polynomial time by a machine which provides an oracle for *H*. Again, a problem is *NP*-hard if each *NP* problem can be reduced to this problem. One of the still open questions is whether P = NP or not, i.e., can a *NP* problem (and hence any of the class) be solved in polynomial time? Were that be the case, then Fig. 4.1 would degenerate as depicted in Fig. 4.2.

Even if it is thought that the answer is negative, the problem is still without a formal solution. The interested reader should consider [25] for a detailed analysis about complexity issues.

An example of a *NP* hard problem of interest here is the following: Given a Lebesgue measurable function $u(\psi) \in [0, 1]$, $\psi \in \Psi \subset \mathbb{R}^l$ and a value $\gamma \in [0, 1]$, does inequality $u(\psi) \leq \gamma$ hold for any $\psi \in \Psi$? The problem, which models the situation where we wonder about the level of satisfaction of a constraint, is surely computationally intractable for a generic $u(\cdot)$ function, since we should query the oracle at each ψ_i and ask the question "is $u(\psi_i)$ below γ?" Even if the oracle responds in a single time step (polynomial time response), the number of queries needed to solve the whole problem is not polynomial for a continuous space Ψ.

We will see in subsequent sections that some hard problems can be addressed and solved by resorting to probability. Such problems are known in the literature as belonging to the class of Randomized Polynomial time (RP) problems.

4.2 Monte Carlo

Monte Carlo methods constitute a class of algorithms that use a repeated random sampling approach and a probabilistic framework to compute the requested output. Due to the possibly large sampling required to provide accurate results, their full effectiveness became available thanks to advances in the computational power exposed by current processors and supercomputers (even if the history of the method dates back to the Manhattan project and it has been formalized in a seminal paper by Metropolis and Ulam [8] already in 1949).

Monte Carlo is an effective tool for addressing problems which can hardly be solved analytically for the mathematical complexity of the involved functions (e.g., integro-differential equations coming from physics and chemistry). It should be noted that Monte Carlo is a set of methods more than a method, each of which personalized to solve a specific class of applications. For instance, we have a method with its own mathematical results to address the integration problem, another for dealing with optimization or computational mathematics. The interested reader can refer to [12, 13] for a comprehensive analysis and further advances. As mentioned above, the core idea is that of sampling from a space and observing the satisfaction of a property or generating an estimate based on the sample ensemble; results are then aggregated to provide an approximated solution to the original problem. In the following, we present at first the idea behind Monte Carlo and, then, the main results the theory provides.

4.2.1 The Idea Behind Monte Carlo

To present the Monte Carlo method with a straight and widely used example: the estimation of π.

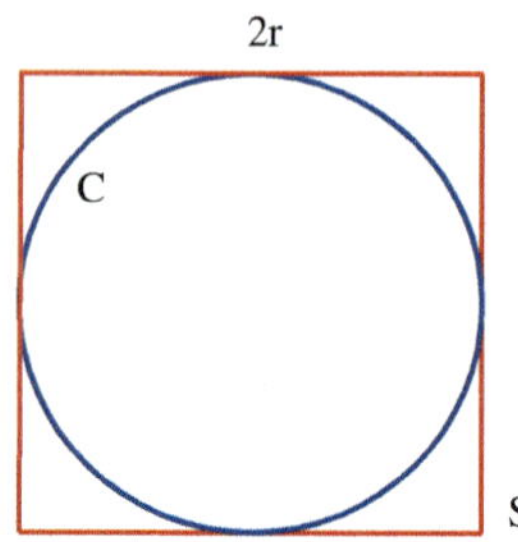

Fig. 4.3 Circle C is inscribed in *square* S representing the sampling world. Defined as $\Pr_C$ the probability of extracting a point belonging to the *circle*, then $\pi = 4\Pr_C$

Example 1: a probabilistic estimate for π

Consider a square S of side length $2r$ and a circle C inscribed within the square (see Fig. 4.3). Assume that a uniform distribution is induced on the square so that each sample drawn from there is equiprobable. Draw then n points inside the square and observe, for each point, whether it belongs to the circle or not. In doing this a straight question would be to ask which is the probability $\Pr_C$ of extracting a point belonging to the circle.

The answer is that such a probability is simply the ratio between the area of the circle and that of the square, i.e., its value is $\frac{\pi}{4}$. Then, $4\Pr_C$ is exactly π: we found a way to compute π with a probabilistic approach.

The issue now becomes that of computing $\Pr_C$ which, a priori, is unknown. We solve the problem with randomization by extracting n samples $s_1, \ldots, s_n$ from S according to the uniform distribution, and evaluating the number of samples n_C falling within the circle and computing the empirical probability

$$\hat{p}_n = \frac{n_C}{n}.$$

The procedure is formalized as follows: Consider the indicator function I_C

$$I_C(s_i) = \begin{cases} 1 \text{ if } s_i \in C \\ 0 \text{ if } s_i \notin C \end{cases}$$

The empirical probability $\hat{p}_n$ can be computed as

$$\hat{p}_n = \frac{1}{n} \sum_{i=1}^{n} I_C(s_i)$$

and represents an approximation of $\Pr_C$. Having an estimate for $\Pr_C$, we generate an estimate for π as

$$\hat{\pi}_n = 4\hat{p}_n = \frac{4}{n} \sum_{i=1}^{n} I_C(s_i).$$

How good is the approximation $\hat{\pi}_n$ of π ? It is intuitive to believe that the larger the number of samples n the better the estimate (the smaller the error $e(n) = |\hat{\pi}_n - \pi|$).

As such, we should consider a "sufficiently large" n to obtain a good approximation according to some predefined accuracy level. Such aspect will be addressed in Sect. 4.3. Instead, the convergence issue of $\hat{\pi}_n$ to π will be studied in Sect. 4.2.2. A high level algorithm for the Monte Carlo method is given in Algorithm 3.

Example 2: a different probabilistic approach to estimate π

Let us consider a different approach to estimate π with randomization. Consider the equation of a sector of circumference $y = f(x), x, y \in [0, 1]$

$$y = \sqrt{1 - x^2}$$

and observe that π can be obtained as

$$\pi = 4 \int_0^1 \sqrt{1 - x^2} dx.$$

If we induce a uniform distribution f_x on the input domain [0, 1] we have that π can also be intended as the expected value of y

$$\pi = 4E_x[y(x)] = 4 \int_0^1 \sqrt{1 - x^2} dx.$$

We comment that the variance σ_y^2 of y is bound

$$\sigma_y^2 = \int_0^1 (y(x) - E_x[y(x)])^2 \, dx \leq 1.$$

Extract then n samples x_i from [0, 1] and evaluate the sample mean

$$\hat{E}_n(y(x)) = \frac{1}{n} \sum_{i=1}^{n} y(x_i).$$

Then, by invoking the Tchebychev inequality in the form

$$\Pr\left(|z - \mu| \geq \alpha\right) \leq \frac{\sigma^2}{\alpha^2}$$

where z is an i.i.d random variable of mean μ, variance σ^2, and α is a positive number [2], we have that

Algorithm 3: The Monte Carlo algorithm

1- Identify the input space D of the algorithm and a random variable s, with probability density function f_s over D;
2- Extract n samples $S_n = \{s_1, \ldots, s_n\}$ from D according to f_s;
3- Evaluate the algorithm on S_n;
4- Generate an estimate of the algorithm output.

$$\Pr\left(|\hat{E}_n(y(x)) - E_x[y(x)]| \geq \varepsilon\right) \leq \frac{\sigma_y^2}{n\varepsilon^2} \leq \frac{1}{n\varepsilon^2}.$$

where the variance of the estimator is $\mathrm{Var}(\hat{E}_n(y(x))) = \frac{\sigma_y^2}{n}$. We can then select the confidence δ as

$$\frac{1}{n\varepsilon^2} < \delta \Longrightarrow n > \frac{1}{\delta\varepsilon^2}.$$

This says that, if we choose $n \geq \frac{1}{\delta\varepsilon^2}$, then

$$\Pr\left(|\hat{E}_n(y(x)) - E_x[y(x)]| \leq \varepsilon\right) \geq 1 - \delta$$

holds with probability $1 - \delta$ and we can estimate π as

$$\Pr\left(|4\hat{E}_n(y(x)) - 4E_x[y(x)]| \leq 4\varepsilon\right) = \Pr\left(|\hat{\pi}_n - \pi| \leq 4\varepsilon\right) \geq 1 - \delta$$

from which we derive the number of points needed to estimate π at a given tolerated level. For instance, if we select $\varepsilon = 0.025$ and $\delta = 0.01$ we need $n \geq 1600$. We extracted $n = 1600$ samples from a uniform distribution f_x and obtained the estimate $\hat{\pi}_n = 3.148$ for which $|\hat{\pi}_n - \pi| = 0.006 \leq 0.1 = 4\varepsilon$.

In the previous experiments, we have implicitly assumed that sampling is associated with a continuous random variable. However, similar results hold by sampling over a discrete space (e.g., a regular grid over $(0, 1)^k$, $k \in \mathbb{N}$).

Interestingly, the second solution proposed to estimate π provides as well the minimum number of samples satisfying a given accuracy and a confidence level. In the sequel, we will be interested in this latter approach by improving bounds so as to reduce the number of samples needed to solve a specific problem after having investigated the asymptotic behavior of the estimate.

4.2.2 Weak and Strong Laws of Large Numbers

In Sect. 4.2.1 we have seen that, by extracting n samples from S it is possible to build sequence $\hat{\pi}_1, \hat{\pi}_2, \ldots, \hat{\pi}_n$; it would be appreciable to discover that such a sequence converges to the expected value π as the second example showed provided that $n \to \infty$. This main result is known in the literature as the *Law of large numbers*; the interested reader can refer to [14] for its proof.

4.2.2.1 Weak Law of Large Numbers

Let $x \in D$ be a continuous scalar random variable of finite expectation μ and finite variance σ_x^2 and $x_1, \ldots, x_n$ a set of n independent and identically distributed samples drawn from D (e.g., $D = \mathbb{R}$) according to the continuous probability density function f_D. Generate the empirical mean $\hat{\mu}_n = \frac{1}{n}\sum_{i=1}^{n} x_i$. Then, for any $\varepsilon \in D$, the weak law of large numbers guarantees that

$$\lim_{n \to +\infty} \Pr(|\hat{\mu}_n - \mu| \geq \varepsilon) = 0.$$

An identical result also holds for the discrete random variable case.

4.2.2.2 Strong Law of Large Numbers

Let $x \in D$ be a continuous random scalar variable of finite expectation μ and finite variance σ_x^2 and $x_1, \ldots, x_n$ a set of n independent and identically distributed samples drawn from D (e.g., $D = \mathbb{R}$) according to the continuous probability density function f_D. Generate the empirical mean $\hat{\mu}_n = \frac{1}{n}\sum_{i=1}^{n} x_i$. Then, the strong law of large numbers guarantees that relationship

$$\lim_{n \to +\infty} \hat{\mu}_n = \mu$$

holds with probability one.

Comments

The difference between the strong and the weak formulation of the laws of large numbers is in the convergence modality. In the weak case, the probability of generating an estimate $\hat{\mu}_n$ so that $|\hat{\mu}_n - \mu| \geq \varepsilon$ decreases as the number of samples increases. Differently, the strong law of large numbers implies that the sequence $\hat{\mu}_n$ converges to μ with probability one.

When we apply the laws of large numbers to the Monte Carlo method, we have that $\hat{\mu}_n$ converges to μ (and $\hat{\pi}_n$ to π).

The assumption of finite variance is not truly necessary but makes the proof easier. In fact, a large or infinite variance negatively affects the convergence rate. However, the variance must exist. When this assumption does not hold, as it happens in example 3, the laws of large numbers cannot be applied.

Example 3: breaking the law of large numbers

Let $x \in \mathbb{R}$ be a continuous random variable characterized by the Cauchy density function

$$f_x = \frac{1}{\pi(1+x^2)}.$$

Then the expectation $E[x]$ does not exist, because the integral

$$\int_{-\infty}^{+\infty} \frac{x}{\pi(1+x^2)} dx$$

diverges; likewise the variance does not exist, hence violating the assumptions requested by the laws of large numbers. If we compute the sample mean by using n samples drawn from the Cauchy density it can be proved that the average is still ruled by a Cauchy probability density function [26].

A main consequence is that if the noise affecting measurements is ruled by a Cauchy density and we average over a number of measurements (think of the estimation module we introduced in Sect. 2.1.1) to mitigate the presence of uncertainty, then the average cannot be expected to be more accurate than any individual measurement!

4.2.3 Some Convergence Results

The laws of large numbers are rather general and can be applied to several interesting cases among which those related to probability and expected value estimation.

Define the real function $u(\psi)$, $\psi \in \Psi \subseteq \mathbb{R}^l$ to be measurable according to Lebesgue in Ψ and denote by f_ψ the probability density function of a random variable ψ with support on the input space Ψ. Assume that ψ has finite mean and variance.

4.2.3.1 Probability Function Estimation

The problem can be formalized as follows: Given a generic value $\gamma \in \mathbb{R}$, evaluate the probability $p(\gamma)$ for which $u(\psi)$ is below γ when ψ spans Ψ, i.e., compute

$$p(\gamma) = \Pr(u(\psi) \leq \gamma).$$

In other terms, we are asking if the embedded system is satisfying a given constraint γ given performance function $u(\psi)$. Formulation of probability $p(\gamma)$ in a closed form can be achieved only in particular cases, e.g., for very specific choices of $u(\psi)$ and f_ψ. However, the problem can be addressed and solved by resorting

Algorithm 4: Estimating the probability that a requested performance value is attained

1- Extract n independent and identically distributed samples $Z_n = \{\psi_1, \ldots, \psi_n\}$ from Ψ according to f_ψ;

2- Evaluate, for the i-th sample ψ_i, the indicator function

$$I(\psi_i) = \begin{cases} 1 \text{ if } u(\psi_i) \leq \bar{\gamma} \\ 0 \text{ if } u(\psi_i) > \bar{\gamma}. \end{cases}$$

3- Construct the estimate $\hat{p}(\bar{\gamma})$ of $p(\bar{\gamma})$ as

$$\hat{p}_n(\bar{\gamma}) = \frac{1}{n} \sum_{i=1}^{n} I(\psi_i)$$

---

to randomization. In the following, we aim at solving the problem with the laws of large numbers and, to this end, we assume at first that γ is given and assumes value $\bar{\gamma}$. However, obtained results are valid for any $\bar{\gamma}$.

Extract n independent and identically distributed samples $Z_n = \{\psi_1, \ldots, \psi_n\}$ from $\psi \in \Psi$ according to f_ψ and evaluate the indicator function

$$I(\psi_i) = \begin{cases} 1 \text{ if } u(\psi_i) \leq \bar{\gamma} \\ 0 \text{ if } u(\psi_i) > \bar{\gamma} \end{cases}$$

The estimate $\hat{p}_n(\bar{\gamma})$ of $p(\bar{\gamma})$ is

$$\hat{p}_n(\bar{\gamma}) = \frac{1}{n} \sum_{i=1}^{n} I(\psi_i)$$

Algorithm 4 summarizes the needed steps to provide an estimate $\hat{p}_n(\bar{\gamma})$ of $p(\bar{\gamma})$.

The laws of large numbers hold under the respective hypotheses and, for any $\varepsilon \in (0, 1)$ we have that

weak law of large numbers

$$\lim_{n \to +\infty} \Pr(|\hat{p}_n(\bar{\gamma}) - p(\bar{\gamma})| \geq \varepsilon) = 0$$

strong law of large numbers

$$\lim_{n \to +\infty} \hat{p}_n(\bar{\gamma}) = p(\bar{\gamma})$$

with probability one.

In other terms $\hat{p}_n(\bar{\gamma})$ converges to $p(\bar{\gamma})$. The obtained results, evaluated for a given $\bar{\gamma}$ value, can now be extended to deal with any given γ value (different

γs will experience different convergence rates). We can then write, for an arbitrary given γ that

weak law of large numbers

$$\lim_{n \to +\infty} \Pr(|\hat{p}_n(\gamma) - p(\gamma)| \geq \varepsilon) = 0, \quad \forall \gamma \in \mathbb{R}$$

strong law of large numbers

$$\lim_{n \to +\infty} \hat{p}_n(\gamma) = p(\gamma), \quad \forall \gamma \in \mathbb{R}$$

with probability one.

4.2.3.2 Expected Value Estimation

Another interesting case, which can be immediately derived from the theory, refers to the problem of evaluating the expected value

$$E_{\Psi}[u(\psi)] = \int_{\Psi} u(\psi) f_{\psi} d\psi$$

through the empirical mean

$$\hat{E}_n(u(\psi)) = \frac{1}{n} \sum_{i=1}^{n} u(\psi_i).$$

where ψ_is have been extracted according to f_{ψ}.

In this case, we wish to evaluate some expected performance the embedded system should have based on measured instances telling us how the system performs for a given input.

Convergence of $\hat{E}_n(u(\psi))$ to $E_{\Psi}[u(\psi)]$ is granted under the assumptions of the laws of large numbers.

weak law

$$\lim_{n \to +\infty} \Pr(|\hat{E}_n(u(\psi)) - E_{\Psi}[u(\psi)]| \geq \varepsilon) = 0$$

strong law

$$\lim_{n \to +\infty} \hat{E}_n(u(\psi)) = E_{\Psi}[u(\psi)]$$

with probability one.

The goodness of the estimate can be evaluated by taking expectation with respect to the sequence of n samples in Z_n. In particular, it can be proved, e.g., by referring to [22], that the variance of the estimate is

$$\mathrm{Var}\left(\hat{E}_n(u(\psi))\right) = E_{Z_n}\left[\left(E_\Psi[u(\psi)] - \hat{E}_n(u(\psi))\right)^2\right] = \frac{\mathrm{Var}(u(\psi))}{n}.$$

The result has a main conceptual impact and states that the variance of the estimate is the variance of function $u(\psi)$ scaled by n^{-1}. The above expression states that if $\mathrm{Var}(u(\psi))$ and $\mathrm{Var}(\hat{E}_n(u(\psi)))$ are bound, we can estimate a priori the number of samples needed to obtain a required accuracy in the estimate. In fact, if we know the variance $\mathrm{Var}(u(\psi))$ (or it is possible to provide a bound for it) and we set $\mathrm{Var}(\hat{E}_n(u(\psi)))$ at a tolerated level c, then the number of samples to be drawn is

$$n \geq \frac{\mathrm{Var}\,(u(\psi))}{c}.$$

4.2.4 *The Curse of Dimensionality and Monte Carlo*

The *Curse of dimensionality* refers to the bad scaling of the number of points n needed to explore a space as its dimension d increases. Consider the segment $\Psi = [0, 1)$ and subdivide it into $N = 10$ points so that each segment has resolution of 0.1. It comes out that, if we wish to keep the same grid resolution for a d dimensional space, the number of points we need to consider to "explore" the space is $n = N^d$. Such an exploration of the space grows exponentially with d and, soon, becomes computationally prohibitive.

The "curse of dimensionality" represents a major problem every time we need to sample a space and take future actions, e.g., if our task is to estimate the function $E_\Psi[u(\psi)]$ through $\hat{E}_n(u(\psi))$.

However, as nicely pointed out in [2] the mean square error of the Monte Carlo estimate of the expected value does not depend on the dimension d of the space which, somehow, breaks the "curse of dimensionality." As it will be clear in Sect. 4.3 this is a consequence of the fact that we associated a probability density function to Ψ: instead of exploring Ψ with a uniformly-spaced grid we do that by extracting the due number of points according to f_ψ. In other words, the curse of dimensionality can be avoided if we move our analysis from a strictly deterministic to a probabilistic framework.

4.3 Bounds on the Number of Samples

With Monte Carlo, we have seen that it is difficult to estimate the number of samples n we should consider to solve a given problem. Results, e.g., see [15–17], exploit some trial tests or a priori information about the specific problem to decide when stopping the sampling procedure. In other cases, e.g., as it happens in Example 2, we were able to identify the minimum number of points required to satisfy the accuracy and confidence requirements.

However, this cannot be granted for a generic application, characterized by a generic Lebesgue measurable function. Moreover, since we are looking for generality so as to cover a large set of applications, a pdf-free approach must be considered. The price we have to pay in a pdf-free framework is the a priori larger number of samples needed to solve our problem compared with that we would need by knowing the probability density function.

Several improved bounds on the number of samples n have been presented in the literature to solve large classes of problems through randomization. We will review such bounds starting from Bernoulli's one.

The theoretical framework is that of a Bernoulli process where the random variable x assumes value 1 with probability p and value 0 with probability $1-p$. The expected value is $E[x] = p$ and the variance $\mathrm{Var}(x) = p(1-p)$. Denote by $x_1, \ldots, x_n$ the sequence of n independent samples drawn from x and compute the empirical mean

$$\hat{E}_n = \frac{1}{n}\sum_{i=1}^{n} x_i$$

which represents the estimate of the probability that $x = 1$ in the n trials. $\hat{E}_n$ is a binomially distributed variable with expected value $E[\hat{E}_n] = p$ and variance $\mathrm{Var}(\hat{E}_n) = \frac{p(1-p)}{n}$.

4.3.1 The Bernoulli Bound

Inequality

$$\Pr\left(|\hat{E}_n - E[\hat{E}_n]| < \varepsilon\right) > 1 - \delta$$

holds for any accuracy level $\varepsilon \in (0, 1)$ *and confidence* $1 - \delta, \delta \in (0, 1)$ *provided that at least* $n \geq \frac{1}{4\delta\varepsilon^2}$ *independent and identically distributed samples are drawn.*

The proof follows by recalling the Tchebychev theorem in the form

$$\Pr\left(|x - \mu| \geq \alpha\right) \leq \frac{\sigma^2}{\alpha^2}$$

where x is the random variable of mean μ, variance σ^2, and α is a positive number. By substituting x with $\hat{E}_n$ and α with the accuracy variable ε, we obtain

$$\Pr\left(|\hat{E}_n - E[\hat{E}_n]| \geq \varepsilon\right) \leq \frac{p(1-p)}{n\varepsilon^2}. \tag{4.1}$$

Since $p(1-p)$ is maximized by $\frac{1}{4}$, we can be finally bound (4.1) as

$$\Pr\left(|\hat{E}_n - E[\hat{E}_n]| \geq \varepsilon\right) \leq \frac{1}{4n\varepsilon^2}. \tag{4.2}$$

By introducing a confidence value $\delta \in (0, 1)$, we can rewrite (4.2) as

$$\Pr\left(|\hat{E}_n - E[\hat{E}_n]| < \varepsilon\right) \geq 1 - \delta. \tag{4.3}$$

By setting

$$\frac{1}{4n\varepsilon^2} \leq \delta$$

we derive the number of samples granting (4.3) to hold.

$$n \geq \frac{1}{4\delta\varepsilon^2} \tag{4.4}$$

Comments

The Bernoulli bound shows that the number of required samples grows quadratically (inversely proportional) with the requested accuracy for the estimate ε and linearly (inversely proportional) with the requested confidence δ. We can obtain a good estimate of $\hat{E}_n$ with a polynomial sampling of the space. For instance, with the choice $\varepsilon = 0.05$, $\delta = 0.01$ we need to extract at least $n = 10000$ samples; with the choice $\varepsilon = 0.02$, $\delta = 0.01$ we need to extract at least $n = 62500$ samples. Figure 4.4 shows how the Bernoulli bound scales with δ and ε. We recall we shall consider small values for δ and ε to have enough confidence and accuracy.

The cost of sampling is the drawback we have to pay for generality (i.e., any p or application). Fortunately, the Bernoulli bound can be tightened with the Chernoff's one.

4.3.2 The Chernoff Bound

The Chernoff bound [1] largely improves over the Bernoulli's bound by reducing the number of samples to be drawn. We study at first the case where variable x is a Bernoulli random variable.

4.3.2.1 The Bernoulli Case

In the Bernoulli case, the main result states that

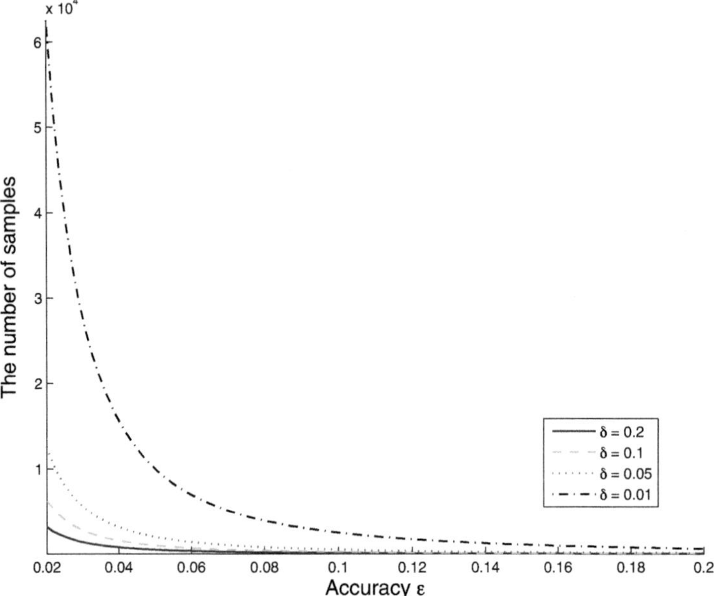

Fig. 4.4 The number of samples requested by the Bernoulli bound

Inequality

$$\Pr\left(|\hat{E}_n - E[\hat{E}_n]| < \varepsilon\right) > 1 - \delta$$

holds for any accuracy level $\varepsilon \in (0, 1)$ *and confidence* $1 - \delta, \delta \in (0, 1)$ *provided that at least*

$$n \geq \frac{1}{2\varepsilon^2} \ln \frac{2}{\delta}$$

independent and identically distributed samples x are drawn.

To prove the bound we recall that $E[\hat{E}_n] = p$ and

$$\Pr\left(|\hat{E}_n - E[\hat{E}_n]| < \varepsilon\right) = \Pr\left(|\hat{E}_n - p| < \varepsilon\right) \leq$$

$$\Pr\left(\hat{E}_n < p + \varepsilon\right) + \Pr\left(\hat{E}_n > p - \varepsilon\right).$$

By relying on the Binomial distribution, we can derive analytically those probabilities

$$\Pr\left(\hat{E}_n > p+\varepsilon\right) = \Pr\left(n\hat{E}_n > n(p+\varepsilon)\right) = \sum_{k>n(p+\varepsilon)}^{n} \binom{n}{k} p^k (1-p)^{n-k}$$

and

$$\Pr\left(\hat{E}_n < p-\varepsilon\right) = \Pr\left(n\hat{E}_n < n(p-\varepsilon)\right) = \sum_{k=0}^{k\leq n(p-\varepsilon)} \binom{n}{k} p^k (1-p)^{n-k}.$$

From those expression it is possible to derive the smallest n such that the sum of the two probabilities is greater than $1-\delta$, but no close form solution is known for the problem. Chernoff provided a bound for each of the above terms. In its additive form, we have that

$$\Pr\left(\hat{E}_n \geq p+\varepsilon\right) \leq e^{-2n\varepsilon^2}$$

and

$$\Pr\left(\hat{E}_n \leq p-\varepsilon\right) \leq e^{-2n\varepsilon^2}.$$

thus

$$\Pr\left(|\hat{E}_n - p| \geq \varepsilon\right) \leq 2e^{-2n\varepsilon^2}$$

i.e.,

$$\Pr\left(|\hat{E}_n - E[\hat{E}_n]| < \varepsilon\right) > 1 - 2e^{-2n\varepsilon^2}.$$

It comes out that

$$\Pr\left(|\hat{E}_n - E[\hat{E}_n]| < \varepsilon\right) > 1 - \delta$$

holds if we extract at least n samples so that $\delta \leq 2e^{-2n\varepsilon^2}$. This happens if we select

$$n \geq \frac{1}{2\varepsilon^2} \ln \frac{2}{\delta}.$$

Results, obtained in the case of x distributed as a Bernoulli variable, can be extended to cover the continuous case where the distribution is generic.

4.3.2.2 The General Case: The Hoeffding Inequality

The Chernoff bound for a generic probability density function and continuous variable ψ can be derived from the Hoeffding inequality [18]

Hoeffding inequality

Let $x_1, \ldots x_n$ be a sequence of independent random variables so that each x_i is almost surely bounded by the interval $[a_i, b_i]$, i.e., $\Pr(x_i \in [a_i, b_i]) = 1$. Then, defined the empirical mean $\hat{E}_n = \frac{1}{n}\sum_{i=1}^{n} x_i$, we have that for any ε value inequality

$$\Pr\left(|\hat{E}_n - E[\hat{E}_n]| \geq \varepsilon\right) \leq 2e^{\frac{-2\varepsilon^2 n^2}{\sum_{i=1}^{n}(b_i - a_i)^2}} \tag{4.5}$$

holds.

Under the above assumptions, we can rewrite (4.5) as

$$\Pr\left(|\hat{E}_n - E[\hat{E}_n]| < \varepsilon\right) > 1 - 2e^{\frac{-2\varepsilon^2}{\sum_{i=1}^{n}(b_i - a_i)^2}}. \tag{4.6}$$

In the interesting case where $\hat{E}_n$ represents the estimate $\hat{p}_n(\gamma)$ of a probability, e.g., $p(\gamma) = \Pr(u(\psi) \leq \gamma)$ for a given positive scalar γ (but any other event applies), we have that for a generic random variable ψ_i the indicator function

$$I\left(u(\psi_i) \leq \gamma\right) = \begin{cases} 1 \text{ if } u(\psi_i) \leq \gamma \\ 0 \text{ if } u(\psi_i) > \gamma \end{cases}$$

I assumes values in $\{0, 1\}$. As a consequence, $a_i = 0, b_i = 1$ and (4.6) becomes

$$\Pr\left(|\hat{E}_n - E[\hat{E}_n]| < \varepsilon\right) > 1 - 2e^{-2n\varepsilon^2}.$$

Since, $\hat{p}_n(\gamma) = \hat{E}_n$ and $E(\hat{p}_n(\gamma)) = p(\gamma)$ the expression becomes

$$\Pr\left(|\hat{p}_n(\gamma) - p(\gamma)| < \varepsilon\right) > 1 - 2e^{-2n\varepsilon^2}. \tag{4.7}$$

from which we derive the Chernoff bound by requesting $\delta \leq 2e^{-2n\varepsilon^2}$

$$n \geq \frac{1}{2\varepsilon^2} \ln \frac{2}{\delta}. \tag{4.8}$$

The Hoeffding inequality plays a major role since it allows us to

- derive the Chernoff bound of (4.8) that will be used to determine the number of samples needed to estimate the probability of performance satisfaction;
- derive the Chernoff bound formally identical to that of (4.8) granting the empirical mean to converge to its expected value with given accuracy and confidence levels;
- derive a set of bounds for estimating the maximum/minimum value of a function within a probabilistic framework.

Figure 4.5 presents the number of samples as function of δ and ε requested by the Chernoff bound.

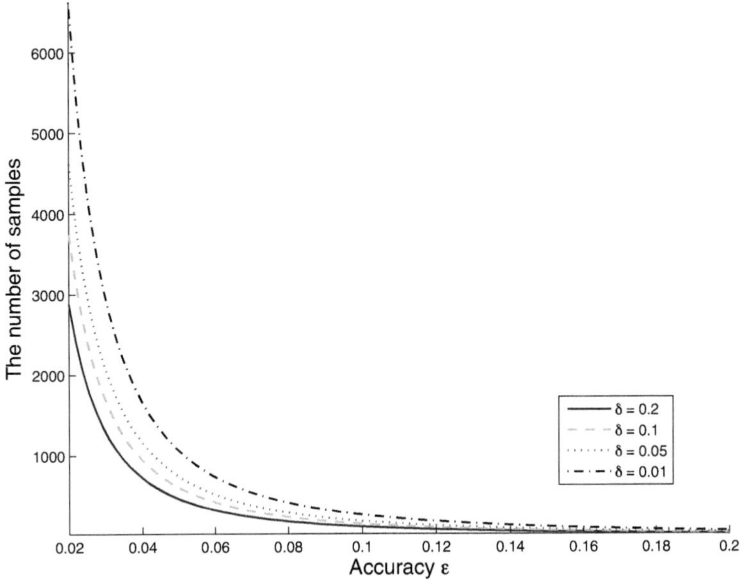

Fig. 4.5 The number of samples requested by the Chernoff bound as function of confidence δ and accuracy ε

Table 4.1 The number of samples $n = n(\varepsilon, \delta)$

Bound	$\varepsilon = 0.05, \delta = 0.02$	$\varepsilon = 0.05, \delta = 0.01$	$\varepsilon = 0.02, \delta = 0.01$	$\varepsilon = 0.01, \delta = 0.01$
Bernoulli	5000	10000	62500	250000
Chernoff	922	1060	6623	26492

Comments

The Chernoff bound shows that the number of required samples grows quadratically (inversely proportional) with the requested accuracy of the estimate ε but logarithmically with the confidence δ. Even if it might appear as a limited gain in reality it is not and represents a true achievement. In fact, if we refer to Table 4.1 we appreciate the significant improvement of the Chernoff bound over the Bernoulli one.

Interestingly, it appears that accuracy is more sampling demanding than confidence since the former is ruled by a quadratic term whereas the latter is bound by a linear one. Figure 4.6 compares the Bernoulli and the Chernoff bound. When δ and ε assume small values, as generally requested by applications since we wish to get high confidence and accuracy, the Chernoff bound significantly improves over the Bernoulli one with a gain $n_c = 2\delta \ln \frac{2}{\delta} n_b$ where n_c and n_b represent the number of samples requested by Chernoff and Bernoulli, respectively.

Other interesting bounds can be obtained by assuming some a priori information about p. For instance, the Chernoff-Okamoto bound [4] is tighter than the Chernoff

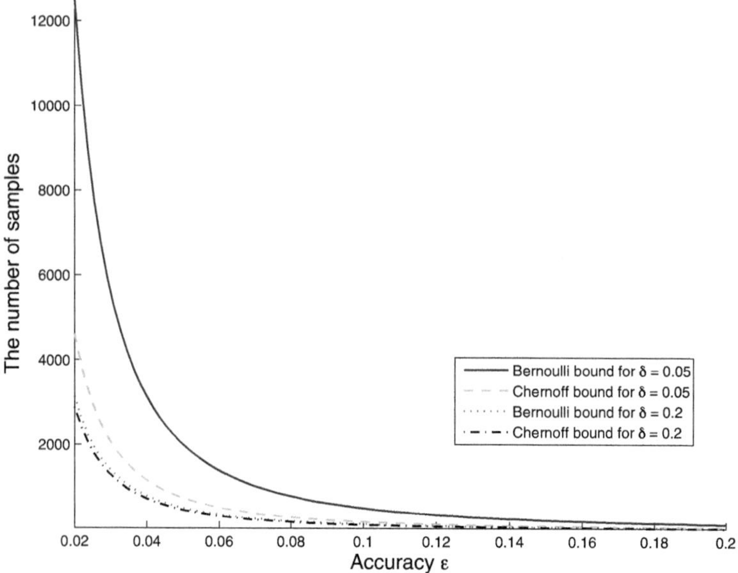

Fig. 4.6 The number of samples requested by the Bernoulli and the Chernoff bounds as function of confidence δ and accuracy ε. Chernoff largely improves over Bernoulli provided that δ and ε assume small values, as requested by applications

one but assumes that $p \leq 0.5$. Other bounds use only one side of the Chernoff bound and can be used to deal with special cases. The interested reader can refer to [2, 4].

As it will be clear in Sect. 4.3.3 the Chernoff bound is one of those main results which make the use of randomized algorithms viable.

4.3.3 A Bound on Samples to Estimate the Maximum Value of a Function

Sections 4.3.1 and 4.3.2 have shown how it is possible to derive bounds on the number of samples needed to guarantee convergence of the empirical mean to its expectation. We show here that many problems such as the verification of a constraint satisfaction problem can be modeled as a realization of a Bernoulli process; at the same time many problems can be reduced to the evaluation of the empirical mean of a quantity.

In this section, we aim at using a sampling technique (randomization) to estimate the maximum value of a function (and, of course, its minimum by changing the sign of the function). Say that we wish to maximize function $u(\psi) \in \mathbb{U} \subset \mathbb{R}$, $\psi \in \Psi \subseteq \mathbb{R}^l$ by identifying the maximum value $u_{\max}$

$$u_{\max} = \max_{\psi \in \Psi} u(\psi).$$

There exists a very large literature addressing the function optimization problem. Different techniques exploit a priori information about the function to be optimized, e.g., as it happens with gradient descent techniques where differentiability is requested. Some techniques explore the search space by looking for regularity and building blocks such as in the case of genetic algorithms; others, explore the search space with a probabilistic approach as in simulated annealing or introduce a blind search strategy as it happens with Monte Carlo. It can be proven that under mild hypotheses on the function to be optimized, all the above techniques converge in probability to the maximum value, also in the case of a blind random search exploration of the parameter space [19]. Different methods either differ in performance accuracy or convergence rate.

Consider the case where random variable ψ, with probability density function f_ψ, is defined over Ψ and generate the estimate

$$\hat{u}_{\max} = \max_{i=1,\ldots,n} u(\psi_i)$$

after having drawn n random samples $\{\psi_1, \ldots, \psi_n\}$. To move back to embedded systems consider $u(\psi)$ as a performance function and ask which is the maximum (minimum) value the function assumes given the fact we can only provide n measurements $u(\psi_i)$. That said, how good is the estimate $\hat{u}_{\max}$? The answer is given by the laws of large numbers.

4.3.3.1 Weak and Strong Laws of Large Numbers for Empirical Maximum

Assume that $u(\psi)$ is continuous in $\psi_{\max} = \text{argmax}_{\psi \in \Psi} u(\psi)$ and that f_ψ assigns a non-null probability to every neighborhood of $\psi_{\max}$.

Then, for any $\varepsilon > 0$ we have that

weak law of large numbers

$$\lim_{n \to +\infty} \Pr(u_{\max} - \hat{u}_{\max} \geq \varepsilon) = 0$$

strong law of large numbers

$$\lim_{n \to +\infty} \hat{u}_{\max} = u_{\max}$$

with probability one.

Since asymptotic results are of scarce utility in real applications we determine a bound on the number of samples granting $\hat{u}_{\max}$ and $u_{\max}$ to be close in probabilistic terms [2].

4.3.3.2 A Bound for a Probabilistic Estimate of the Maximum of a Function

The problem can be simply solved by noting that the determination of the maximum of a function is related to the probability estimation problem addressed in Sect. 4.3.2 and, in particular, Eq. (4.7):

$$\Pr\left(|\hat{p}_n(\gamma) - p(\gamma)| < \varepsilon\right) > 1 - 2e^{-2n\varepsilon^2}. \tag{4.9}$$

In fact, if we set $\gamma = \hat{u}_{\max}$ we have that

$$p(\gamma) = \Pr\left(u(\psi) \le \hat{u}_{\max}\right) = 1 - \Pr\left(u(\psi) > \hat{u}_{\max}\right)$$

and

$$\hat{p}_n(\gamma) = 1$$

since all taken samples satisfy inequality $u(\psi) \le \hat{u}_{\max}$ by construction. Therefore, from (4.9)

$$\Pr\left(|\hat{p}_n(\gamma) - p(\gamma)| < \varepsilon\right) = \Pr\left(\Pr\left(u(\psi) > \hat{u}_{\max}\right) < \varepsilon\right) > 1 - 2e^{-2n\varepsilon^2}$$

which holds by selecting n according to the Chernoff bound. However, the bound can be improved as shown in [2] and leads to the final result:

Inequality

$$\Pr\left(\Pr\left(u(\psi) > \hat{u}_{\max}\right) \le \varepsilon\right) \ge 1 - \delta$$

holds for any accuracy level $\varepsilon \in (0, 1)$ *and confidence* $1 - \delta, \delta \in (0, 1)$ *provided that at least*

$$n \ge \frac{\ln \delta}{\ln(1 - \varepsilon)} \tag{4.10}$$

independent and identically distributed samples are drawn.

Other results about convergence exist, but are outside the goal of this book. The interested reader can refer to [14] where a more complete analysis is carried out. Derived results will be used in Sect. 4.4.2.

4.4 Randomized Algorithms

Consider a problem influenced by some variables grouped in vector ψ with a pdf f_ψ over the space Ψ. Randomized algorithms are algorithms that, by sampling from space Ψ according to f_ψ, provide results valid in probability. The method is general

Algorithm 5: The algorithm behind randomized algorithms

1- Transform the deterministic problem into a probabilistic problem;
2- Identify the input space Ψ of the algorithm and define a random variable ψ, with probability density function f_ψ over Ψ;
3- Identify the accuracy and the confidence levels and, then, the number of samples n required by the randomization process;
4- Draw n samples $S_n = \{s_1, \ldots, s_n\}$ from Ψ according to f_ψ;
5- Evaluate the algorithm on samples in S_n;
6- Provide the probabilistic outcome of the algorithm.

and can be applied to a very large class of functions, namely those Lebesgue measurable: a filter bank, a Fast Fourier Transform (FFT), a discrete cosine transform, wavelets transform, and a generic circuit response function are some very simple examples of Lebesgue measurable functions.

At a very high abstraction level, the procedure behind a randomized algorithm is given in Algorithm 5.

In the following, we will apply randomized algorithm to an interesting class of problems. In Chaps. 5 and 7 results will be applied to the robustness problem and to characterize the level of approximate computation, respectively. Randomized algorithms will also be used to assess the performance of embedded applications as well as evaluate the level of constraints satisfaction within a noise-affected environment.

4.4.1 The Algorithm Verification Problem

The algorithm verification problem aims at evaluating the satisfaction level of an inequality. Even though solving this problem might appear strange, we will see that it constitutes the core of many problems.

Consider function $u(\psi) \in \mathbb{U} \subset \mathbb{R}$, $\psi \in \Psi \subseteq \mathbb{R}^l$ Lebesgue measurable over Ψ onto which a random variable ψ is defined, with pdf f_ψ over Ψ, and a given, but generic, $\gamma \in \mathbb{R}$ scalar. As we already pointed out the problem models the case where we wish to determine the level of satisfaction of performance function $u(\psi)$ given a constant value γ, generally acting as a tolerated performance. Without loss of generality we study here and in next sections a scalar performance function. However, the simultaneous attainment of several scalar performance functions may be easily handled with the introduced techniques. The problem can be finalized as:

Verify the level of satisfaction of inequality

$$u(\psi) \le \gamma, \forall \psi \in \Psi.$$

In other words, we wish to determine the "percentage" of points of Ψ satisfying the inequality. Such a value is simply the ratio

$$n_{u(\psi)\leq\gamma} = \frac{\int_{u(\psi)\leq\gamma,\psi\in\Psi} d\psi}{\int_{\Psi} d\psi}.$$

Determination of $n_{u(\psi)\leq\gamma}$ is surely a computationally hard problem for a generic $u(\psi)$ function and cannot be computed in a closed form unless $u(\cdot)$ presents a form that makes the mathematics amenable. Differently, the problem can be solved with a randomized algorithm by transforming the deterministic problem into a probabilistic one. By relying on the previously mentioned probability density function f_ψ defined over Ψ, we are able to evaluate the probability

$$p(\gamma) = \frac{\int_{u(\psi)\leq\gamma,\psi\in\Psi} f_\psi(\psi)d\psi}{\int_{\Psi} f_\psi(\psi)d\psi} = \Pr\left(u(\psi)\leq\gamma\right), \forall\psi\in\Psi.$$

We have seen in Sect. 4.2.3 that $p(\gamma)$ can be evaluated through randomization and that, given a γ value, the event

$$u(\psi)\leq\gamma$$

is associated with the Bernoulli variable

$$\psi\in\Psi : I\left(u(\psi)\leq\gamma\right) = \begin{cases} 1 \text{ if } u(\psi)\leq\gamma \\ 0 \text{ if } u(\psi)>\gamma \end{cases}$$

and by sampling n i.i.d. realizations $\{\psi_1,\ldots,\psi_n\}$ from ψ

$$\hat{p}_n(\gamma) = \frac{1}{n}\sum_{i=1}^{n} I\left(u(\psi_i)\leq\gamma\right).$$

We invoke the Chernoff inequality with $\hat{E}_n = \hat{p}_n(\gamma)$, and $E[\hat{E}_n] = p(\gamma)$ and provide the main result

Performance verification problem

Let $u(\psi)\in\mathbb{U}\subset\mathbb{R}$ be a performance function measurable according to Lebesgue on its input domain $\Psi\subseteq\mathbb{R}^l$ and ψ be a random variable, with probability density function f_ψ over Ψ. *Define*

$$p(\gamma) = \Pr\left(u(\psi)\leq\gamma\right)$$

and evaluate the estimate $\hat{p}_n$ from the n i.i.d. samples $\psi_1,\ldots,\psi_n$. Then,

$$\Pr\left(|\hat{p}_n(\gamma)-p(\gamma)|\leq\varepsilon\right)\geq 1-\delta$$

holds for any accuracy level $\varepsilon\in(0,1)$, confidence $\delta\in(0,1)$ and $\forall\gamma\in\mathbb{R}$ provided that

$$n\geq\frac{1}{2\varepsilon^2}\ln\frac{2}{\delta}.$$

Algorithm 6: Randomized algorithms for the algorithm performance verification problem: the given performance loss case $\overline{\gamma}$

1- The probabilistic problem requires evaluation of $p(\gamma) = \Pr\left(u(\psi) \leq \overline{\gamma}\right)$ for a given $\overline{\gamma}$;
2- Identify the input space Ψ and a random variable ψ, with density function f_ψ over Ψ;
3- Select accuracy ε and confidence δ;
4- Draw $n \geq \frac{1}{2\varepsilon^2} \ln \frac{2}{\delta}$ samples $\psi_1, \ldots, \psi_n$ from ψ;
5- Estimate

$$\hat{p}_n(\overline{\gamma}) = \frac{1}{n} \sum_{i=1}^{n} I\left(u(\psi_i) \leq \overline{\gamma}\right), \qquad I\left(u(\psi_i) \leq \overline{\gamma}\right) = \begin{cases} 1 \text{ if } u(\psi_i) \leq \overline{\gamma} \\ 0 \text{ if } u(\psi_i) > \overline{\gamma} \end{cases}$$

6- use $\hat{p}_n(\overline{\gamma})$;

Value $\hat{p}_n(\gamma)$ is the probabilistic outcome of the algorithm.

By using the algorithm given in Algorithm 6 we estimate $p(\gamma)$ for a given $\bar{\gamma}$ so as to solve the problem of determine the level of satisfaction for the inequality, i.e.,

$$\Pr\left(u(\psi) \leq \bar{\gamma}\right), \forall \psi \in \Psi.$$

In other applications, we could be interested in constructing function $p(\gamma)$ for an arbitrary large but given and finite set of γs. The natural solution to this problem is to provide a decomposition of the feasible interval of γ, $[a_\gamma, b_\gamma]$ (e.g., with an equally spaced grid) and obtain for each $\gamma \in \Gamma = \{\gamma_1, \ldots, \gamma_k\}$ an estimate $\hat{p}_n(\gamma_i)$ by invoking Algorithm 6 for $i \in \{1, \ldots K\}$. In such a case the algorithm can be extended as in Algorithm 7.

Comments

Randomization has allowed us to solve the algorithm verification problem by transforming the deterministic problem in a probabilistic one. At the same time the Chernoff bound has provided the number of samples satisfying it a given accuracy ε and confidence δ.

Having provided a first complete algorithm based on randomization it is worth to shed light on some operational aspects somehow hidden within the theory.

Here, ε represents the accuracy of estimating $p(\gamma)$, given γ, with $\hat{p}_n(\gamma)$, that is to say it represents an upper bound for the error $|\hat{p}_n(\gamma) - p(\gamma)|$. If ε is small then we can confuse $\hat{p}_n(\gamma)$ with $p(\gamma)$ in our subsequent use of $p(\gamma)$. At the same time we shall note that $|\hat{p}_n(\gamma) - p(\gamma)|$ is a random variable depending on the particular realization of the sampling set. A different sampling set would have provided a different estimate $\hat{p}_n(\gamma)$.

Then one should ask how credible the statement $|\hat{p}_n(\gamma) - p(\gamma)| \leq \varepsilon$ is $\forall \psi \in \Psi$; the answer is that the statement holds with probability $1 - \delta$. This means that we

Algorithm 7: Randomized algorithms for solving the algorithm verification problem

1- The probabilistic problem requires evaluation of $p(\gamma) = \Pr(u(\psi) \le \gamma)$ for any γ belonging to a finite set of arbitrary γ values;
2- Identify the input space Ψ and a random variable ψ, with density function f_ψ over Ψ;
3- Select accuracy ε and confidence δ;
4- Identify the interested performance level set $\Gamma = \{\gamma_1, \dots, \gamma_k\}$;
5- $\hat{p}_{n,\Gamma}(\gamma)$ = verification-problem $(\Psi, f_\psi, u(\psi), \Gamma, \varepsilon, \delta)$;
6- use $\hat{p}_{n,\Gamma}(\gamma)$;

function verification-problem $(\Psi, f_\psi, u(\psi), \Gamma, \varepsilon, \delta)$
Draw $n \ge \frac{1}{2\varepsilon^2} \ln \frac{2}{\delta}$ samples $\psi_1, \dots, \psi_n$ from ψ;
For each $\gamma \in \Gamma$ estimate

$$\hat{p}_n(\gamma) = \frac{1}{n} \sum_{i=1}^{n} I(u(\psi_i) \le \gamma), \qquad I(u(\psi_i) \le \gamma) = \begin{cases} 1 \text{ if } u(\psi_i) \le \gamma \\ 0 \text{ if } u(\psi_i) > \gamma \end{cases}$$

Group all $\hat{p}_n(\gamma)$s in vector $\hat{p}_{n,\Gamma}$;
Return $\hat{p}_{n,\Gamma}$

could extract a sequence of points for which the inequality $|\hat{p}_n(\gamma) - p(\gamma)| \le \varepsilon$ is not verified but this happens with probability δ, which needs to be kept small.

As a last note we observe that the sampling space is $\mathbb{R}^l$: the Chernoff bound is independent from the dimension l of the input sampling space. A *small dimension or a large dimension requires the same number of samples*: again we find that randomization has somehow broken the "curse of dimensionality."

4.4.2 The Maximum Value Estimation Problem

The maximum value estimation problem, also known in the literature as worst-case analysis, aims at estimating the maximum value a function can assume.

Consider a $u(\psi) \in \mathbb{U} \subset \mathbb{R}$ function which is Lebesgue measurable over $\Psi \subseteq \mathbb{R}^l$. The problem can be cast in the canonical form requesting the evaluation of

$$u_{\max} = \max_{\psi \in \Psi} u(\psi). \tag{4.11}$$

Analytical determination of $u_{\max}$ is impossible for a large class of functions as the Lebesgue measurable one is and its evaluation might be a computational hard problem.

As we did for the verification case, we generate a probabilistic version of the problem. Observe that the (4.11) can be reformulated as searching for that value $u_{\max}$ of $u(\psi)$ for which

$$u(\psi) \leq u_{\max}, \quad \forall \psi \in \Psi. \tag{4.12}$$

Now we resort to probability by relaxing the deterministic approach intrinsic with (4.12). In particular, we are looking for an estimate $\hat{u}_{\max}$ of $u_{\max}$ and say that the estimate is good if the probability of receiving a ψ for which $u(\psi) > \hat{u}_{\max}$ is small, say assumes value τ.

In other words we are requesting that

$$\Pr\left(u(\psi) > \hat{u}_{\max}\right) \leq \tau. \tag{4.13}$$

Assume that a random variable ψ, with probability density function f_ψ, is defined over Ψ and draw n i.i.d. samples $\psi_1, \ldots, \psi_n$ from ψ. Construct estimate $\hat{u}_{\max}$ as

$$\hat{u}_{\max} = \max_{i=1,\ldots,n} u(\psi_i).$$

As we have seen in Sect. 4.3.3 the weak and strong laws of large numbers grant convergence of $\hat{u}_{\max}$ to $u_{\max}$ in probability.

Unfortunately, solution of (4.13) requires a number of points which is exponential in the dimension of the input space and, as such, the problem solution is computationally hard [6]. To solve this issue we note that (4.13) is again a random variable since different realizations of the sampling set would provide different estimates of $\hat{u}_{\max}$. To address this last aspect, we introduce a confidence value δ and use a second level of probability. Since we have reformulated our problem in a canonical form, we immediately use the bound given in (4.10).

Maximum value estimation problem

Let $u(\psi) \in \mathbb{U} \subset \mathbb{R}$ be a performance function measurable according to Lebesgue on its input domain $\Psi \subseteq \mathbb{R}^l$ onto which is defined a random variable ψ with probability density function f_ψ. Define value $u_{\max}$ to be the maximum value function $u(\psi)$ assumes, i.e.,

$$u(\psi) \leq u_{\max}, \quad \forall \psi \in \Psi.$$

Draw n i.i.d. samples $\psi_1, \ldots, \psi_n$ according to f_ψ and generate the estimate $\hat{u}_{\max}$

$$\hat{u}_{\max} = \max_{i=1,\ldots,n} u(\psi_i)$$

then,

$$\Pr\left(\Pr\left(u(\psi) \geq \hat{u}_{\max}\right) \leq \varepsilon\right) \geq 1 - \delta$$

holds for any accuracy level $\varepsilon \in (0, 1)$, confidence $\delta \in (0, 1)$ and $\forall \psi \in \Psi$ *provided that*

$$n \geq \frac{\ln \delta}{\ln(1 - \varepsilon)}$$

Algorithm 8: Randomized algorithm to estimate the maximum value of a function

1- The probabilistic problem requires evaluation of $\hat{u}_{\max}$;
2- Identify the input space Ψ and a random variable ψ with pdf f_ψ over Ψ;
3- Select the accuracy ε and the confidence δ levels;
4- $\hat{u}_{\max}$ = Max-estimate ($\Psi, f_\psi, u(\psi), \varepsilon, \delta$);
5- use $\hat{u}_{\max}$;

Max-estimate ($\Psi, f_\psi, u(\psi), \varepsilon, \delta$)
Draw $n \geq \frac{\ln \delta}{\ln(1-\varepsilon)}$ samples $\psi_1, \ldots, \psi_n$ from ψ according to f_ψ ;
Compute $\hat{u}_{\max} = \max_{i=1,\ldots,n} u(\psi_i)$;
Return $\hat{u}_{\max}$

Table 4.2 The number of samples $n = n(\varepsilon, \delta)$

	$\varepsilon = 0.05, \delta = 0.02$	$\varepsilon = 0.05, \delta = 0.01$	$\varepsilon = 0.02, \delta = 0.01$	$\varepsilon = 0.01, \delta = 0.01$
n	77	90	228	459

Value $\hat{u}_{\max}$ *is the probabilistic outcome of the algorithm.*

The algorithm solving the maximum value estimation problem, i.e., the probabilistic version of the worst case analysis, is given in Algorithm 8.

Comments

As it can be seen from Table 4.2, the required number of samples $n \geq \frac{\ln \delta}{\ln(1-\varepsilon)}$ is well below the one requested by Chernoff to solve the performance verification problem. In fact, for a sufficiently small ε, $\ln(1-\varepsilon) \simeq -\varepsilon$: the number of samples scales as $\frac{1}{\varepsilon^2}$ with Chernoff and $\frac{1}{\varepsilon}$ for the above.

Figure 4.7 compares the number of samples requested by Chernoff with those requested to solve the maximum value estimation problem. We appreciate the fact that the latter bound significantly improves over the former with a gain set by $\frac{1}{\varepsilon}$.

However, since there does not exist a free lunch, the price we have to pay is that our estimate requires two levels of probability

$$\Pr\left(\Pr\left(u(\psi) \geq \hat{u}_{\max}\right) \leq \varepsilon\right) \geq 1-\delta.$$

The inner inequality $\Pr\left(u(\psi) \geq \hat{u}_{\max}\right) \leq \varepsilon$ states that we are requesting an estimate which is good not in terms of classic accuracy but according to Lebesgue. In other terms the inequality requires that, at least with probability $1-\delta$, the probability of encountering points whose $u(\psi)$ is larger than $\hat{u}_{\max}$ is below ε.

Figure 4.8 shows the situation. Function $u(\psi)$ is given and $\hat{u}_{\max}$ determined as discussed above. Those points $u(\psi) \geq \hat{u}_{\max}$ belong to two intervals Ψ_1, Ψ_2 so that

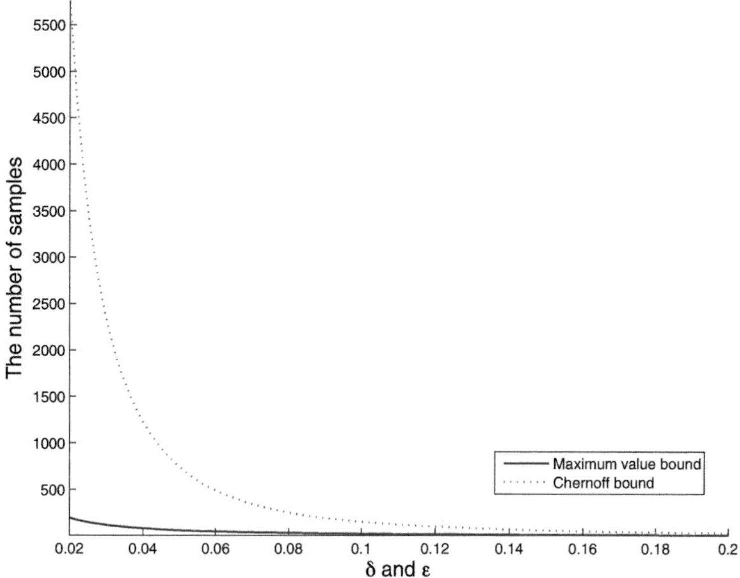

Fig. 4.7 The number of samples requested by the Chernoff bound and that requested to solve the maximum value estimation problem. ε and δ assume the same values to ease the comparison

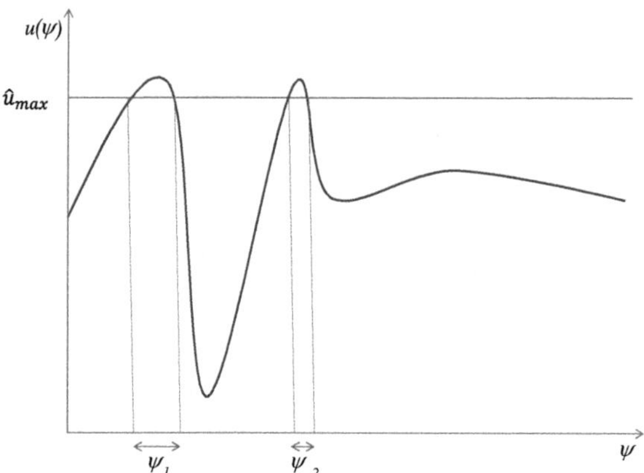

Fig. 4.8 The maximum estimated value for function $u(\psi)$ is $\hat{u}_{\max}$. The probability of having points $u(\psi) \geq \hat{u}_{\max}$ is associated with two supports Ψ_1, Ψ_2, for which $\Pr\left(u(\psi)|_{\psi\in\Psi_1} \geq \hat{u}_{\max}\right) \leq \varepsilon_1$, $\Pr\left(u(\psi)|_{\psi\in\Psi_2} \geq \hat{u}_{\max}\right) \leq \varepsilon_2$ and sum $\varepsilon_1 + \varepsilon_2 \leq \varepsilon$

$\Pr\left(u(\psi)|_{\psi\in\Psi_1} \geq \hat{u}_{\max}\right) \leq \varepsilon_1$ and $\Pr\left(u(\psi)|_{\psi\in\Psi_2} \geq \hat{u}_{\max}\right) \leq \varepsilon_2$, respectively. However, the sum $\varepsilon_1 + \varepsilon_2 \leq \varepsilon$ at least with confidence $1 - \delta$.

There might even be an infinity of points ψ for which $u(\psi)$ is larger than the estimated $\hat{u}_{\max}$ but the probability of encountering such points is no more than ε. This note should be carefully recalled when using the obtained estimates.

It can be proved that the bound is tight under regularization and smoothness hypotheses on the probability function of the random variable $u(\psi)$, for instance continuity (e.g., refer to [7]).

4.4.3 The Expectation Estimation Problem

In many applications it is crucial to be able to estimate the expected value of a given function $u(\psi)$, operation generally carried out by estimating the empirical mean. Again, the problem is to identify the minimum number of samples granting an arbitrary level of accuracy and confidence.

Consider a $u(\psi) \in [0, 1]$ function which is Lebesgue measurable over $\Psi \subseteq \mathbb{R}^l$ and let f_ψ be the probability density function of a random variable ψ defined over Ψ. Expectation estimation requires evaluation of

$$E[u(\psi)] = \int_\Psi u(\psi) f_\psi(\psi) d\psi. \tag{4.14}$$

As in other problems, evaluation of (4.14) is computationally hard for a generic u function and the empirical mean

$$\hat{E}_n(u(\psi)) = \frac{1}{n} \sum_{i=1}^{n} u(\psi_i) \tag{4.15}$$

is constructed instead based on the n i.i.d. samples $\psi_1, \ldots, \psi_i, \ldots, \psi_n$ drawn from ψ according to f_ψ. Of course, $\hat{E}_n(u(\psi))$ is a random variable depending on the particular realization of the n samples. By invoking the Hoeffding inequality (4.5) where $a_i = 0, b_i = 1, i \in \{1, \ldots, n\}$

$$\Pr\left(|\hat{E}_n(u(\psi)) - E[u(\psi)]| \geq \varepsilon\right) \leq 2e^{-2\varepsilon^2 n} \tag{4.16}$$

we derive the Chernoff bound

$$n \geq \frac{1}{2\varepsilon^2} \ln \frac{2}{\delta}. \tag{4.17}$$

Expectation estimation problem

Let $u(\psi) \in [0, 1]$ be a performance function measurable according to Lebesgue on its input domain $\Psi \subseteq \mathbb{R}^l$ onto which is defined a random variable ψ with probability density function f_ψ. Define $E[u(\psi)]$ to be the expectation of function $u(\psi)$.

Draw n i.i.d. samples $\psi_1, \ldots, \psi_n$ according to f_ψ and generate the estimate

$$\hat{E}_n(u(\psi)) = \frac{1}{n} \sum_{i=1}^{n} u(\psi_i)$$

then,

$$\Pr\left(|\hat{E}_n(u(\psi)) - E[u(\psi)]| \leq \varepsilon\right) \geq 1 - \delta$$

holds for any accuracy level $\varepsilon \in (0, 1)$, confidence $\delta \in (0, 1)$

Algorithm 9: Randomized algorithm to estimate the expected value of a function

1- The probabilistic problem requires evaluation of $E[u(\psi)]$;
2- Identify the input space Ψ and a random variable ψ with pdf f_ψ over Ψ;
3- Select the accuracy ε and the confidence δ levels;
4- Draw $n \geq \frac{1}{2\varepsilon^2} \ln \frac{2}{\delta}$ samples $\psi_1, \ldots, \psi_n$ from ψ according to f_ψ;
5- Compute $\hat{E}_n(u(\psi)) = \frac{1}{n} \sum_{i=1}^{n} u(\psi_i)$;
6- use $\hat{E}_n(u(\psi))$;

$$n \geq \frac{1}{2\varepsilon^2} \ln \frac{2}{\delta}.$$

Value $\hat{E}_n(u(\psi))$ is the probabilistic outcome of the randomized algorithm.

The randomized algorithm to estimate the expected value of a function is given in Algorithm 9.

Comments

Interestingly, the determination of the expected value problem can be addressed with the same number of samples (Chernoff bound) used to address the probability estimation problem. The structural difference is in the use of the empirical sum in one case and the indicator function in the other. Even if their derivations came from a different perspective, both cases are a special case of the Hoeffding inequality (which leads to the Chernoff bound). As a consequence, the request that $u(\psi) \in [0, 1]$ is only made to ease the derivation of the bound through the Hoeffding's inequality. In general, it is enough to require $u(\psi_i)$ bound, e.g., to the same $a_i = a, b_i = b, i = 1, \ldots, n$. As a consequence, the bound on the number of samples would become

$$n \geq \frac{(b-a)^2}{2\varepsilon^2} \ln \frac{2}{\delta} \tag{4.18}$$

Another aspect which should be addressed is the relationship between the number of needed samples as per the Chernoff bound and that which could be derived by applying the central limit theorem. In fact, if $f_{u(\psi)} = f_{u(\psi)}(\mu, \sigma^2)$ the central limit theorem states that as n increases the distribution of $\hat{E}_n(u(\psi))$ approaches the normal distribution with mean value $E[u(\psi)] = \mu$ and variance $\frac{\sigma^2}{n}$ irrespective of $f_{u(\psi)}$. Said that, we can write that

$$\Pr\left(|\hat{E}_n(u(\psi)) - \mu| \leq \lambda \frac{\sigma}{\sqrt{n}}\right) = \text{erf}\left(\frac{\lambda}{\sqrt{2}}\right) \tag{4.19}$$

If we select $\varepsilon > 0$ so that $\varepsilon = \lambda \frac{\sigma}{\sqrt{n}}$, then the implicit relationship between ε, δ, and n is

$$\delta = 1 - \text{erf}\left(\frac{\varepsilon\sqrt{n}}{\sigma\sqrt{2}}\right)$$

since for $x > 0$ we can provide the Chernoff-Rabin bound

$$\frac{1}{2}\left(1 - \text{erf}\left(\frac{x}{\sqrt{2}}\right)\right) \leq \frac{1}{2}e^{\frac{-x^2}{2}}$$

then, being $x = \frac{\varepsilon\sqrt{n}}{\sigma}$ we can write that

$$\delta \leq e^{\frac{-\varepsilon^2 n}{2\sigma^2}}$$

from which

$$n \geq \frac{2\sigma^2}{\varepsilon^2} \ln \frac{1}{\delta}. \tag{4.20}$$

We recall that the Chernoff bound requires $u(\psi) \in [0, 1]$ as a working hypothesis but we commented that results can be extended provided that the variable is bounded. The variance σ^2 might be small. In such a case the bound (4.20) provided by the central limit theorem could slightly improve over the Chernoff bound (the opposite holds). That said, the Chernoff bound should always be preferred independently of the value assumed by σ^2. In fact, (4.20) relies on the assumption that the distribution of the empirical mean is Gaussian which is only true asymptotically with the increasing n and its convergence ratio depends on σ. Differently, the (4.17) is general and does not require any particular assumption on the distribution.

As an example, let us assume that $u(\psi)$ is uniformly distributed in interval $[a, b] = [0, 1]$. Then, the central limit theorem (using Eq. 4.20) would lead to

$$n \geq \frac{2(b-a)^2}{12\varepsilon^2} \ln \frac{1}{\delta} = \frac{1}{6\varepsilon^2} \ln \frac{1}{\delta}$$

against the bound derived from the Hoeffding inequality (Eq. 4.18)

$$n \geq \frac{(b-a)^2}{2\varepsilon^2} \ln \frac{2}{\delta} = \frac{1}{2\varepsilon^2} \ln \frac{2}{\delta}$$

Figure 4.9 presents the bound set by the central limit theorem against the Chernoff one for the choice $\delta = \varepsilon$. As we see the CLT, by taking advantage of the fact the distribution of the empirical mean is Gaussian (when it is only asymptotically), improves over Chernoff that is not assuming any particular distribution.

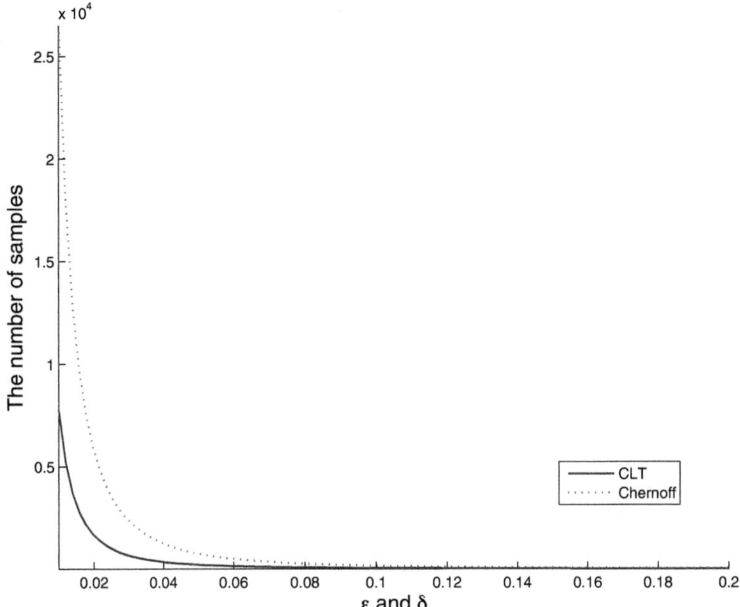

Fig. 4.9 The number of samples requested by the Chernoff bound and the Central limit theorem as function of confidence and accuracy $\delta = \varepsilon$

4.4.4 The Minimum (Maximum) Expectation Problem

The minimum (maximum) expectation problem aims at estimating the minimum (maximum) value of the expectation of a function. Without any loss in generality, we consider here the minimization problem by keeping the same structure given in [2].

Consider the Lebesgue measurable function $u(\psi, \Delta) \in [0, 1]$, $\psi \in \Psi \subseteq \mathbb{R}^l$ and $\Delta \in D \subseteq \mathbb{R}^k$. Define f_Ψ and f_Δ to be the probability density functions associated to random variables ψ and Δ defined over Ψ and D, respectively. The problem requires minimization either of function

$$u_{\min} = \min_{\psi \in \Psi} E_\Delta[u(\psi, \Delta)] \tag{4.21}$$

or

$$u_{\min} = \min_{\Delta \in D} E_\Psi[u(\psi, \Delta)].$$

The two problems are structurally equivalent; as such we consider the first one and the other follows immediately. The problem can then be described by the system

$$\begin{cases} \phi(\psi) = E_\Delta[u(\psi, \Delta)] \\ u_{\min} = \min_{\psi \in \Psi} \phi(\psi). \end{cases}$$

In Sect. 4.4.3, we have seen how the empirical mean converges to its expectation if we draw a number of samples satisfying the Chernoff bound. Let us then consider a given value $\bar{\psi}$ and estimate the expected value $E_\Delta[u(\bar{\psi}, \Delta)]$ with its empirical mean

$$\hat{E}_n(u(\bar{\psi})) = \frac{1}{n} \sum_{j=1}^{n} u(\bar{\psi}, \Delta_j) \tag{4.22}$$

based on n i.i.d. samples $\Delta_1, \ldots, \Delta_n$. The Hoeffding inequality can then be applied and leads to

$$\Pr\left(|\hat{E}_n(u(\bar{\psi})] - E_\Delta[u(\bar{\psi}, \Delta)]| \geq \varepsilon \right) \leq 2e^{-2n\varepsilon^2} \tag{4.23}$$

from which we derived the Chernoff bound (4.23) holds for $\bar{\psi}$ but it also independently holds *for any* finite sequence of $\bar{\psi} \in \{\psi_1, \ldots, \psi_m\}$ drawn from ψ.

Moreover, we can interpret $u(\bar{\psi}, \Delta)$, as a set of functions parameterized in $\bar{\psi}$ composing the function family A.

We would appreciate the actual mean evaluated on the generic i-th sample $\hat{E}_n(u(\psi_i))$ to be close to the expected value $E_\Delta[u(\psi_i, \Delta)]$ for any ψ_i, $i = 1, \ldots, m$.

In other words, we would like the empirical mean to converge to its expectation *uniformly* as n goes to infinity and for each element of the family $A = \{u(\psi_i, \Delta), i = 1, \ldots, m\}$. When this holds we say that function family A satisfies the *Uniform Convergence of Empirical Mean* (UCEM) property. If the family A is finite (say composed of m functions) then, by repeated application of the Hoeffding inequality, we have that

$$\Pr\left(\sup_{u \in A} |\hat{E}_n(u(\psi)) - E_\Delta[u(\psi, \Delta)]| > \varepsilon \right) \leq 2me^{-2n\varepsilon^2} \tag{4.24}$$

and, when $n \to \infty$, (4.24) goes to zero. The UCEM property then holds for any finite function family. However, the property might hold also for an infinite function family, e.g. $A = \{u(\psi, \Delta), \psi \in \Psi\}$. It can be proved that the UCEM property holds for all those families for which the Pollard dimension d_P of A is finite [4].

4.4.4.1 The Pollard Dimension

Let Ψ be a measurable space and $F \subseteq [0, 1]^k$ a family of measurable functions. A set of points $\psi_1, \ldots, \psi_n$ is said to be *P-shattered* by F if there exists a real vector $c \in [0, 1]^n$ such that, for every binary vector $b \in \{0, 1\}^n$, there exists a function $f_b \in F$ such that

$$\begin{cases} f_b(\psi_i) < c_i & \text{if } b_i = 0 \\ f_b(\psi_i) \geq c_i & \text{if } b_i = 1 \end{cases}$$

The Pollard dimension d_P of F is the largest integer n for which there exists a set of cardinality n P-shattered by F [4].

To better understand the concept of P-shattered consider a real vector $c \in [0, 1]^n$ and the generic point ψ_i. For each function $f \in F$ we have that $f(\psi_i)$ can be larger (or equal) or smaller than value c_i. Then there are 2^n possible behaviors as f varies in F. Set $\psi_1, \ldots, \psi_n$ is said to be *P-shattered* by F if *each* of the possible 2^n behaviors is realized by some $f \in F$.

The d_P is a generalization of the Vapnik–Chervonenkis (VC) dimension defined on binary valued functions F. Moreover, for binary valued functions $d_P = d_{VC}$ where d_{VC} is the VC-dimension.

When the Pollard's dimension is known, we can state the important Corollary [2]:

The minimum expectation problem. Corollary:

Let $u(\psi, \Delta) \in [0, 1]$ be a performance function measurable according to Lebesgue on its domains $\Psi \subseteq \mathbb{R}^l$ and $D \subseteq \mathbb{R}^k$, onto which are defined the random variables ψ and Δ, respectively, with probability density functions f_Ψ and f_Δ. Let d_P of function $u(\cdot)$ be finite.
Draw m i.i.d. samples $\psi_1, \ldots, \psi_i, \ldots, \psi_m$ from ψ and n i.i.d. samples $\Delta_1, \ldots, \Delta_j, \ldots \Delta_n$ from Δ and compute

$$\hat{E}_n(u(\psi)) = \frac{1}{n}\sum_{j=1}^{n} u(\psi, \Delta_j)$$

$$\hat{u}_{\min} = \min_{i=1,\ldots,m} E_n[u(\psi_i)]$$

then,

$$\Pr\left(\Pr\left(E_\Delta[u(\psi, \Delta)] \le \hat{u}_{\min} - \varepsilon_1\right) \le \varepsilon_2\right) \ge 1 - \delta$$

holds for any accuracy level $\varepsilon_1, \varepsilon_2 \in (0, 1)$, confidence $\delta \in (0, 1)$ provided that

$$m \ge \frac{\ln \frac{2}{\delta}}{\ln(\frac{1}{1-\varepsilon_2})}$$

and

$$n \ge \frac{32}{\varepsilon_1^2}\left[\ln \frac{16}{\delta} + d_P \left(\ln \frac{16e}{\varepsilon_1} + \ln \frac{16e}{\varepsilon_1}\right)\right]$$

Value $\hat{u}_{\min}$ is the probabilistic outcome of the algorithm.

Instead, when the Pollard dimension is not know we can use the main result given in the following theorem [2]

Table 4.3 The number of samples n, $m = g(\varepsilon, \delta)$

$\varepsilon_1 = \varepsilon_2 = \varepsilon$	$\varepsilon = 0.05, \delta = 0.02$	$\varepsilon = 0.05, \delta = 0.01$	$\varepsilon = 0.02, \delta = 0.01$	$\varepsilon = 0.01, \delta = 0.01$
(m, n)	(89, 1960)	(104, 2126)	(263, 14451)	(528, 61296)

The minimum expectation problem. Theorem:

Let $u(\psi, \Delta) \in [0, 1]$ be a performance function measurable according to Lebesgue on its input domains $\Psi \subseteq \mathbb{R}^l$ and $D \subseteq \mathbb{R}^k$, onto which are defined the random variables ψ and Δ, respectively, with probability density functions f_Ψ and f_Δ.
Draw m i.i.d. samples $\psi_1, \ldots, \psi_i, \ldots, \psi_m$ from ψ and n i.i.d. samples $\Delta_1, \ldots, \Delta_j$, $\ldots \Delta_n$ from Δ, compute

$$\hat{E}_n(u(\psi)) = \frac{1}{n}\sum_{j=1}^{n} u(\psi, \Delta_j)$$

$$\hat{u}_{\min} = \min_{i=1,\ldots,m} E[u(\psi_i)]$$

then,

$$\Pr\left(\Pr\left(E_\Delta[u(\psi, \Delta)] \leq \hat{u}_{\min} - \varepsilon_1\right) \leq \varepsilon_2\right) \geq 1 - \delta$$

holds for any accuracy level $\varepsilon_1, \varepsilon_2 \in (0, 1)$, confidence $\delta \in (0, 1)$ provided that

$$m \geq \frac{\ln \frac{2}{\delta}}{\ln(\frac{1}{1-\varepsilon_2})}$$

and

$$n \geq \frac{1}{2\varepsilon_1^2} \ln \frac{4m}{\delta}$$

Value $\hat{u}_{\min}$ is the probabilistic outcome of the algorithm.

Comments

We see from Table 4.3 that the required number of samples can be very high depending on the selected accuracy and confidence levels since the number of samples n is function of the number of samples m, yet through alogarithm.

However, the number of samples required by the corollary is significantly higher than those requested by the theorem. For instance, if we choose $\varepsilon_1 = \varepsilon_2 = \varepsilon = 0.02$

Algorithm 10: Randomized algorithm for the minimum expectation problem

1- The probabilistic problem requires to estimate $\min E[u(\psi, \Delta)]$;
2- Identify the input spaces Ψ, D and random variables ψ and Δ with probability density function f_ψ over Ψ and f_Δ over D, respectively;
3- Select the accuracy ε and the confidence δ levels;
4- Draw $m \geq \frac{\ln \frac{2}{\delta}}{\ln(\frac{1}{1-\varepsilon})}$ i.i.d. samples $\psi_1, \ldots, \psi_i, \ldots, \psi_m$ from ψ;
5- Draw $n \geq \frac{1}{2\varepsilon^2} \ln \frac{4m}{\delta}$ i.i.d. samples $\Delta_1, \ldots, \Delta_j, \ldots \Delta_n$ from Δ according to f_Δ;
6- Compute $\hat{u}_{\min}(\psi_i) = \frac{1}{n} \sum_{j=1}^{n} u(\psi_i, \Delta_j)$ for each i;
7- use $\hat{u}_{\min} = \min_{i=1,\ldots,m} \hat{u}_{\min}(\psi_i)$ and $\hat{\psi} = arg \min_{i=1,\ldots,m} \hat{u}_{\min}(\psi_i)$;

and $\delta = 0.01$ then $m = 263, n = 14,451$ from the Theorem and $m = 263, n = 1,367,851$ from the Corollary with the easiest (yet unlikely) dimension $d_P = 1$. For this reason, we surely use the Theorem's results in the Randomized algorithm framework, mostly with the choice $\varepsilon_1 = \varepsilon_2 = \varepsilon$.

The randomized algorithm for solving the minimum expectation problem is finally summarized in Algorithm 10.

4.5 Controlling the Statistical Volume of the Sampling Space

Randomization requests to sample from a given space Ψ and a random variable with probability density function f_ψ defined over Ψ. By acting on some controlling parameter of f_ψ, we can tune the statistical volume Ψ defined as

$$\text{Vol}(\Psi) = \int_\Psi f_\psi d\Psi$$

which is a very useful operation in many applications. For instance, if we wish to control the space of uncertainty affecting a computation we find useful to introduce a control parameter that allows the shrinkage/enlargement of the space. A norm applied to the vector is a first element that can control it. Another possibility—which can be related to the norm—is the introduction of a mechanism controlling the scattering of points in the space. For their nature, the variance for a scalar and the covariance matrix for a vector can control effectively the statistical volume of a space: the larger the scattering index the larger the embraced volume.

If $\Psi \subset \mathbb{R}^l$, it is common to describe it either in terms of a controllable hypercube or a controllable ball onto which ϕ is defined with pdf f_ψ (both situations can be managed by introducing the concept of norm). In the former case, a common description is such that each component $\psi(i)$ of ψ belongs to a bounded interval, i.e., $\psi(i) \in [a_i, b_i]$. Here, the control of the volume is on a_i and b_i. If we set identical and symmetrical values for a_i and b_i so that $a_i = -\rho, b_i = \rho$, then we have that each edge of the hypercube has length 2ρ and Ψ can be controlled in expansion and contraction with the single parameter ρ and $\Psi = \Psi(\rho)$.

Algorithm 11: The algorithm for extracting vectors according to a uniform distribution from a l_p norm-ball

1- Generate l independent random real scalars ξ_i distributed according to the generalized gamma density function

$$G(x) = \frac{p}{\Gamma(\frac{1}{p})} e^{\xi^p}, \xi \geq 0,$$

where Γ is the gamma function and p the norm value.

2- Construct random vector $x \in \mathbb{R}^l$ of components $x_i = s_i \xi_i$ where s_i is a random sign. Random vector $y = \frac{x}{\|x\|_p}$ is uniformly distributed on the boundary of $\mathbb{B}_\rho$.

3- Return $\psi = \rho y w^{\frac{1}{l}}$, where w is a random variable uniformly distributed in $[0, 1]$

We recall that if we have a uniform distribution defined in the $[-\rho, \rho]$ interval, the variance is $\frac{\rho^2}{3}$, and the control of ρ implies a control in variance. This situation is formalized by the $\|\psi\|_\infty$ norm

$$\|\psi\|_\infty = \max \{|\psi(1)|, |\psi(2)|, \cdots, |\psi(l)|\}$$

being $\psi(i)$ the i-th component of vector ψ. Following the definition, $\|\psi\|_\infty = \rho$ induces a hypercube of edge 2ρ. In the latter case, e.g., the norm-ball controlled case, ψ is restricted within $\Psi(\rho)$ described in terms of norm-bounded balls of radius ρ

$$\Psi(\rho) = \{\|\psi\|_p \leq \rho\}$$

where

$$\|\psi\|_p = \left(\sum_{i=1}^{l} |\psi(i)|^p\right)^{\frac{1}{p}}.$$

In general, the L^2-norm is used but other norms can be considered to bound Ψ and having it controlled as $\Psi(\rho)$. Interestingly, the maximum norm $\|\psi\|_\infty$ is the limit of the $\|\psi\|_p$ norm when $p \to \infty$.

Though a uniform distribution sample extraction algorithm is immediate for a $\|\psi\|_\infty$ norm where we simply need to uniformly sample from each axis, the problem is more complex if we wish to generate a uniform sampling from a norm-bounded ball. Clearly, verifying the appurtenance of a sample to the ball as we did when estimating π with the square-circle mechanism of Sect. 4.2.1, instead of a hypercube is not an effective solution. Fortunately, [3] provides a simple algorithm that returns a sample ψ belonging to a ball $\mathbb{B}_\rho$

$$\mathbb{B}_\rho = \Psi(\rho) = \{\psi \in \Psi : \|\psi\|_p \leq \rho\}.$$

The algorithm is given in Algorithm 11. Interestingly, if we arrest the algorithm to the second step, we obtain a sample that is uniformly distributed on the boundary $\|\psi\|_p = \rho$.

If $\Psi = \mathbb{R}^l$ and a multivariate probability density function f_ψ is defined for ψ, say Gaussian, then we can control the statistical volume by acting on the covariance matrix C_ψ. The interested reader can refer to [2] for a deeper investigation.

Chapter 5
Robustness Analysis

Robustness is the property of being strong and healthy in constitution. When we transpose it to a system, it refers to the ability of tolerating perturbations that might affect the system's functional body. As such, a robust system will be able to somehow resist to a set of perturbations by providing a graceful loss in performance. When we consider an embedded system, perturbations are associated with either a physical realization of the device (e.g., fluctuations introduced by the production process in an analog implementation) or finite precision representation of structural parameters defining the computational flow (e.g., truncation operators and lookup tables in a digital implementation). Faults and aging phenomena represent other examples of perturbations. If the application is robust enough then its porting on the embedded system will be effective and the ability to guarantee a quality of service over some time is granted despite the presence of aging effects; in both cases the performance loss is kept within a tolerated margin.

In the following we formalize at first the robustness analysis problem. We then study the effect of perturbations affecting a computational flow and quantify the induced impact on the chosen figure of merit. In particular, the *Robustness in the small* problem is addressed, where it is assumed that perturbations are small in magnitude. The "small" magnitude request is hard to be verified since small or large depends on the specific problem. However, the assumption allows us to provide results in a closed form thanks to the more amenable mathematics. Conversely, we say that we address a *Robustness in the large* problem when no assumptions are made about the magnitude of perturbations, which can be either small or large.

C. Alippi, *Intelligence for Embedded Systems*, DOI: 10.1007/978-3-319-05278-6_5,

5.1 Problem Formalization

Consider a system/application described by a Lebesgue measurable function $g(\theta, x) \in \mathbb{R}$ depending on the column parameter vector $\theta \in \Theta \subset \mathbb{R}^d$ and the column input vector $x \in X \subset \mathbb{R}^l$ and its perturbed version $g(\theta, \delta\theta, x) \in \mathbb{R}$ depending also on the perturbation $\delta\theta \in \Delta \subseteq \Theta$.[1]

5.1.1 Robustness

We say that $g(\theta, x)$ is robust with respect to perturbations $\delta\theta \in \Delta \subseteq \Theta$ at level $\gamma \in \mathbb{R}^+$ when, given a discrepancy function $u\,(g(\theta, x), g(\theta, \delta\theta, x)) \in \mathbb{U} \subset \mathbb{R}$ the system experiences a degradation in performance within γ i.e.,

$$u\,(g(\theta, x), g(\theta, \delta\theta, x)) \leq \gamma, \quad \forall \delta\theta \in \Delta, \forall x \in X. \tag{5.1}$$

In some relevant cases the set X is the discrete one $X = \tilde{X}$ and contains a finite number of input instances. For instance, this happens when we have a limited number of data or signals and we wish to estimate the robustness level of the application constrained to available set $\tilde{X}$. When this is the case the (5.1) becomes

$$u\,(g(\theta, x), g(\theta, \delta\theta, x)) \leq \gamma, \quad \forall \delta\theta \in \Delta, \forall x \in \tilde{X}. \tag{5.2}$$

We commented above that perturbations on parameters can be intended as uncertainty affecting parameters, e.g., introduced by their analog implementation within an embedded hardware. At the same time, we might wish to port an algorithm designed with parameters represented in double precision to a fixed point representation microprocessor. Although this is not a robustness problem (the perturbation is in fact fixed for a given architecture) and the problem should be addressed with the PACC framework of Chap. 7, we have that if the application is designed to be robust then the application porting will be successful provided that the induced loss on the figure of merit is below a tolerated value. In other words, a robust enough computation will be able to address also the specific perturbation introduced by the limited precision architecture.

The issue is much more relevant than it might appear. In fact, if the application is not robust, perturbations affecting the parameters, say of a nonlinear model, though small, can introduce a significant change in the function's behavior as well as a drastic fall in performance. All scholars having trained a neural network model with a high precision machine and ported to an embedded-less precise-hardware have experienced the dramatic loss in accuracy associated with this operation unless the

[1] We should consider that the perturbed parameter vector also belongs to Θ. From now on we assume that this condition is also satisfied.

network was trained to be robust. Evaluation of the robustness level possessed by an application is requested in a vast scenario of applications and research areas, from model estimation [113] to robust computation [114, 115] and control [2, 116] to cite a few examples. When electronic devices are taken into account we have different sources of perturbations. Without pretending to be exhaustive, we encounter stochastic variations in the production process, e.g., [119], permanent and transient faults affecting the electronics, finite precision data representation, and processing in digital devices [37, 117, 118] and aging effects in analog ones [120].

In Sect. 4.5 we show how it is possible to control the space of uncertainty affecting a computation by acting on a suitable norm for Δ. Likewise, we can use the same framework to describe and control the perturbation space $\Delta = \Delta(\rho)$ either in terms of a hypercube whose edge length is modulated by a single positive real parameter ρ or by shaping $\Delta(\rho)$ through a norm-bounded ball of radius ρ. By operating on parameter ρ we enlarge/shrink the volume of the perturbation space. When $\rho = 0$ the perturbation space disappears and the system degenerates into its nominal perturbation-free description.

A robustness analysis aims at evaluating different aspects:

- **The performance loss verification problem** A perturbation space Δ or $\Delta(\rho)$ is given as well as a tolerated performance loss level γ. We wish to verify if (5.1) or (5.2) is satisfied for function $g(\theta, x)$. Whenever they are not satisfied we might be interested in determining the level of satisfaction by computing the percentage of points in X (or $\tilde{X}$) satisfying the relationship, i.e.,

$$\frac{\int_{\Delta,X} I(\delta, x) \mathrm{d}\delta \mathrm{d}x}{\int_{\Delta,X} \mathrm{d}\delta \mathrm{d}x}$$

  where $I(\delta, x)$ is the indicator function

$$I(\delta, x) = \begin{cases} 1 & \text{if} \quad u\left(g(\theta, x), g(\theta, \delta\theta, x)\right) \leq \gamma \\ 0 & \text{otherwise} \end{cases}$$

- **The evaluation of the robustness level problem** A perturbation space Δ is given and we wish to determine the minimum γ granting (5.1) or (5.2) to hold. In this case, γ provides the robustness index of the application.
- **The robustness function problem** Not rarely, we iterate the robustness level problem by acting on $\Delta(\rho)$ through ρ and evaluate function $\gamma(\rho)$ which provides the robustness profile for the application.

5.1.2 Robustness at the Computational Flow Level

It should be commented that the formalization given in (5.1) and (5.2) addresses a more complete problem than that of evaluating the effects of perturbations solely

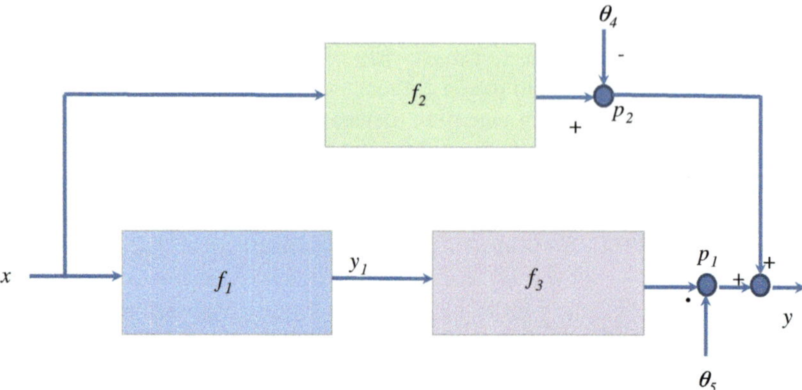

Fig. 5.1 Decomposition of function $g(\theta, x)$. Parameter vector θ of function $g(\cdot, \cdot)$ contains both parameters defining the operational modality of sub-functions (here f_1, f_2, f_3) and those controlling the subsequent insurgence of perturbations, i.e., θ_4 and θ_5. The nominal perturbation-free case is characterized by the choice of functional parameters $\theta_4 = 0$ (the perturbation operates with an additive model on the computational flow) and $\theta_5 = 1$ (the perturbation, once activated, operates according to a multiplicative model)

affecting the parameters of a function (e.g., the weights of a neural network or the parameters of a linear time invariant dynamic model in a predictive form). An appropriate use of parameters can, in fact, allow us to deal with perturbations acting on a computational flow. To ease the understanding let us consider the functional flow depicted in Fig. 5.1 where function $g(\theta, x)$

$$g(\theta, x) = f_2(\theta_2, x) - \theta_4 + f_3\left(\theta_3, f_1(\theta_1, x)\right) \theta_5$$

is partitioned in sub-functions $y_1 = f_1(\theta_1, x)$, $f_2(\theta_2, x)$, $f_3(\theta_3, y_1)$ (inputs and partial outcomes of the processing must be intended as defined in the appropriate real spaces). The partitioning in subsystems depends on the nature of the computational flow, on design issues, and where the robustness analysis must be carried out. In the example, there are two perturbation injection points p_1 and p_2 controlled by parametric values θ_4, θ_5 defined in a suitable space and initially set to zero and one so that their influence on the computation is null. The vector θ is then $\theta = [\theta_1, \theta_2, \theta_3, \theta_4, \theta_5]$. The introduction of parameters θ_4, θ_5 allows us to address sources of perturbations in the computational flow which are silent during the nominal perturbation-free case and become active when perturbations are introduced. Initial values for those parameters must be set so that their contribution is neutral in the nominal conditions. For instance, if we are interested in evaluating the robustness level of the computation w.r.t point p_1 we need to solve the (5.1) by selecting a perturbation $\delta\theta$ affecting parameter θ_4 only. Then, the perturbation must have the structure $\delta\theta = [0, 0, 0, \delta\theta_4, 0]$ and $\Delta(\rho)$ is derived accordingly.

5.2 Robustness in the Small

This section addresses the robustness in the small problem. The operational framework is that of Sect. 5.1 where we also assume that $g(\theta, x)$ is differentiable twice w.r.t. column parameter vector θ and that the magnitude of the perturbation $\delta\theta$ is small.

5.2.1 *Evaluating the Impact of Small Perturbations at the Function Output*

At first we wish to evaluate the pointwise effect of a small perturbation affecting the parameter vector on the function output. If $g(\theta, \delta\theta, x)$ is the value of the perturbed function, then the small perturbation hypothesis allows us to determine its impact on the perturbed output $u(g(\theta, x), g(\theta, \delta\theta, x))$. For instance, consider the additive perturbation mechanism and $u(g(\theta, x), g(\theta, \delta\theta, x)) = g(\theta+\delta\theta, x) - g(\theta, x) = \delta g$. By expanding with Taylor the perturbed function $g(\theta + \delta\theta, x)$ around the perturbation-free parameter vector θ, we obtain

$$g(\theta + \delta\theta, x) = g(\theta, x) + \left.\frac{\partial g}{\partial \theta}^T\right|_\theta \delta\theta + \frac{1}{2}\delta\theta^T \left.\frac{\partial^2 g}{\partial \theta^2}\right|_\theta \delta\theta + O(\delta\theta^T \delta\theta).$$

By neglecting higher order terms, the induced perturbation at the function output becomes

$$\delta g = \left.\frac{\partial g}{\partial \theta}^T\right|_\theta \delta\theta + \frac{1}{2}\delta\theta^T \left.\frac{\partial^2 g}{\partial \theta^2}\right|_\theta \delta\theta. \tag{5.3}$$

As expected, the perturbation depends on the local geometry of the parameter space through the gradient $\left.\frac{\partial g}{\partial \theta}\right|_\theta$ and the hessian matrix $\left.\frac{\partial^2 g}{\partial \theta^2}\right|_\theta$. No more can be said at this level unless a priori information or extra assumptions are taken into account about the nature of the perturbation or the chosen function $g(\cdot, \cdot)$. For instance, if we keep only the linear term in the expansion, we can repeat the derivations carried out in Chap. 3 when assessing the stochastic properties (mean value, variance, and pdf where applicable) of uncertainty affecting the inputs of a linear system. Clearly, here the random variable to be considered is $\delta\theta$ (ruled by a probability density function $f_{\delta\theta}$ and defined over Δ) and x (to which is associated pdf f_x over X). This is an easy exercise we leave to the reader.

5.2.2 *Perturbations at the Empirical Risk Level*

The case where the parameters of the function $g(\theta, x)$ are derived from a learning process carried out by adopting an effective gradient-based learning algorithm is of particular relevance. We might need to run the algorithm several times due to the presence of local minima in the empirical risk function V_N before obtaining a good approximating model $f(\hat{\theta}, x)$, (see Sect. 3.4.1). We know that the parameter vector $\hat{\theta}$ belongs to the neighborhood of the unknown local optimal parameter θ° as seen in Chap. 3.

If we now select function $g(\theta, x)$ to be the empirical risk defined on the training set Z_N, an additive perturbation and $u\,(g(\theta, x), g(\theta + \delta\theta, x)) = \delta V_N$, then we can evaluate the effects that a perturbation affecting the parameters $\hat{\theta}$ induce on the empirical risk. Since V_N is based on Z_N, space X is constrained so that xs to be considered are those in Z_N, i.e., $\tilde{X} = Z_N$. We have that the linear term

$$\left.\frac{\partial V_N}{\partial \theta}^T\right|_{\hat{\theta}} \delta\theta$$

in (5.3) is null (the training algorithm converged to a minimum of V_N). The variation induced at the function output δg depends then only on the quadratic form

$$\delta V_N = \frac{1}{2}\delta\theta^T \left.\frac{\partial^2 V_N}{\partial \theta^2}\right|_{\hat{\theta}} \delta\theta. \tag{5.4}$$

Moreover, if we assume V_N to be the Mean Squared Error (MSE) $V_N = \frac{1}{N}\sum_{i=1}^{N}(y_i - f(\theta, x_i))^2$ and defined $e(x) = y - f(\theta, x)$, then term $\frac{\partial^2 V_N}{\partial \theta^2}|_{\hat{\theta}}$ of (5.4) becomes

$$\left.\frac{\partial^2 V_N}{\partial \theta^2}\right|_{\hat{\theta}} = \frac{1}{N}\sum_{i=1}^{N} \left.\frac{\partial^2 e(x_i)^2}{\partial \theta^2}\right|_{\hat{\theta}}$$

with

$$\left.\frac{\partial^2 e(x)^2}{\partial \theta^2}\right|_{\hat{\theta}} = 2\left.\frac{\partial f(\theta, x)}{\partial \theta}\right|_{\hat{\theta}} \frac{\partial f(\theta, x)^T}{\partial \theta}|_{\hat{\theta}} - 2e(x)\left.\frac{\partial^2 f(\theta, x)^T}{\partial \theta^2}\right|_{\hat{\theta}}.$$

We now introduce the quasi-Newton approximation stating that term

$$e(x)\left.\frac{\partial^2 f(\theta, x)^T}{\partial \theta^2}\right|_{\hat{\theta}}$$

is negligible. This happens when the pointwise error is very small for all the x in the training set ($e(x) \simeq 0$) or the local curvature of $\frac{\partial^2 V_N}{\partial \theta^2}|_{\hat{\theta}}$ around the minimum $\hat{\theta}$ can be nicely approximated with a quadratic semidefinite positive form. Under the quasi-Newton assumption δV_N degenerates to the quadratic form

$$\delta V_N = \delta\theta^T H \delta\theta = \text{trace}\left(H\delta\theta\delta\theta^T\right)$$

where trace is the trace operator and H is the semidefinite positive matrix

$$H = \frac{1}{N}\sum_{i=1}^{N} \frac{\partial f(\theta, x_i)}{\partial \theta}\bigg|_{\hat{\theta}} \frac{\partial f(\theta, x_i)^T}{\partial \theta}\bigg|_{\hat{\theta}}.$$

Since we have a quadratic semidefinite positive form by construction, *any* perturbation $\delta\theta$ affecting the parameter vector $\hat{\theta}$ will introduce a loss in approximation performance on Z_N and V_N will not decrease. This means that the training error will not necessarily decrease following a perturbation and is likely to increase.

This is exactly the situation scholars found when they wanted to port a neural network configured in a high precision platform to an embedded system characterized by lower precision.

By recalling that here $\tilde{X} = Z_N$, and given the above, problem (5.2) can be rewritten as

$$\delta\theta^T H \delta\theta \leq \gamma, \quad \forall \delta\theta \in \Delta, \forall x \in Z_N. \tag{5.5}$$

Since the performance loss verification problem aims at verifying the level of satisfaction of (5.5) given a perturbation space $\Delta(\rho)$ and a given tolerated performance loss level γ, its solution can be easily obtained by invoking randomized algorithms according to Algorithm 6. We recall that if the pdf $f_{\delta\theta}$ is unknown, we can consider the uniform distribution for its worst case characteristics.

We now solve the "evaluation of the robustness level problem." Again perturbation space Δ is given and we wish to determine the minimum γ granting the (5.5) to hold.

This problem can be reformulated by looking for the maximum value the perturbed empirical risk can assume following an arbitrary perturbation $\delta\theta \in \Delta$. The maximum value is obtained when the perturbation $\delta\theta$ is a vector parallel to the eigenvector of the matrix H associated with the largest eigenvalue $\lambda_{\max}(H)$

$$\max(\delta V_N) = \|(H)\|_2 \max(\|\delta\theta\|^2) = \lambda_{\max}(H)\rho^2$$

the maximum error depends again on the geometry of the space and the intensity of the perturbation ($\|\delta\theta\|^2$ is the squared magnitude of the perturbation). Note that $\|\delta\theta\|^2 = \rho^2$ both for $\|\delta\theta\|_2$ and $\|\delta\theta\|_\infty$ norms.

The hard problem associated with the evaluation of the maximum error can also be solved by considering the probabilistic solution of the maximum value estimation problem with randomized algorithms according to Algorithm 8.

Another problem we might be interested in requires the evaluation of the expected value the perturbed empirical risk assumes, i.e., $u\left(g(\theta, x), g(\theta + \delta\theta, x)\right) = E_{\delta\theta}[\delta V_N]$. Although the problem can be solved with randomized algorithms by invoking Algorithm 9, we can solve it in a closed form under some hypotheses. Let us assume that perturbation $\delta\theta$ is an i.i.d. random variable of zero mean with all components characterized by identical variance $\sigma^2_{\delta\theta}$. This is exactly the case we encounter when we implement function $f(\hat{\theta}, x)$ with an analog representation where each parameter is subject to a fluctuation induced by the production process. For instance, if we implement a generic parameter with a resistor, the parameter value would be chosen as the nominal value. However, the production process will generate resistors for such a parameter ruled by a Gaussian distribution of mean value centered in the nominal one and a standard deviation set, e.g., to one-third of the component tolerance (refer also to Sect. 2.1.4 where the tolerance of sensors was defined).

Following this comment we can study the behavior of a family of functions $f(\hat{\theta}, x)$ generated by the production process. Since

$$\delta V_N = \delta\theta^T H \delta\theta = \text{trace}\left(H \delta\theta\delta\theta^T\right)$$

by taking expectation w.r.t. $\delta\theta$ to account for randomness on $\delta\theta$, we have that

$$E_{\delta\theta}[\delta V_N] = E_{\delta\theta}[\delta\theta^T H \delta\theta] = \text{trace}\left(H E_{\delta\theta}\left[\delta\theta\delta\theta^T\right]\right).$$

If we assume that components of $\delta\theta$ are i.i.d. with the same variance $\sigma^2_{\delta\theta}$, then

$$E_{\delta\theta}[\delta V_N] = \sigma^2_{\delta\theta}\,\text{trace}(H) = \sigma^2_{\delta\theta} \sum_{i=1}^{d} \lambda_i(H) \quad (5.6)$$

where $\lambda_i(H)$ is the i-th eigenvalue of the H matrix. The expected value of the increment in the empirical risk $E_{\delta\theta}[\delta V_N]$ is function of the intensity of the perturbation $\sigma^2_{\delta\theta}$ and the local geometry of the empirical risk around $\hat{\theta}$.

Comments

We learned from Sect. 3.4.1 that the ultimate goal of learning is to provide a model that not only fits past data but also provides good performance on unseen patterns. In this subsection we have addressed the case where the figure of merit we are looking at is the empirical risk V_N and not the structural risk $\bar{V}(\theta)$ (situation addressed in the next subsection). In fact, in real applications, it is not rare the case where we have all the data needed to learn a nonlinear function for which an explicit relationship does not exist. Having many data for training means that N tends to the asymptotes and the theory tells us that the empirical risk becomes a good estimate of the structural one. Then, the identified model is the one to be used in our application and implemented in the embedded system. V_N is providing a good performance estimate that, basically, coincides with the inherent risk (the approximation and the estimation risk are negligible here provided that the training procedure is effective and a universal approximation neural network is considered).

Porting the model to the embedded system requires a robustness analysis, exactly the case we investigated in this section.

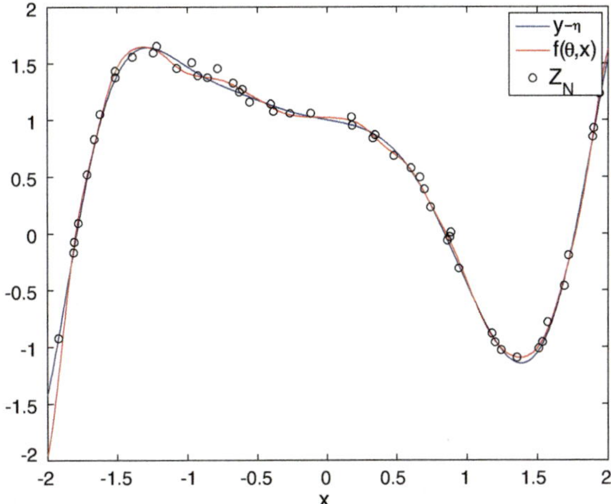

Fig. 5.2 The training data Z_N (*circles*), the noise-free function $y - \eta$, and the learned function $f(\theta, x)$. The hidden layer has 10 hidden units

5.2.2.1 Example (1:4): Learning a Neural Network

Consider the problem of learning an unknown nonlinear regression function from data and port it to an embedded system. Only the parameters of the function are affected by perturbations. The unknown reference function generating the data is

$$y = -x \sin(x^2) + \frac{e^{-0.23x}}{1 + x^4} + \eta$$

where $\eta \sim \mathcal{N}(0, 0.05) \cdot N = 50$ training data have been drawn from interval $X = [-2, 2]$ according to a uniform distribution. The y values were evaluated accordingly.

A feedforward neural network with three layers (the input, the hidden nonlinear one with neurons characterized by a hyperbolic tangent activation function, and the linear output layer) was consider to learn the function based on the available training set Z_N. Training was carried out with a Levenberg-Marquardt algorithm and led to $f(\hat{\theta}, x)$. Figure 5.2 shows the noise-free and the neural network output as well as the considered training data. We appreciate the fact that the learned neural network approximates well the unknown function.

The perturbation in the small analysis was carried out by perturbing the neural network parameter vector according to an additive model $\theta + \delta\theta$ affecting both weights and biases. The generic i-th component of vector $\delta\theta$ is drawn from a uniform distribution $U(-\rho, \rho)$, being ρ a positive real scalar controlling the intensity of the perturbation. We consider the expected value $E_{\delta\theta}[\delta V_N]$ of (5.6) to be the figure of merit chosen to evaluate the robustness property for the neural network

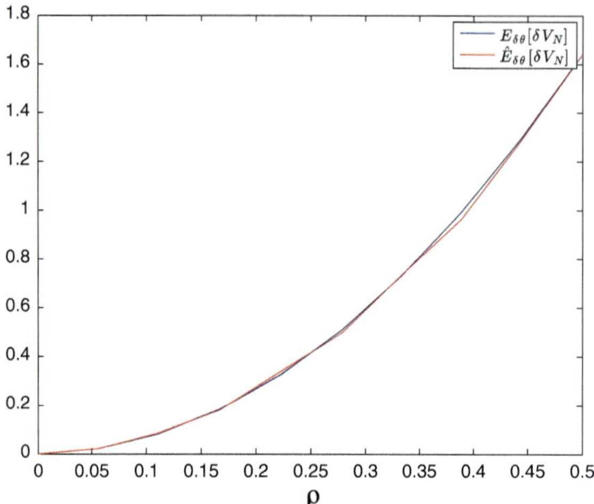

Fig. 5.3 $E_{\delta\theta}[\delta V_N]$ estimated according to (5.7) compared with $\hat{E}_{\delta\theta}[\delta V_N]$ estimated according to (5.8). The number of samples granting convergence of $\hat{E}_{\delta\theta}[\delta V_N]$ to $E_{\delta\theta}[\delta V_N]$ was chosen to satisfy the Chernoff bound

$$E_{\delta\theta}[\delta V_N] = \frac{\rho^2}{3} \sum_{i=1}^{d} \lambda_i(H) \tag{5.7}$$

since $\sigma^2_{\delta\theta} = \frac{\rho^2}{3}$. Equation (5.7) tells us that the expected increase in V_N is linear with the variance of the perturbation and quadratic in ρ, the parameter controlling the intensity of the perturbation.

Figure 5.3 shows the $E_{\delta\theta}[\delta V_N]$ estimated according to (5.7) and

$$\hat{E}_{\delta\theta}(\delta V_N) = \frac{1}{n} \sum_{j=1}^{n} \left(V_N(\theta + \delta\theta_j) - V_N(\theta) \right). \tag{5.8}$$

The number of samples $n = 4612$ was chosen according to the Chernoff bound ($\delta = 0.05$, $\varepsilon = 0.02$, Algorithm 9) to grant that discrepancy between $|E_{\delta\theta}[\delta V_N] - \hat{E}_{\delta\theta}[\delta V_N]| < \varepsilon$ with probability $1 - \delta = 0.95$.

The two curves basically coincide until $\rho = 0.5$ and start diverging around $\rho = 2.7$ implying that the small perturbation hypothesis is no more valid from thereon. Two comments can be made. The first refers to the fact that after $\rho = 2.7$ V_N assumes very high values, e.g., 50 and above. In such a circumstance it does not even make sense to speak about loss in performance since the network behaves as a totally different one. In addition, Fig. 5.3 evidences how the theory represented by Eq. (5.7) well estimates the real behavior (5.8). The figure shows that the loss in performance on V_N ($V_N = 1.2 \times 10^{-4}$ after training) in the nominal error-free condition is acceptable only for very small values of ρ, say $\rho < 0.05$. For larger values V_N increases with a behavior quadratic in ρ as pointed out by (5.7).

5.2.3 Perturbations at the Structural Risk Level

Despite the small-perturbation hypothesis, investigation of the effect of perturbations on the structural risk is a difficult issue that can be solved in a closed form only by assuming some additional hypotheses. Although this might appear an unnatural exercise, we anticipate this is not. In fact, even if the assumed assumptions might be strong they allow us to open a view over the rather complex theory behind the robustness problem: results obtained in closed form showing the relationship between accuracy performance and perturbations set the basis for future research.

In the following, we assume that our application can be modeled as the unknown nonlinear function $g(\theta^\circ, x)$ belonging to a known model family $f(\theta, x)$ so that there exists a unique θ° for which $g(\theta^\circ, x) = f(\theta^\circ, x)$. Moreover, we assume that the system model provides data to be used for model configuration according to the additive signal plus noise model $y = f(\theta^\circ, x) + \zeta$, where ζ is a scalar i.i.d. random variable of zero mean and unknown σ_ζ^2 variance.

The assumption that the system model is complete, i.e., the process generating the data belongs to the envisaged model family used for model identification, permits us to forget about the existence of the approximation and estimation errors contributing to the structural risk and focus instead on the relationship between perturbations and accuracy.

We know that the completeness assumption is a strong one if we have a limited data set, but becomes reasonable when the number of data increases and function family $f(\theta, x)$ is a universal function approximator.

By following the approach delineated in [50] we consider a squared loss figure of merit for the structural risk and a mean squared one for empirical one

$$\bar{V}(\theta) = \frac{1}{2} E\left[(y - f(\theta, x))^2\right]$$

$$V_N(\theta, Z_N) = \frac{1}{2N} \sum_{i=1}^{N} (y_i - f(\theta, x_i))^2.$$

Note that this formulation is slightly different from the one used in Sect. 3.4.1 for the introduction of scaling factor $\frac{1}{2}$ which simplifies subsequent derivations.

Denote by $\hat{\theta}$ the parameter vector obtained by minimizing $V_N(\theta, Z_N)$ with an efficient gradient-based algorithm. We know from Chap. 3 that θ° is the value minimizing the structural risk and assume that $\hat{\theta}$ belongs to its neighborhoods.

By considering a Taylor expansion for both structural and empirical risks, assuming that inputs and noise are independent variables, neglecting higher order terms and adopting a quasi-Newton approximation for the Hessian, we obtain that [50, 122, 123],

$$E\left[V_N(\hat{\theta}, Z_N)\right] = \bar{V}(\theta^\circ) - \frac{\sigma_\xi^2 \bar{p}}{2N} \tag{5.9}$$

$$E\left[\bar{V}(\hat{\theta})\right] = \bar{V}(\theta^\circ) + \frac{\sigma_\xi^2 \bar{p}}{2N}. \tag{5.10}$$

Expectation is here taken with respect to all possible sets Z_N we could generate with N supervised couples.

$$\bar{p} = \text{rank}\left(\left.\frac{\partial^2 \bar{V}(\theta)}{\partial \theta^2}\right|_{\theta^\circ}\right)$$

is the effective number of parameters used by the model to solve the learning problem and does not necessarily coincide with the sum of available parameters. When the model is linear and complete, $\bar{p}$ coincides with the VC dimension minus one. It is expected that $\bar{p}$ is somehow related to the VC dimension also in the nonlinear case but no results are available yet.

The above relationships have an intriguing meaning worth further discussion. We comment that $\bar{V}(\theta^\circ)$ is the structural risk, whereas $E\left[\bar{V}(\hat{\theta})\right]$ represents the expected accuracy performance associated with all possible Z_N sets, each of which is evaluated on the obtained training parameter vector. $E\left[V_N(\hat{\theta}, Z_N)\right]$ represents instead the expected empirical risk evaluated over all the possible realizations of set Z_N. We have that the expected validation error is larger than the structural risk of term $\frac{\sigma_\xi^2 \bar{p}}{2N}$, whereas the expected empirical risk is lower than the structural risk of the same term. In other words, (5.9) and (5.10) state that the expected training error is an optimistic estimate of the true test error.

By estimating the variance of the noise, we obtain

$$E\left[\bar{V}(\hat{\theta})\right] \simeq E\left[V_N(\hat{\theta}, Z_N)\right] \frac{N + \bar{p}}{N - \bar{p}}. \tag{5.11}$$

The expected validation error equalizes the expected training error amplified by factor $\frac{N+\bar{p}}{N-\bar{p}}$.

It can be proved [123] that, by removing the expectation since we have only a data set Z_N and given the estimate $\hat{p} = \text{rank}\left(\left.\frac{\partial^2 \bar{V}(\theta)}{\partial \theta^2}\right|_{\hat{\theta}}\right)$ of $\bar{p}$, Eq. (5.11) reduces to

$$\bar{V}(\hat{\theta}) \simeq V_N(\hat{\theta}, Z_N) \frac{N + \hat{p}}{N - \hat{p}} + l\left(\bar{V}(\theta^\circ) - V_N(\theta^\circ)\right) \tag{5.12}$$

where $l\left(\bar{V}(\theta^\circ) - V_N(\theta^\circ)\right)$ is an unknown constant depending on the given data set Z_N. In other terms, the generalization performance of the model is equal to the training one amplified by term $\frac{N+\hat{p}}{N-\hat{p}}$: the test error is larger than the training one or, which is the same, the training error is an optimistic estimate of the generalization one. As expected, when $N \to \infty$, $V_N(\hat{\theta}, Z_N) \to \bar{V}(\hat{\theta})$, i.e., the empirical risk tends to the structural one and $\hat{\theta}$ to θ°.

Let us now introduce perturbation $\delta\theta$ additive to the parameters and wonder which is the induced perturbation at the test error $\delta\bar{V}(\hat{\theta}) = \bar{V}(\hat{\theta} + \delta\theta) - \bar{V}(\hat{\theta})$. We must pay attention to the fact that the perturbation might change the rank of $\frac{\partial^2 V_N(\theta)}{\partial\theta^2}\big|_{\hat{\theta}}$, and hence, the effective number of parameters $\hat{p}$ of a value δp. From [50], it can be proved that

$$\delta\bar{V}(\hat{\theta}) \simeq \delta\theta^T H \delta\theta \frac{N + \hat{p} + \delta p}{N - \hat{p} - \delta p} + \frac{\hat{\sigma}_\zeta^2 \delta p}{N - \hat{p} - \delta p} \tag{5.13}$$

where $\hat{\sigma}_\zeta^2 = \frac{2N V_N(\hat{\theta})}{N-\hat{p}}$. H is the quasi-Newton approximation of the quadratic form $\frac{\partial^2 V_N(\theta)}{\partial\theta^2}\big|_{\hat{\theta}}$. Note that term $l(\cdot)$ vanishes being a constant. From (5.13) the variation in generalization performance is the sum of two contributions. The first term is related to the sensitivity of the model and is characterized by a quadratic form, the second term is related to the intensity of the noise and its sign depends on δp. Interestingly, if $\delta p < 0$ and $\delta\bar{V}(\hat{\theta}) < 0$ then the perturbation improves the performance of the available model! This is in line with the principal component pruning method proposed in [124] for neural networks, where authors identified those weights whose removal increases the accuracy performance of the model (weights removal can be seen as a particular type of perturbation that sets the weight values to zero). Further details can be found in [50].

We defined in Chap. 3 a perturbation to be acute when it does not change the rank of the matrix. Then, if $\delta\theta$ is an acute perturbation w.r.t. the H matrix the rank of H does not change, $\delta p = 0$ and

$$\delta\bar{V}(\hat{\theta}) \simeq \delta\theta^T H \delta\theta \frac{N + \hat{p}}{N - \hat{p}}. \tag{5.14}$$

We are in a case similar to the δV_N one with the unique difference that there is the amplification term $\frac{N+\hat{p}}{N-\hat{p}}$ taking into account the fact we have the structural risk: results valid for the empirical risk also hold for the structural one under the assumed hypotheses.

We should then ask which is the probability of encountering not acute perturbations, which are the unique perturbations that can improve the generalization ability of the model. Reference [50] answers the question for well-conditioned static neural networks by proving that a continuous perturbation $\delta\theta$, i.e., $\Pr(\delta\theta = \delta\bar{\theta}) = 0$, is acute with the probability one. In other words, non-acute perturbations are extremely rare events and, if we wish to obtain them we need to design ad hoc experiments. For

Algorithm 12: The randomized algorithm evaluating the probability that a generic perturbation is not acute

1- Select the perturbation space Δ of $\delta\theta$ and assign a uniform probability density function $f_{\delta\theta}$ over Δ;
2- Identify the accuracy ε and the confidence δ;
3- Set the number of samples $n \geq \frac{1}{2\varepsilon^2} \ln \frac{2}{\delta}$;
4- Draw n samples $\{\delta\theta_1, \ldots, \delta\theta_n\}$ from $\delta\theta$ according to $f_{\delta\theta}$;
5- Compute the indicator function for each sample

$$I(\delta\theta_i) = \begin{cases} 1 \text{ if } \delta p \neq 0 \\ 0 \text{ if } \delta p = 0 \end{cases}$$

6- The estimated probability of having non-acute perturbations in the application is

$$\hat{p}_{\mathrm{na}} = \frac{1}{n} \sum_{i=1}^{n} I(\delta\theta_i)$$

a generic application we expect perturbations to be acute, even though the property should be evaluated case by case.

Of course, we can use randomized algorithms to estimate which is the probability p_a of having acute perturbations in our application. Simply, we have to extract $\delta\theta$ from the perturbation space Δ, say according to a uniform distribution, and evaluate the probability that

$$p_a = \Pr(\delta\theta | \delta p = 0)$$

i.e., the probability that the perturbation does not change the rank of the H matrix of our model. The randomized algorithm procedure for estimating the probability of not encountering acute perturbations is given in Algorithm 12 for its educational value. We recall that the estimate is provided with accuracy ε and confidence δ.

5.2.3.1 Example Continued (2:4): Learning a Neural Network

We continue the experiment initiated in the previous section by considering here the case where we are interested in changes in generalization accuracy following the perturbations. The experimental setup is that defined at the beginning of the subsection, i.e., the neural network learned in the previous experiment becomes function $g(\theta^\circ, x)$ that, additively affected by Gaussian noise ζ of zero mean and $\sigma_\zeta^2 = 0.05$ variance, generates the training and the test data sets. A second neural network $f(\theta, x)$ belonging to the same family set by $g(\theta^\circ, x)$ was trained on the training set Z_N ($N = 50$) and provided model $f(\hat{\theta}, x)$. In this way model completeness is granted, i.e., $g(\theta^\circ, x) = f(\theta^\circ, x)$. The $f(\theta^\circ, x)$ we refer to is this until the end of the chapter.

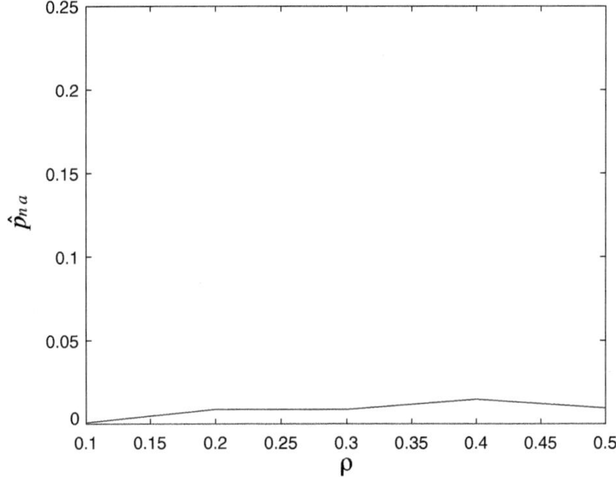

Fig. 5.4 The expected value $\hat{p}_{\text{na}}$ as function of ρ. $p_{\text{na}} = 1 - p_a$ should be null from the theory since continuous perturbations are expected to be acute. The experiments show that the estimate $\hat{p}_{\text{na}}$ is almost null due to the fact we are evaluating it by introducing the quasi-Newton approximation for matrix H

Evaluate, at first, an estimate $\hat{p}_{\text{na}}$ of the probability of encountering a non-acute perturbation $p_{\text{na}} = 1 - p_a$ by relying on randomized algorithms (Algorithm 12) where we set $\varepsilon = 0.04$ and $\delta = 0.05$ requiring $n = 1153$ samples drawn from space $\Delta(\rho)$, $\rho \in [0, 0.5]$. We should expect the estimated $\hat{p}_{\text{na}}$ to be null to confirm that finding non acute perturbations is a very rare event. This is true in the ideal case where we are able to obtain the exact value p. However, in real cases we have only the estimate $\hat{p}$ obtained from the quasi-Newton approximation for H (operation that introduces uncertainty on $\hat{p}$). This situation can be observed in Fig. 5.4 where we plotted the curve $\hat{p}_{\text{na}}$ as function of ρ. We see that $\hat{p}_{\text{na}}$ is basically zero, the discrepancy from it due to the above comments.

Having fixed the non-acute perturbation issue we move further and evaluate the estimated $\hat{p}$. Figure 5.5 shows the expected evolution of $\hat{p}$ as function of ρ. The plot shows the maximum and minimum value bars associated with a given ρ. We comment that the neural network is characterized by 10 hidden units and, therefore, makes available 31 potential degrees of freedom (i.e., the sum of the weights and neuron biases). From the figure, we see that the neural network uses all available degrees of freedom to solve the function approximation task (even though, many eigenvalues are very small but significantly higher than the resolution of the 64 bits machine and, hence, should be considered here).

We comment that the fluctuation in $\hat{p}$ is contained and the minimum value it assumes is 30. Although we should consider (5.13) to evaluate the accuracy performance loss associated with the structural risk, given the small fluctuations around $\hat{p}$ we can consider δp to be null and consider the degenerate form (5.14) instead. That done, we can now compute the real expected change in accuracy performance of the

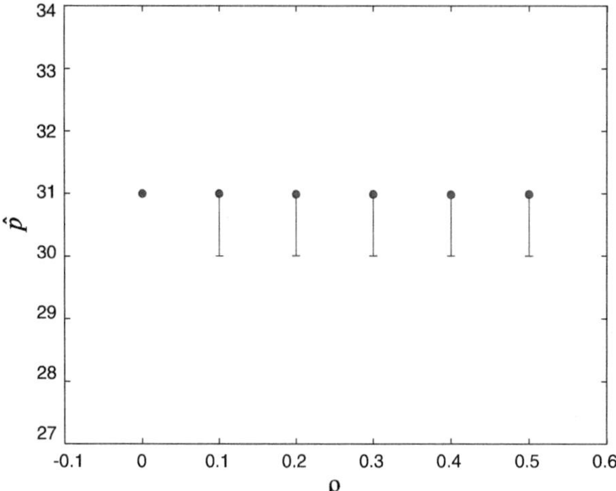

Fig. 5.5 The expected value $\hat{p}$ as function of ρ with the maximum and minimum *bars*. We appreciate the fact that the fluctuation around the expected value is very small. This is a consequence of the fact that the perturbations are acute

obtained model in correspondence to a new data set under perturbations uniformly extracted from $\Delta(\rho)$ as in the previous subsection and compare it with the expected one

$$E[\delta\bar{V}(\hat{\theta})] = \frac{\rho^2}{3}\frac{N+\hat{p}}{N-\hat{p}}\sum_{i=1}^{d}\lambda_i(H). \tag{5.15}$$

Equation (5.15) can be derived from (5.14) by following those derivations that led to derive the (5.7).

To test the accuracy of (5.15) compare it with the expected real change in test performance evaluated on a test data set of cardinality $n_V = 30$. The pointwise change in test can be evaluated as the discrepancy between the test error following the perturbation and the perturbation-free one

$$\delta V_{n_V} = \frac{1}{2n_V}\sum_{j=1}^{n_V}(y_j - f(\hat{\theta}+\delta\theta, x_j))^2 - (y_j - f(\hat{\theta}, x_j))^2.$$

Since we need to compare the estimate provided by (5.15) with $E_{\delta\theta}[\delta V_{n_V}]$, which is unknown, we resort again to randomization to estimate the expected value of a function by selecting $\varepsilon = 0.04$, $\delta = 0.05$. Figure 5.6 compares the two curves.

We can see that the performance loss estimated according to (5.15) is larger than the effective one. The outcome should not be a surprise for two reasons. At first, we do prefer to have more conservative bounds on the real performance of our function. Secondly, we recall that $N = 50$ and $\hat{p} = 31$ (we have full rank within an

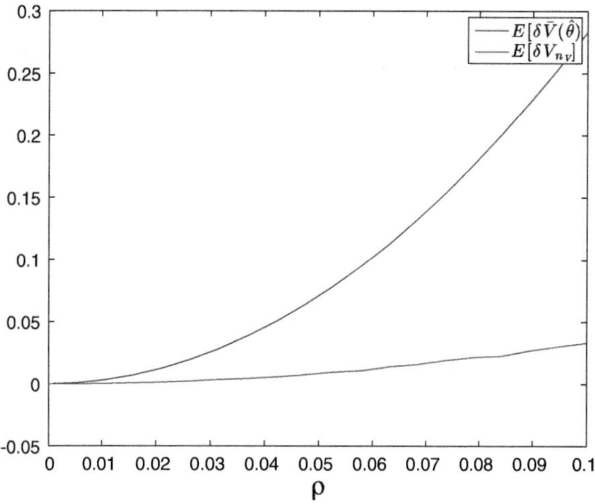

Fig. 5.6 The expected value $E[\delta\bar{V}(\hat{\theta})]$ estimated according to (5.15) and the expected one on the test set $E_{\delta\theta}[\delta V_{n_V}]$ obtained from simulations

acute framework). The number of available data is not significantly larger than the estimated degrees of freedom, a situation that introduces high uncertainty: a more conservative estimate is, in fact, provided by the correction term

$$\frac{N+\hat{p}}{N-\hat{p}} = \frac{1+\frac{\hat{p}}{N}}{1-\frac{\hat{p}}{N}} \tag{5.16}$$

that penalizes "the small data compared to the effective number of degrees of freedom" situation. Ultimately, small or large data w.r.t. $\hat{p}$ depends on the ratio $\frac{\hat{p}}{N}$: in a learning procedure we have few or many data depending on the number of degrees of freedom that need to be estimated. Ideally, we would like $\frac{\hat{p}}{N} \to 0$, statement that, in practice, requires $\frac{\hat{p}}{N} \ll 1$ to which is associated an almost unitary correction term in (5.16). Given the above setting for N and $\hat{p}$ the ratio is 0.62, far from being "much smaller than one." This not optimal learning framework is penalized by the correction term in (5.16) that assumes value 4.26.

5.2.4 Theory Highlights on Robustness

The small perturbation assumption allows us to open some views on the hidden mechanism that rules the robustness-performance relationship.

We discovered that, by introducing the concept of acute perturbations, the empirical risk and structural risk share the same behavior, the latter coinciding with the former amplified by term $\frac{N+\hat{p}}{N-\hat{p}}$, constant for a given application. It is expected that continuous perturbations are acute with probability one in many well-posed applications. In particular, we have acute perturbations when considering well trained static neural networks for nonlinear regression and recall it is a misconception to believe that neural networks are intrinsically robust.

Both in the expected and the maximum perturbation case, we see that the local geometry of the space (the H matrix) plays a relevant role in defining the robustness degree of an application whose parameters have been configured. Obviously, the intensity of the perturbation is another ingredient to this recipe and the amplification term $\frac{N+\hat{p}}{N-\hat{p}}$ must be considered to address the structural risk.

From the above sections we can propose two indexes to evaluate the goodness in robustness for the available solution and decide, in the case we have several equivalent solutions, which one we should consider if robustness is a figure of merit we are interested in. The first index, R_{WCA}, refers to the worst-case analysis, the second, R_{MCA}, to the mean case analysis

$$R_{\text{WCA}} = \lambda_{\max}(V_N''(\hat{\theta}))$$

$$R_{\text{MCA}} = \sum_{i=1}^{d} \lambda_i(V_N''(\hat{\theta})),$$

and $V_N''(\theta)$ can be approximated with the quasi-Newton matrix H whenever computation becomes an issue. Having two equivalent solutions we can contrast them by evaluating the most appropriate figure of merit for robustness (either R_{WCA} or R_{MCA}) and select the solution that minimizes the robustness index.

An interesting followup, which naturally comes from the above derivations, is based on the observation that we can integrate a penalty term directly in the training phase so as to guide the search toward solutions that provide an improved robustness ability. In other terms, the training phase will naturally look for solutions finding a balance between approximation performance and robustness ability. The figures of merit we need to consider for the training are

$$V_{N,R} = V_N + \tau\lambda_{\max}\left(V_N''(\theta)\right)$$

if we wish to improve the worst-case scenario and

$$V_{N,R} = V_N + \tau\sum_{i=1}^{d} \lambda_i\left(V_N''(\theta)\right)$$

for the average case one. $\tau > 0$ is penalty term weighting the relevance of accuracy versus robustness. The above penalty-based framework is known in the literature as Tikhonov regularization and is used to integrate some desired property the solution must possess by suitably guiding the learning process, see [121].

It is evident that evaluation of $V_N''(\theta)$ is a critical aspect from the computational point of view, being high resources demanding. However, the problem can be mitigated by considering the quasi-Newton approximation for $V_N''(\theta)$ that leads to the formulation

$$V_{N,R} = V_N + \tau \lambda_{\max} \left(\frac{1}{N} \sum_{i=1}^{N} \frac{\partial f(\theta, x_i)}{\partial \theta} \frac{\partial f(\theta, x_i)}{\partial \theta}^T \right)$$

for the worst-case scenario and

$$\begin{aligned} V_{N,R} &= V_N + \tau \sum_{i=1}^{d} \lambda_i \left(\frac{1}{N} \sum_{i=1}^{N} \frac{\partial f(\theta, x)}{\partial \theta} \frac{\partial f(\theta, x)}{\partial \theta}^T \right) \\ &= V_N + \tau \text{ trace} \left(\frac{1}{N} \sum_{i=1}^{N} \frac{\partial f(\theta, x_i)}{\partial \theta} \frac{\partial f(\theta, x_i)}{\partial \theta}^T \right) \\ &= V_N + \frac{\tau}{N} \sum_{j=1}^{d} \sum_{i=1}^{N} \left(\frac{\partial f(\theta_j, x_i)}{\partial \theta_j} \right)^2 \\ &= V_N + \frac{\tau}{N} \sum_{i=1}^{N} \frac{\partial f(\theta, x_i)^T}{\partial \theta} \frac{\partial f(\theta, x_i)}{\partial \theta} \end{aligned}$$

for the average case one.

Clearly, pursuit of the worst-case scenario implies also the average case one, whereas the opposite does not hold a priori. Conversely, the average case scenario is much easier to be implemented and significantly less time-consuming than the worst case. In fact, the extra computation reduces to add a penalty term function of the magnitude of the gradient vector: no eigenvalues need to be computed. We comment that the gradient information is freely available in a gradient descent algorithm. If τ is a function of the learning iterations we can also guide the learning process to differently weigh the robustness issue over time with a profile favoring more accuracy at the beginning of the training phase and smoothly integrate the robustness constraint later.

A different approach for integrating robustness directly in the training phase requires to add perturbations $\delta\theta$ affecting parameters θ. The idea is that by perturbing the training process the final network will become less sensitive to perturbations affecting parameters. In practice, if we consider a gradient descent algorithm, then we need to update parameters during the learning process and add, at each iteration time t, a realization $\delta\theta_t$ of perturbation random variable $\delta\theta \in \Delta$

$$\theta_{t+1} = \theta_t - \tau \frac{\partial f(\theta, x)}{\partial \theta} |_{\theta_t} + \delta\theta_t.$$

However, in order for methodology to be effective, the learning procedure should not converge too rapidly otherwise the Δ space will be badly explored. A simple gradient-based learning procedure, e.g., back-propagation, should be preferred instead of more effective quadratic-based solutions taking into account also an estimate of the local Hessian, e.g., the Levenberg–Marquardt algorithm [125].

An interesting problem that naturally derives from the analysis of the structural risk has been addressed in [128]. There the authors evaluate the degradation in performance according to the structural risk associated with linear models and identify sufficient conditions for structural redundancy at the model level. Moreover, it is shown that the intuitive idea of considering a fully linear neural network to improve the robustness index of the linear application by spreading the information over more degrees of freedoms is an insane operation. In fact, although the solution improves the intrinsic robustness of the model, the gain is achieved at the expense of a high computational load: structural redundancy methods should be considered instead, being characterized by a lower computational cost.

5.3 Robustness in the Large

It is clear from previous derivations that, despite the elegant structure, the "perturbation in the small" approach cannot be always envisaged to address practical applications. However, such a method has the striking potentiality of providing results in closed form, hence shedding light on the hidden relationships among accuracy, robustness, and application at the theory level. One might object that many assumptions have been made. Although that is true, there are situations where the assumptions hold and results are valid. We surely have a limited view of the robustness mechanism but it is available and usable given the closed analytical form.

If we want to keep generality at the application level and, at the same time provide estimates for the robustness index associated with the "robustness in the large" framework, we cannot do anything else but leave a deterministic approach and move to a probabilistic one. We will see that the hard problem associated with the evaluation of the robustness level possessed by an application can be solved by resorting to randomized algorithms once the robustness problem has been suitably formalized.

5.3.1 Problem Definition: The $u(\delta\theta)$ Case

The operative framework is that set in Sect. 5.1.1. Consider an $u(\delta\theta) \in \mathbb{U} \subset \mathbb{R}$ function which is Lebesgue measurable over subset $\Delta \subset \mathbb{R}^l$ and a given, but arbitrary, $\gamma \in \mathbb{R}$ positive scalar. We comment that, as it is, $u(\delta\theta)$ has no explicit function in the inputs in the sense that if inputs are there they are finite in number and fixed and belong to set $\tilde{X}$ of (5.2).

For instance, the $u(\delta\theta)$ function can either represent the perturbation impact on the empirical risk $\delta V_N(\hat{\theta})$ or the structural one $\delta\bar{V}(\hat{\theta})$ as introduced in Sects. 5.2.3 and 5.2.2, respectively. We recall that, there, figures of merit have been evaluated over the Z_N data set. The more general case closer to the robustness problem set in (5.1) is treated in Sect. 5.3.3.

We introduce two fundamental definitions of robustness in the large, the first based on a deterministic approach, the second, derived from the first one, characterized by a probabilistic framework.

Definition: Deterministic Robustness

We say that a computation is robust at level $\bar{\gamma}$ on perturbation space Δ when $\bar{\gamma}$ is the smallest value for which $u(\delta\theta) \leq \bar{\gamma}$, $\forall \delta\theta \in \Delta$.

The deterministic definition of robustness is very strong and requires determination of the minimum value of $\bar{\gamma}$ satisfying the inequality. We saw cases in previous sections where such a value was obtainable under very strong assumptions. Determination of $\bar{\gamma}$ cannot be obtained in closed form for a generic application and, surely, a point-by-point investigation of the property satisfaction would be computationally intractable.

We relax the above definition by resorting to a dual probabilistic problem.

Definition 1: Probabilistic Robustness

We say that a computation is robust at level $\bar{\gamma}$ for perturbation space Δ with probability $1-\eta$ it when *$\bar{\gamma}$ is the smallest value for which* $\Pr(u(\delta\theta) \leq \bar{\gamma}) \geq 1-\eta$, $\forall \delta\theta \in \Delta$.

η represents a small positive value defined in the [0, 1] interval and $1 - \eta$ can be intended as a confidence level.

The probabilistic characterization of the problem tolerates the existence of perturbations not satisfying the inequality. The proportion of such points is

$$\frac{\text{Vol}\,(\delta\theta | u(\delta\theta) > \bar{\gamma})}{\text{Vol}(\Delta)} = \eta$$

where Vol is the volume operator. Of course, we would like to have $\eta = 0$ so that, with probability one, all perturbations would satisfy inequality $u(\delta\theta) \leq \bar{\gamma}$ and the two definitions would become close. By investigating all points in Δ the definition assumes that perturbations are equiprobable, implicitly stating that there exists a uniform pdf $f_{\delta\theta}$ associated to $\delta\theta$ over Δ. The given definition automatically considers also the case where a generic $f_{\delta\theta}$ is considered. In the following we will consider such a framework and complete the definition as

Definition 2: Probabilistic Robustness

We say that a computation is robust at level $\bar{\gamma}$ *with probability* $1-\eta$ *for the perturbation space* Δ *when, given a pdf* $f_{\delta\theta}$ *associated to random* $\delta\theta$, $\bar{\gamma}$ *is the smallest value granting* $\Pr(u(\delta\theta) \leq \bar{\gamma}) \geq 1-\eta,\ \forall \delta\theta \in \Delta$.

Let us compare the deterministic D and probabilistic P problems

$$\begin{cases} D : \bar{\gamma}_D = \operatorname{argmin}_\gamma |u(\delta\theta) \leq \gamma, \forall \delta\theta \in \Delta \\ P : \bar{\gamma}_P = \operatorname{argmin}_\gamma | \Pr(u(\delta\theta) \leq \gamma) \geq 1-\eta, \forall \delta\theta \in \Delta \end{cases}$$

The goal of robustness is hence to estimate $\bar{\gamma}_P$ for the probabilistic problem and $\bar{\gamma}_D$ for the deterministic one.

When η is small, say zero, we have that all $\delta\theta$s satisfy the inequality with the probability one. However, we recall that this means that inequality might not hold for a set of perturbations Ω whose Lebesgue measurability is null: it is null the probability that, by sampling from a continuous $f_{\delta\theta}$, we obtain a perturbation not satisfying the inequality in correspondence with $\bar{\gamma}_P$. Differently, if problem P holds with probability $1-\eta$, then we have a set of Lebesgue measure up to η of points not satisfying the inequality (100η % of the total probabilistic volume). When the u function is continuous over Δ we have that perturbations not satisfying the inequality lie close to the ones which satisfy it [126] and, hence, the estimate of $\bar{\gamma}_P$ of $\bar{\gamma}_D$ is good even if the Ω set is not empty.

An estimate for $\bar{\gamma}_P$ can be obtained by resorting to randomization as also suggested in [127].

5.3.2 Randomized Algorithms and Robustness: The $u(\delta\theta)$ Case

Consider the probabilistic problem of robustness in the large

$$P : \bar{\gamma}_P = \operatorname{argmin}_\gamma | \Pr(u(\delta\theta) \leq \gamma) \geq 1-\eta, \forall \delta\theta \in \Delta$$

and define $p(\gamma)$ to be the probability that $u(\delta\theta) \leq \gamma$ for an arbitrary but given γ value:

$$p(\gamma) = \Pr(u(\delta\theta) \leq \gamma)$$

we wish to estimate such a probability at first.

This is exactly the algorithm performance verification problem designed in Sect. 4.4.1 that we re-propose in Algorithm 13 specialized to the robustness problem.

We now know that, for each value $\gamma \in \Gamma$ and chosen the respective value $\hat{p}_n(\gamma)$ from $\hat{p}_{n,\Gamma}$ provided by Algorithm 13, relationship

Algorithm 13: Randomized Algorithms to solve the probabilistic robustness evaluation problem

1- Identify the perturbation space Δ and the random variable $\delta\theta$ with pdf $f_{\delta\theta}$ over Δ;
2- Select the accuracy ε and confidence δ ;
3- Identify the interested performance level set $\Gamma = \{\gamma_1, \ldots, \gamma_k\}$;
4- $\hat{p}_{n,\Gamma}(\gamma)$= verification-problem($\Delta$, $f_{\delta\theta}, u(\delta\theta), \Gamma, \varepsilon, \delta$);
5- use $\hat{p}_{n,\Gamma}(\gamma)$;

function verification-problem($\Delta, f_{\delta\theta}, u(\delta\theta), \Gamma, \varepsilon, \delta$*)*
Draw $n \geq \frac{1}{2\varepsilon^2} \ln \frac{2}{\delta}$ samples $\delta\theta_1, \ldots, \delta\theta_n$ from $\delta\theta$ according to $f_{\delta\theta}$;
For each $\gamma \in \Gamma$ estimate

$$\hat{p}_n(\gamma) = \frac{1}{n}\sum_{i=1}^{n} I\left(u(\delta\theta_i) \leq \gamma\right), \qquad I\left(u(\delta\theta_i) \leq \gamma\right) = \begin{cases} 1 \text{ if } u(\delta\theta_i) \leq \gamma \\ 0 \text{ if } u(\delta\theta_i) > \gamma \end{cases}$$

Return $\hat{p}_{n,\Gamma}$

$$\Pr\left(|\hat{p}_n(\gamma) - p(\gamma)| \leq \varepsilon\right) \geq 1 - \delta$$

holds with accuracy ε and accuracy δ. Define as $\hat{\gamma}$ the smallest value of $\gamma \in \Gamma$ for which $\hat{p}_n(\gamma) = 1, \forall \gamma \geq \hat{\gamma}$. Then, since

$$|\hat{p}_n(\hat{\gamma}) - p(\hat{\gamma})| \leq \varepsilon$$

we can write that

$$\hat{p}_n(\hat{\gamma}) - \varepsilon \leq p(\hat{\gamma}) \leq 1$$

namely,

$$p(\hat{\gamma}) \geq 1 - \varepsilon.$$

By selecting η of the deterministic problem equal to ε we solve the probabilistic robustness problem and the estimate for $\bar{\gamma}_P$ is $\hat{\bar{\gamma}}_P = \hat{\gamma}$.

The intractable problem associated with the deterministic problem is solved by resorting to probability.

5.3.2.1 Example Continued (3:4): Learning a Neural Network

We can complete the analysis of robustness for the neural network learned in part (2:4) of the experiment by investigating how it behaves when affected by large perturbations. In particular, we explore how perturbations affecting the parameters of $f(\hat{\theta}, x)$ influence the empirical risk evaluated on Z_N. The chosen figure of merit is then $u(\delta\theta) = V_N(\hat{\theta} + \delta\theta) - V_N(\hat{\theta})$, $\Delta = \Delta(\rho)$ is parametric in ρ, e.g., according to the multiplicative model so that, for the generic ith component θ_i,

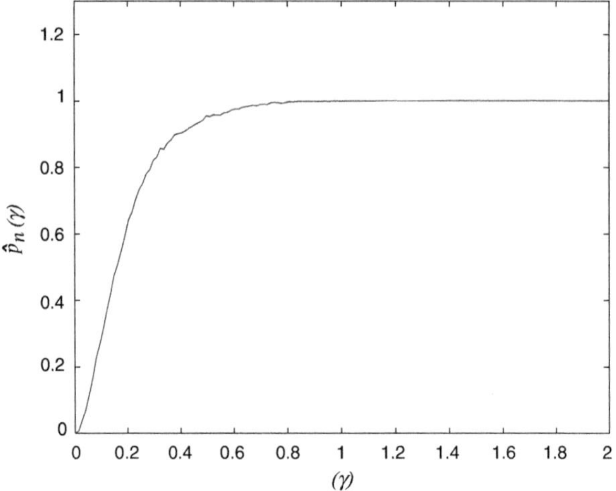

Fig. 5.7 The *curve* $\hat{p}_n(\gamma)$ estimated with Algorithm 13 for $\rho = 9.83$. The corresponding value $\hat{\gamma}$ is $\hat{\gamma} = 1.2$

$$\theta_i + \delta\theta = \theta_i(1 + \delta'\theta) \ \text{ s.t. } \delta'\theta \text{ is uniform in}[-\rho, \rho].$$

In this way the perturbation is affecting each bias and weight of the neural network with a percentage impact bounded by $\pm 100\rho\,\%$. The experiment allows us to investigate in more detail the interesting steps we undertook from the theory point of view when studying the perturbation in the large case. In the following, $\varepsilon = 0.02$, $\delta = 0.05$.

At first we set $\rho = 9.83$ and evaluate the associated probabilistic robustness problem by determining the function $\hat{p}_n(\gamma)$ curve estimated with Algorithm 13.

The outcome is given in Fig. 5.7 and provides $\hat{\gamma} = 1.2$.

Figure 5.8 compares three $\hat{p}_n(\gamma)$ functions associated with increasing ρ values. As expected, a larger ρ (a stronger perturbation) induces a larger $\hat{\gamma}$ (a larger performance loss).

Exploration of different perturbation spaces provides several $\hat{p}_n(\gamma)$ functions, each of which is associated with its $\hat{\gamma}$ value. The robustness in the large curve $\bar{\gamma}_P(\rho)$ provides the global performance loss of the perturbed function in correspondence with different perturbation spaces and solves the robustness in the large problem. Solution to the specific problem requires identification of the perturbation space we expect for our application and sees if the performance loss is below a tolerated value.

The robustness in the large curve for this experiment is given in Fig. 5.9 where $\bar{\gamma}_P$ is estimated with $\hat{\gamma}$.

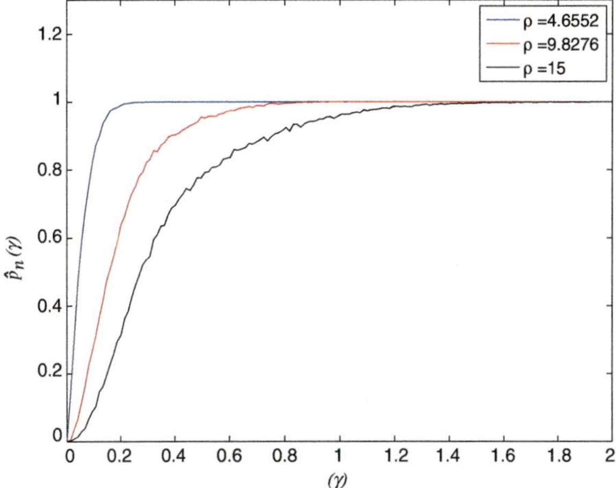

Fig. 5.8 Three *curves* $\hat{p}_n(\gamma)$ estimated with Algorithm 13 for $\rho = 4.66, 9.83, 15$. The larger the perturbation space controlled by ρ the larger the performance loss $\hat{\gamma}$

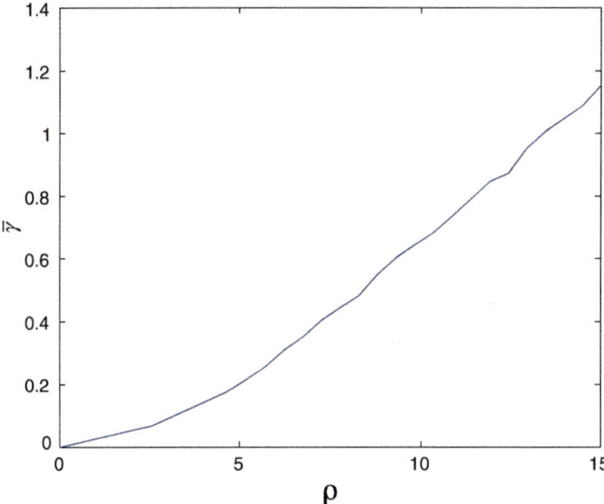

Fig. 5.9 The *curve* $\bar{\gamma}_P$ as a function of ρ and estimated with $\hat{\gamma}$. As expected, by increasing the strength of the perturbations affecting the neural network parameters the empirical risk increases. A multiplicative model has been considered for the parameters so that the maximum percentage impact is bounded by $\pm 100\rho\,\%$

5.3.3 The Maximum Expectation Problem

Equation (5.1) is characterized by a problem, the evaluation of the inequality satisfaction $u\,(g(\theta, x), g(\theta, \delta\theta, x)) \leq \gamma$, that requires a double level of space exploration, involving both the Δ and the X spaces. The problem is clearly computationally hard

Algorithm 14: Randomized algorithm to compute a probabilistic estimate $\hat{u}_{\max}$ of $u_{\max}$ associated with the problem defined in (5.18)

1- The probabilistic problem requires to estimate $\max_{\delta\theta \in \Delta} E_x[u(\delta\theta, x)]$;
2- Identify the input spaces X, Δ and define random variable x and $\delta\theta$ with pdf f_x over X and $f_{\delta\theta}$ over Δ, respectively;
3- Select the accuracy ε and the confidence δ levels;
4- Draw $m \geq \frac{\ln \frac{2}{\delta}}{\ln(\frac{1}{1-\varepsilon})}$ i.i.d. samples $\delta\theta_1, \ldots, \delta\theta_i, \ldots, \delta\theta_m$ from $\delta\theta$;
5- Draw $n \geq \frac{1}{2\varepsilon^2} \ln \frac{4m}{\delta}$ i.i.d. samples $x_1, \ldots, x_j, \ldots, x_n$ from x according to f_x;
6- Compute $\hat{u}_{\max}(\delta\theta_i) = \frac{1}{n} \sum_{j=1}^{n} u(\delta\theta_i, x_j), \forall i = 1, \ldots, m$
7- use $\hat{u}_{\max} = \max_{i=1,\ldots,m} \hat{u}_{\max}(\delta\theta_i)$;

for a generic function, albeit Lebesgue measurable. In Sects. 5.2.2 and 5.2.3 we "simplified" the problem—which however is still hard—by restricting the exploration of the input space to a finite data set composed of data instances present in the Z_N training set. In other words, we restricted problem (5.1) as in (5.2), i.e.,

$$u\left(g(\theta, x), g(\theta, \delta\theta, x)\right) \leq \gamma, \forall \delta\theta \in \Delta, \forall x \in Z_N \tag{5.17}$$

for a given Z_N. In fact, although in Sect. 5.2.3 we took expectation with respect to Z_N in (5.9) and (5.10), in subsequent derivations (e.g., see (5.12)) we made approximations and confined ourselves to the unique available data set Z_N.

Problem (5.1) cannot be easily addressed in its general form but can be managed in very interesting and common cases where it can be converted into the determination of the maximum value $u_{\max}$ that function $u(\delta\theta, x) = u\left(g(\theta, x), g(\theta, \delta\theta, x)\right) \in [0, 1]$ assumes $\forall \delta\theta \in \Delta, \forall x \in X$. By construction, $u_{\max}$ coincides with γ granting the (5.1) to hold. To keep under control the complexity of the exploration of the input space, expectation is taken with respect to X according to f_x. Finally, the problem we aim at solving requires determination of

$$u_{\max} = \max_{\delta\theta \in \Delta} E_x[u(\delta\theta, x)]. \tag{5.18}$$

Fortunately, the hard problem stated in (5.18) can be solved by resorting to randomized algorithms with derivations given in Sect. 4.4.4 and Algorithm 10 given in Chap. 4 suitably adapted to host the problem in (5.18) as proposed in Algorithm 14.

5.3.3.1 Example Continued (4:4): Testing the Robustness in the Large of a Neural Network

This last experiment aims at investigating the robustness in the large properties of the neural network $f(\hat{\theta}, x)$ learned from the $g(\theta^{\circ}, x)$ one and given in experiment (2 : 4). In particular, the figure of merit we use aims at evaluating the change in

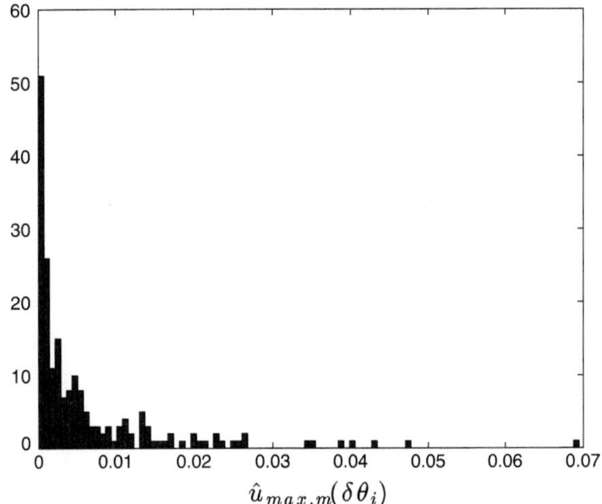

Fig. 5.10 The histogram of values $\hat{u}_{\max}(\delta\theta_i)$ of Algorithm 14 in correspondence with $\rho = 1.58$. $\hat{u}_{\max,m} = 0.07$

discrepancy between the two functions evaluated according to a quadratic figure of merit induced by additive perturbations affecting the parameters of function $f(\hat{\theta}, x)$ leads, following (5.18), to problem

$$u_{\max} = \max_{\delta\theta \in \Delta} E_x[(f(\hat{\theta} + \delta\theta, x) - g(\theta^\circ, x))^2]$$

that can be solved by invoking Algorithm 14. Also in the following experiments the $\delta\theta$ random variable follows uniform distribution governed by ρ such that $\|\delta\theta\|_\infty \le \rho$. Similarly, x is defined with uniform pdf f_x over $X = [-2, 2]$.

At first we wish to plot the histogram of the instances $\hat{u}_{\max}(\delta\theta_i) = \hat{E}_n[u(\delta\theta_i, x)]$ of Algorithm 14 estimating $E_x[u(\delta\theta_i, x)]$ when $x \in X$ for a given $\delta\theta_i$. The operational parameters are $\varepsilon = 0.02$, $\delta = 0.04$ inducing $m = 194$ samples to be taken from the $\Delta(\rho)$ space and $n = 12342$ from $X \cdot \rho = 1.58$. The histogram of the distribution is given in Fig. 5.10. The robustness level of the neural network evaluated according to the chosen figure of merit is $\hat{u}_{\max} = 0.07$.

We wish to investigate now how the estimate $\hat{u}_{\max}$ changes with the intensity of the perturbation (i.e., the enlargement of the perturbation space controlled by $\|\delta\theta\|_\infty = \rho$). Figure 5.11 provides three curves estimating $u_{\max}(\rho)$ for ρ values in interval [0, 2]. The operational parameters are as follows: $\delta = 0.04$; in the $\varepsilon = 0.005$ case we have $m = 781$ and $n = 225315$; in the $\varepsilon = 0.02$ case we have $m = 194$ and $n = 12342$; in the $\varepsilon = 0.04$ one we have $m = 96$ and $n = 2866$. When ε decreases (at the cost of a larger number of samples) the curve becomes more regular as expected and the obtained estimate tends to the ideal $u_{\max}(\rho)$.

It is clear that every time we draw n data from X and m from Δ we generate a realization for the $u_{\max}(\rho)$ curve. Figure 5.12 presents such an ensemble associated

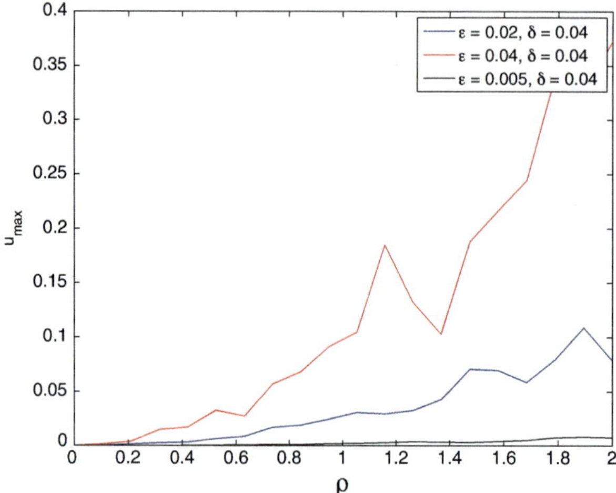

Fig. 5.11 *Curves* estimating $u_{\max}(\rho)$ for $\delta = 0.04$ and $\varepsilon = 0.02, 0.04, 0.05$

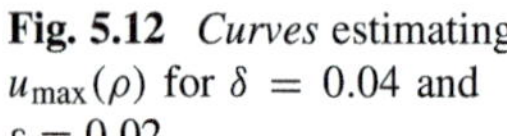

Fig. 5.12 *Curves* estimating $u_{\max}(\rho)$ for $\delta = 0.04$ and $\varepsilon = 0.02$

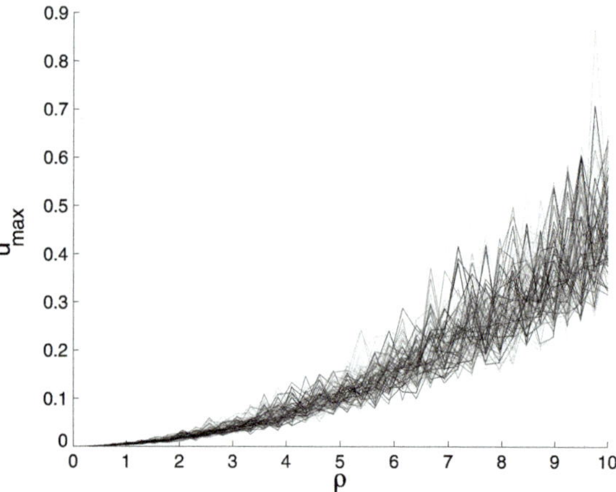

with 100 curves ($\delta = 0.04$ and $\varepsilon = 0.02$). We need to comment that, despite the fact the ensemble is compact as we shall expect since the neural functions are continuous and, then, it is the associated figure of merit, some curves introduce spread values. This should not surprise us since the estimate for the maximum/minimum value of a function is carried out in probability and must be intended as discussed in Sect. 4.4.2.

Chapter 6
Emotional Cognitive Mechanisms for Embedded Systems

A cognitive embedded system is an embedded system that takes advantage of cognitive processes to propose intelligent solutions.

It is strongly perceived that the future of intelligent embedded systems is oriented towards the implementation, either in software or hardware, of cognitive mechanisms. However, the predicted future is not something will happen decades from now. Since adaptation, namely the ability to automatically modifying the system to host a new situation, is a basic form of cognition associated with elementary automatic reactions, we can safely claim that the future of embedded systems has started already. In fact, many embedded solutions present in the market introduce adaptation mechanisms at various levels (see Chap. 8). At the same time, technological advances in the hardware both at the microprocessor, FPGA and Graphics Processing Units (GPU) level make available a computational power that permits the execution of sophisticated embedded solutions also integrating online learning and cognitive mechanisms.

This chapter aims at providing an engineering-oriented perspective of brain functions. Obviously, from a biological view point, most functions are not done by a specific region, but emerge over many regions through natural neurodynamics. Nevertheless, the spirit of this chapter is to apply "lessons" from the brain to embedded systems (and not to duplicate the way the brain works). More specifically, the chapter introduces a functional description of some basic processes of the human brain, with a special focus on emotional and cognitive processing. Emotion processing is, in fact, a complete framework that enables most of the fundamental data processing and storage elements modellable with automatic and controlled processes and, as such, it represents an ideal candidate to be mimicked. Access to the memory to retrieve/store information is a key function here, with information "stored" in different ways, from simple patterns associated with feature instances to more advanced forms of semantic knowledge.

We anticipate that many mechanisms introduced in subsequent chapters, e.g., those related to PACC, adaptation, and learning in a non-stationary and evolving environments can be immediately cast within a cognitive representation of emotions.

C. Alippi, *Intelligence for Embedded Systems*, DOI: 10.1007/978-3-319-05278-6_6,

As such, each of those chapters will have a section showing how the presented methods are modellable as instances of the emotional cognitive processing. However, it is worth outlining that the association "emotional processing-intelligent algorithm" finds affinity in other neural cognitive mechanisms. We leave the interested reader to deepen the investigation in reference textbooks, e.g., see [220] for a recent essay on neural anatomy and information processing and [221] for cognitive aspects.

6.1 Emotional Cognitive Structure

Brain phenomena can be modeled as a hierarchy of subsystems differentiating in time activation and accuracy levels [138–140] that, possibly, rely on a memory containing knowledge suitably stored for decision making and supporting other processes.

Lower levels are generally characterized by fast automatic processes so as to be able to quickly take a decision and/or provide a prompt reaction following the presentation of external stimuli. In the meanwhile, either these processes or the presence of stimuli activate higher levels of knowledge processing characterized by more complex and articulated controlled mechanisms able to assess ongoing performance and abort/modify/complete actions and decisions made by lower levels. Not rarely, higher cognitive levels introduce feedback mechanisms by providing information to lower ones so that learning can be perfected, e.g., through a fine tuning of the existing schemata. Clearly, new situations encountered during lifetime must be included in the processing mechanism and stored in the semantic memory where long-term knowledge is kept.

The joint activity of automatic and controlled processes allow us for modeling the emotional responses in humans. Here, lower levels provide immediate actions following a stimuli pattern presentation. Clearly, promptness in the reply i.e., low latency in pattern classification and decision making, is to be preferred than high accuracy in emotion labeling: if a potential threat is somehow detected we need to immediately intervene instead of waiting for more information to come or the outcome of a more sophisticated, and hence time consuming, processing. Higher processing levels will then be activated either from available stimuli, so as to permit a parallel processing, or be initiated by automatic ones. In both cases, the goal is to assess the action taken by lower levels and intervene whenever appropriate to confirm, abort or correct the decision made.

In the last decades there has been a lot of work on emotional processing, much is known, but a complete understanding of automatic and controlled cognitive mechanisms is still missing.

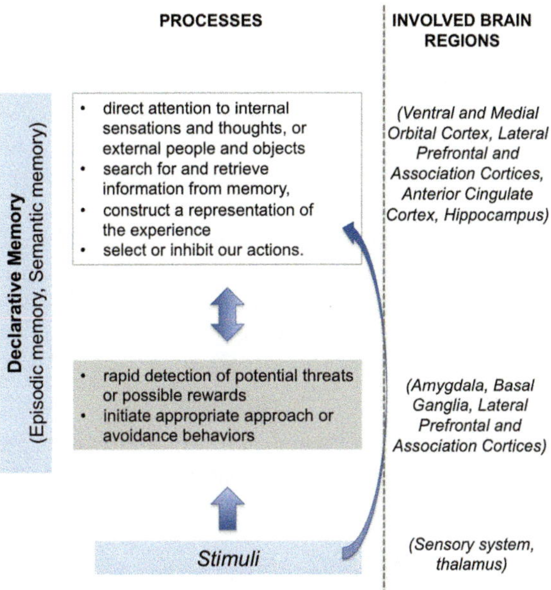

Fig. 6.1 Automatic and controlled processes and involved brain regions

6.2 Automatic and Controlled Processes

Automatic and controlled processes involved in emotional expression processing and response activation are briefly described in the following with Fig. 6.1 representing the reference framework. The interested reader can refer to e.g., [140] for a more detailed analysis of the presented concepts.

6.2.1 Automatic Processes

Automatic emotion processes, that refer to the lowest levels of the hierarchical cognitive system, are characterized by an interesting and relevant detection-reaction mechanism designed to quickly identify potential dangers and the possibility to get a reward and initiate suitable actions/reactions, e.g., by increasing the heartbeat or the respiration rate and releasing stress-hormones following a perceived threat. Here, reduced latency is more important than keeping under control the false positive rate following a conservative primordial principle. In fact, it is much better to react with an unnecessary action following a perceived—possibly new—threat than being insensitive to it.

These processes are meant to be quick and effortless in generating an emotional response (which could be either positive or negative) to external stimuli such as presentation of faces, objects or events [140]. This emotional response is part of the emotion processing and the associated decision making level, which involves the detection of danger or plan and schedule actions to get a reward, the recall of previously acquired information related the situation/event/emotion and the activation of a proper action/reaction. It is worth noting that the emotional response (possibly together with additional environmental information) is then processed to become part of the knowledge (i.e., stored in the memory) to be recalled whenever necessary.

By inspecting Fig. 6.1 we see that automatic processes receive stimuli made available by the sensory system or the thalamus and provide outputs to feed controlled processes, which also influence automatic ones with a feedback mechanism. Clearly, we must also react to a threat by activating the sensory-motor mechanism primarily implemented by the sensory-motor cortices, basal ganglia and cerebellum and not considered here being outside the strict emotion processing analysis.

The main tasks carried out by automatic processes can be summarized as

- Rapid detection of potential threats and activate avoidance behaviours;
- Initiate appropriate approaches oriented towards the achievements of rewards.

6.2.2 Controlled Processes

Consciously, we also direct attention to our personal sensations, construct our emotional background, and select or inhibit actions depending on a lifetime experience. The use of computationally demanding processes in the generation and regulation of emotions is named *controlled emotion processing*. By deliberately monitoring, activating and processing emotions we can re-interpret and alter their meaning (learning mechanism), change the current personal experience and perception of the world as well as the way we interpret emotions and respond to stimuli.

After a preliminary automatic response, cognitive processes are consciously activated by higher levels of the cognitive hierarchical system. The aim of these processes is to integrate, improve and (if necessary) correct the output provided by the automatic processing level. As such, these high-level processes allow the human brain for consciously focusing our attention to a specific stimuli or emotion, recalling events or sensations stored in the memory or modifying the action/reaction pattern activated by lower cognitive levels. Differently from automatic processes, this "validation" step is not automatic, is time and energy consuming and requires consciousness.

Moreover, a relevant activity performed by controlled processes is their capability to create an abstract representation/meaning of experienced events or stimuli. These representations, coded in a suitable way, can then be stored in memory and recalled whenever necessary.

With reference to Fig. 6.1 we see that controlled processes receive information directly from incoming stimuli as well as the output of automatic processes. The declarative memory plays a fundamental role being present both for automatic and controlled processes to provide episodic and semantic memory instances to be retrieved, stored or updated. The main tasks carried out by controlled processes can be summarized as

Table 6.1 Roles, functions and characteristics of brain regions. Refer to [138] for a more complete treatment

Brain regions	Function	Operations	Type of process
Amygdala	Detecting, learning about stimuli	Detects potentially threatening stimuli and associates them with appropriate actions	Automatic
Basal ganglia	Registering rewards, acquiring habits, selective gating of behavior	Mediates selection and initiation of actions, automatizes sequences of behavior and reinforced thoughts	Automatic or controlled
Lateral pre-frontal/association cortices	Retrieving and storing semantic emotion knowledge	Identifies stimuli, differentiates feeling states; attributes emotional qualities to stimuli; repository of regulatory strategies, lay emotion knowledge	Retrieval can be automatic or controlled
Anterior cingulate cortex	Conflict monitoring	Monitors on-going behavior and determines whether a change is necessary or not	Conflicts detected automatically, but making changes takes control
Ventral/Medial orbital frontal cortex	Context- dependent action selection	Inhibits on-going emotional responses based on analyses of context	Controlled
Hippocampus	Long-term strategies	Understanding spatial relations within the environment	Controlled

- Select or inhibit the actions activated by automatic processes;
- Construct a representation of the emotional experience over time and perfect it according to the received external stimuli and the final situation outcome;
- Direct attention to internal sensations and thoughts;
- Search and retrieve information from the declarative memory.

6.3 Basic Functions of the Neural Emotional System

Evidences from multiple domains suggests that emotional processing is carried out by at least six distinct neural subsystems. Amygdala can be associated mainly to automatic processes; Anterior Cingulate Cortex, Ventral and Medial prefrontal, Orbital Cortices and Hippocampus to controlled processes, while Lateral Prefrontal and Association Cortices and Basal Ganglia to both [138–141].

The above subsystems differ both in the mechanisms they rely upon and the carried out tasks: their joint interaction is a primary ability of the human brain in emotion processing. Roles, functions and characteristics of these brain regions are summarized in Table 6.1, derived from [140] and here suitably expanded. The functional description of these brain regions are summarized in the sequel. The interested reader can find a more detailed description e.g., in [140].

6.3.1 Amygdala

The role of the amygdala is quite complex and involves the detection of a threat by inspecting stimuli patterns and activating a reaction. Moreover, it helps in modulating the long-term memory consolidation of stimuli-stimuli and stimuli-response association in the emotional declarative memory.

The capability to quickly identify potentially threatening events is crucial for survival purposes and is preparatory for any subsequent action following the detection. The amygdala is devoted to this aspect by assessing the risk associated with acquired stimuli patterns and, if necessary, it activates the proper reaction. Any novel or ambiguous stimulus might initially be intended as threatening within a conservation of species principle, and thus warrants a response from the amygdala even if later the event would be consciously labeled as a false positive based on further information processing.

Among the areas composing the amygdala, the activation of physiological responses (e.g., increase of the heartbeat or respiration rate and release of stress-hormone) is managed by the central nuclei.

6.3.2 Long-Term Memory

The creation and association of long-term memory to emotional events is another relevant activity carried out by the amygdala. Specifically, as pointed out by several researches, the basolateral complex of the amygdala is deeply involved in the association of threatening stimuli with the long-term declarative memories. Moreover, although these mechanisms are not fully understood yet, the amygdala is thought to have a relevant role in the consolidation of long-term (potentially lifelong) memory suggesting the idea that emotions are involved in the long-term storage of perceived events/stimuli. In fact, recent studies suggest that, while the amygdala is not itself a long-term memory storage site, and learning can occur without it, one of its roles is to regulate memory consolidation in other brain regions. This suggests that the amygdala contributes to the development of semantic knowledge by influencing the information incorporated into the long term semantic memory.

Having the above in mind, the amygdala may be considered as a module processing stimuli patterns to detect threats. Since novel and unknown stimuli are initially considered as threatening, amygdala enters into an "alarm" state following their detection. Afterwards, other mechanisms intervene e.g., the anterior cingulate cortex to provide, if required, a new assessment for the threat label recommended by the amygdala. All these aspects are aligned with change detection tests and will be deeply investigated in Chap. 9.

6.3.3 Basal Ganglia

Research indicates that basal ganglia provides functions related to the voluntary motor control (e.g., eyes control) as well as emotional processing, for instance that leading to learn routine behaviors to attain rewards.

Generation of sequential steps for attaining a goal is of paramount importance to achieve rewards requesting an activity planning. Research demonstrated that this task is not carried out within the amygdala but in the basal ganglia. Specifically, although the basal ganglia are involved in sequences of actions and have been proposed as learning such "chunks" [247–249], the sequential aspect is seen as the main function of basal ganglia. Rather, they facilitate action selection, which is important for both sequential and non-sequential response. The function of actually discovering and configuring action sequences is thought to occur in the cortex—mainly the supplementary motor area of the motor cortex [140]. Once a sequence is found to be important enough to be learned as a single action, it is chunked and then gated through the basal ganglia.

In intelligent embedded systems this activity refers to the ability of modeling time dependent events by means of dynamic systems or machine learning techniques (e.g., Markov processes) as done in Chap. 10. The availability of these models is crucial for forecasting purposes and to define sequences of actions necessary to achieve a long-term goal by means of planning.

6.3.4 Lateral Prefrontal and Association Cortices

The lateral prefrontal and the association cortices (LPAC) provide mechanisms enabling the storage and retrieval of semantic emotion knowledge as well using the memory content to asses the relevance of stimuli and events. As such, the role of this subsystem appears to be that of storing emotional concepts and providing mechanisms to connect different memories characterized by similar emotional associations. Access to this emotional database is automatic during the generation of an emotional state, or when we consciously represent or label emotional states to draw inferences about those emotions we are experiencing. However, it is not sure if these brain areas are responsible for emotion labeling.

Although the neural mechanism behind these behaviors is not fully understood, there is evidence that the lateral prefrontal and the association cortices are involved both in supporting the emotional automatic and controlled processes. In particular, the research has found that this system is part of the automatic generation of an emotion in response to external stimuli and that, over time, the relevant regularities of our episodic memory slowly become incorporated into the database of semantic knowledge. We saw that the emotional contents of the semantic memory are influenced by the amygdala, which facilitates long-term consolidation of episodic memory for significant, arousing events.

The lateral prefrontal and association cortices are closely related to the knowledge acquired during the operational life by recurrent adaptive classifiers discussed in Chap. 9. This knowledge is organized into concepts, each of which represents a memory of the state. Similar concepts can be fused together to improve/integrate the knowledge over time.

6.3.5 Anterior Cingulate Cortex

The role of the Anterior Cingulate Cortex (ACC) is to assess the "congruence" of emotions and feelings that have been generated in response to external stimuli. Moreover, the ACC is involved in forecasting whether external stimuli would induce threats or pain in the future or not. This capability, which is conscious, is crucial in the activity planning to achieve long-term goals.

The ACC activity, fundamental in a complex, high-connected systems such as the human brain, is conceptually very close to the validation procedure of hierarchical change detection tests introduced in Chap. 9: in response to an event, signals are jointly evaluated to determine whether what perceived is associated with a false alarm (congruence among stimuli) or, differently, it represents a true change to be taken into account.

6.3.6 Orbital and Ventral-Medial Prefrontal Cortices

The orbital and ventral-medial prefrontal cortices known as OFC and VM-PFC (both simplified in VM-PFC in the sequel) appear to represent the current, context-specific, emotional value carried by an external stimulus and provide functions that allow us to both alter our emotional responses based on analyses of the current context and generate affective responses based on these analyses. These two functions form the foundation for the active regulation of emotion and emotion-guided behavior.

As such, the role of the VM-PFC is closely related to the active approach modeling the human behavior. Emotions, stimuli and memory patterns automatically generated by lower cognitive levels are integrated and linked to the long-term memory to define a "cognitive" high-level response. This response, which could consider and take into account long-term goals, can integrate (or even substitute) the automatic response taken from lower levels. Interestingly, researches (e.g., [138]) found that the "cognitive" ability of the VM-PFC resides in the capacity to connect the memory systems (which include the working and declarative memory) with emotional systems (where the amygdala comes into play) to evaluate the taken actions and recall associated somatic states.

In the field of decision-making [141], the activity of the VM-PFC has been widely and deeply studied. Research demonstrated that damages in the VM-PFC prevent the human brain from effectively integrating information coming from external stimuli, emotions and memory. The effect of this damage induces extremely poor decision-making abilities.

In the context of intelligent embedded systems, this approach is very close to cognitive analysis in distributed fault diagnosis systems where low level information is integrated and analysed by taking into account also the network topology to be able to distinguish between false alarms, real events associated with faults or model bias associated with the models introduced to describe the physical phenomenon under monitoring, see Chap. 10.

6.3.7 Hippocampus

The hippocampus is an old structure of cerebral cortex that takes part in many declarative memory functions which refer to the memory of facts and events. Encoding and recalling information from the memory are the two main tasks of this fundamental subsystem. Interestingly, the hippocampus interacts with the amygdala in the formation of short-term memory, which is a preliminary step for the storage of long-term information. Research demonstrated that lesions affecting the hippocampus induce errors in the processing of information present in short-term memory solely and do not influence knowledge previously stored in the long-term one.

The role of hippocampus in spatial representations is rather clear in the case of rodents. Differently, in primates, it seems to have roles in declarative memory and perhaps sensory integration (though that seems to happen much more in the parietal cortex).

Thought concept formation is much more likely to happen in other parts of the cortex (especially other parts of the temporal lobe and perhaps in the prefrontal cortex), the hippocampus is needed for memory recall only for some period after the memory is first acquired [250]. In intelligent embedded systems, the role of the hippocampus can be associated with the ability to recall previously acquired concepts whenever necessary (refer to Chap. 9).

6.4 Emotion and Decision-Making

Decision-making involves the orchestration of multiple neural structures and cognitive subsystems, e.g., VM-PFC, amygdala, LPAC, and hippocampus [138]. The key reference for understanding this mechanisms is Fig. 6.2.

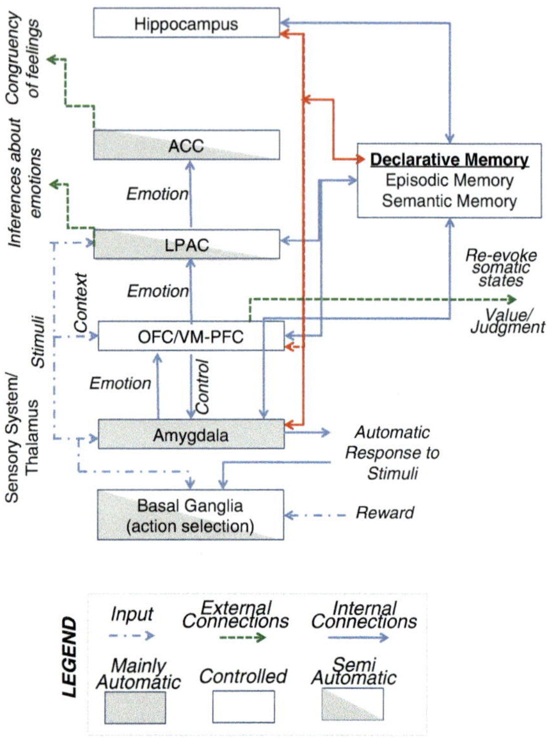

Fig. 6.2 The basic functional elements behind emotion and decision-making in emotion processing

Research has shown that areas such as the VM-PFC, the amygdala, the insula, the somatosensory cortex, the dorsolateral prefrontal cortex and the hippocampus are all involved in various aspects of decision-making [138].

Sensory information is acquired/processed by the sensory system/thalamus and forwarded to amygdala, OFC/VM-PFC and LPAC together with contextual information. Emotions are processed, integrated and abstracted in the processing flow starting from the amygdala and ending with the LPAC, while the congruency of these feelings/emotions is assessed in the ACC. The memory of events/emotions/decisions then come into play through the hippocampus and the VM-PFC together with information about the reward (provided by the basal ganglia) and the value/judgment (provided by the VM-PFC). All these mechanisms cooperate and constitute the basis of the decision-making process.

Interestingly, this complex process is very close to the processing in adaptive classifiers (Chap. 9) where an initial decision is initially taken by considering external stimuli and previously acquired information.

Chapter 7
Performance Estimation and Probably Approximately Correct Computation

The analysis phase of a problem aims at evaluating, given a computation, its performance. Performance can be intended in several ways depending on the specific target problem as well as the abstraction level where it is carried out. For instance, at the device level we have cost, latency, throughput, power, energy, complexity to name some major performance design indexes. At the algorithm level we have accuracy, confidence, energy, and complexity. Not rarely we constrain such indexes and we saw in Chap. 4 how it is possible to evaluate their satisfaction level.

Performance and design indexes are evaluated through suitable figures of merit applied to architectural and functional elements of the embedded system or the algorithm. Despite the fact that the provided methodological analysis is able to address all figures of merit and architectural aspects, we will mainly focus on accuracy as a case study of a performance/design index without any loss in generality. As a consequence, we look at algorithms to be executed on embedded systems whereas the embedded system being characterized by finite resources introduces physical constraints which, in turn, affect the algorithm itself and its performance. At the same time, as already pointed out in the introduction, we discover that it is too expensive, and most of times not necessary, to provide a worst-case analysis for accuracy. In line with approximate computation and probabilistic computation, we are here interested in an algorithm that, mounted in the embedded system, provides an outcome correct in probability.

The chapter introduces at first methods for assessing the accuracy of a computation, then formalizes the concept of Probably Approximately Correct Computation PACC. Finally, it provides techniques for accuracy assessment in terms of PACC and methods for answering to the following questions:

1. Which are the performances of my algorithm?
2. If I simplify my algorithm, which is the introduced performance loss?
3. I have different algorithms solving my problem. Which one is the best?
4. I have different algorithms solving my problem. Which one is the best on a given embedded system?
5. Shall I use a floating point unit or a cheaper fixed point representation suffices?

C. Alippi, *Intelligence for Embedded Systems*, DOI: 10.1007/978-3-319-05278-6_7,

7.1 Accuracy Estimation: Figures of Merit

Different figures of merit can be considered to evaluate the discrepancy between two functions which, here, must be intended as the optimal—ideal—solution for a given problem and the approximated solution proposed as the candidate to be implemented in the embedded system. The discrepancy can be intended as a performance loss, difference in accuracy between an identified model and the true one or, simply, a measure of the "distance" between the two. The figure of merit depends on the specific application and the goodness of a solution also depends on the chosen figure of merit in the sense that the same solution can be more or less good depending on the chosen performance evaluation tool.

Since our goal is to provide a function approximating the real one, in the following we consider assessing the discrepancy between two functions according to figure of merit $u \in \mathbb{R}$. The evaluation is carried out by taking into account the punctual discrepancy of two functions $u(x) = u(y(x), \hat{y}(x)) \in \mathbb{R}$,[1] where we assume $x \in X \subset \mathbb{R}^d$. X is a probability space, whose probability measure μ induces pdf f_x over it, and $y(x)$, $\hat{y}(x) \in \mathbb{R}$, which we assume to be measurable functions according to Lebesgue. By applying an aggregation operator to the punctual discrepancy $u(x)$ to $x \in X$ we obtain the discrepancy u. In the sequel, we indicate as y the *reference* function and $\hat{y}$ the *approximating* one. Two interesting cases arise from the applications:

- Functions $y(x)$ and $\hat{y}(x)$ are given. The figure of merit u evaluates the discrepancy over the whole input space X. Again, if f_x is unknown we shall consider a uniform distribution for its worst case properties.
- Function $y(x)$ is not given, but can be queried, i.e., once a sample x_i is drawn from the input space, function $y(x)$ acts as an oracle and provides value $y(x_i)$, possibly affected by uncertainty. The number of samples can be finite or infinite depending on the nature of the application, thus $x_i \in \tilde{X} \subset X, i \in \mathbb{N}$ are sampled according to the pdf f_x induced by the probability measure over X and are subsequently considered for the discrepancy computation. In this situation also function $\hat{y}(x)$ is not known, therefore it must be at first identified, by considering a suitable model family function $\hat{y}(x, \theta)$ parametric in the vector of parameters $\theta \in \Theta \subset \mathbb{R}^l$. Then, with abuse of notation $\hat{y}(x, \hat{\theta}) = \hat{y}(x)$ once $\hat{\theta}$ has been provided.

The first case arises in all those applications for which the theory provides the optimal solution to the problem and we need to approximate it for several reasons, e.g., because our embedded system is not able to host the high accurate solution for its complexity and an approximation needs to be considered instead. As an example, we have the optimal design of filters and representation of a complex numerical algorithm. The second case is what the theory of system identification and learning is about: starting from a sequence of input/output pairs we determine

[1] With an abuse of notation that eases the understanding, we consider the punctual discrepancy as a function of x, since y and $\hat{y}$ are fixed.

the approximating function. This topic, which is fundamental in this book, has been addressed in Sect. 3.4.1.

In the following we will present, without the intent to be exhaustive, three interesting u figures of merit. We recall that it is the application designer who identifies the right figure of merit for a given problem based on a priori knowledge, experience, and application constraints. However, when one does not know which figure of merit to consider, it is a rather common approach to use the mean squared one.

7.1.1 Squared Error

The Squared Error (SE) is a rather common quadratic figure of merit adopted to quantify the difference between two functions. The corresponding discrepancy u_{SE} is a risk function, corresponding to the expected value of the squared punctual discrepancy $u(x) = (y(x) - \hat{y}(x))^2$

$$u_{SE} = E[u(x)] = E\left[(y(x) - \hat{y}(x))^2\right]$$

which represents the second order moment of the error (considering a zero mean error). It is worth observing that the discrepancy between the two functions is weighted by the probability density function f_x over the input space.

Its empirical version evaluated over a finite number of n points drawn according to f_x (i.e., considering a finite space $\tilde{X} = \{x_1, \ldots, x_n\}$ where x_i is a realization of a random variable $x \in X$ with pdf f_x), provides the empirical estimate of the quadratic error or a Mean Squared Error (MSE)

$$u_{MSE} = \frac{1}{n}\sum_{i=1}^{n} u(x_i) = \frac{1}{n}\sum_{i=1}^{n} \left(y(x_i) - \hat{y}(x_i)\right)^2.$$

We have been using u_{SE} and u_{MSE} widely over the book for their intriguing structure that makes the mathematics amenable. Moreover, despite the fact that the use of a SE loss function has been criticized, e.g., in speech and image applications for its quadratic behavior which amplifies large point-wise discrepancies more than small ones and is not a perception-based figure of merit [41, 42], it is commonly adopted since it is easy to use [40]. Moreover, a quadratic function is a natural way to measure the energy of the discrepancy function $u(x)$ and, thanks to the Parseval theorem, the energy of the signal can be equivalently computed in the signal space or frequency domain. It is clear that, if functions $y(x)$ and $\hat{y}(x)$ are deterministic, then the u_{SE} simply evaluates the approximation risk (model bias), namely the integral of the squared discrepancy. In this case, the smaller the u_{MSE} the better the approximating function. Conversely, if $y(x)$ is affected by noise and $\hat{y}(x)$ is learned as proposed in Sect. 3.4.1, then expectation must be extended to the noise as well. Results presented in Sect. 3.4.1 hold.

7.1.2 Kullback–Leibler

The Kullback–Leibler divergence [43, 44] measures the distance between two probability density functions, which, in the following, we denote as $y(x)$ and $\hat{y}(x)$.

More specifically, if the densities $y(x)$ and $\hat{y}(x)$ exist then the Kullback–Leibler divergence is

$$u_{KL} = \int_X y(x) \log \frac{y(x)}{\hat{y}(x)} \mathrm{d}x \tag{7.1}$$

The figure of merit, also known as information divergence and relative entropy, is not a metric (and this is the reason why it is called divergence), since it does not satisfy the symmetry property, i.e., $u_{KL}\left(y(x), \hat{y}(x)\right) \neq u_{KL}\left(\hat{y}(x), y(x)\right)$. However, $u_{KL}\left(y(x), \hat{y}(x)\right) \geq 0$ and assumes value zero only when $\hat{y}(x) = y(x),\ \forall x \in X$, that is to say when the two distributions are equal.

In the machine learning field, the Kullback–Leibler divergence plays a leading role. For instance, in Bayesian machine learning it is used to approximate an intractable density model [48]. In other application scenarios, the divergence is used for parameter estimation [46], text classification [45] and, again multimedia applications [47] just to name the few.

7.1.3 L^p Norms and Other Figures of Merit

Several other figures of merit can be designed to asses the discrepancy between two functions based on the L^p norms. For instance, if we consider as punctual discrepancy $u(x) = y(x) - \hat{y}(x)$, we can use its L^p norm as figure of merit

$$u_{L^p} = \|u(x)\|_p = \|y(x) - \hat{y}(x)\|_p = \left(\int_X |y(x) - \hat{y}(x)|^p f_x(x)\mathrm{d}x\right)^{\frac{1}{p}}. \tag{7.2}$$

where $f_x(x)\mathrm{d}x = \mathrm{d}\mu(x)$ is the differential of the probability measure over X. Of particular interest are the L^1, L^2 (equivalent to the SE approach, since $u_{SE} = u_{L^2}$) and L^∞ norms.

In some cases, the punctual discrepancy is weighted by a given function $w(x) \geq 0$, $\forall x \in X$ and (7.2) becomes discrepancy induced by the weighted L^p norm

$$u_{L^p,w} = \left(\int_X w(x)|y(x) - \hat{y}(x)|^p f_x(x)\mathrm{d}x\right)^{\frac{1}{p}}$$

Other figures of merits may be derived by taking into account the mutual information [49], cross entropy [49], or maximum likelihood. However, whatever the chosen figure of merit is, the designer has solely to provide a Lebesgue measurable function with domain in a probability space, which is the unique requirement we ask for in subsequent analyses.

7.2 Probably Approximately Correct Computation

Consider given algorithm A associated with function $y(x), x \in X$ measurable according to Lebesgue and X a probability space. Denote as $\hat{y}(x), x \in X$ a given function approximating the $y(x)$ implementing algorithm $\hat{A}$. Let f_x be the probability density function associated with the measure over X. As previously mentioned we can relax the assumption by assuming that function $y(x)$ is not known but operates as an oracle providing value $y(x_i)$ once queried on x_i.

Definition *We say that function $\hat{y}(x)$ is a Probably Approximately Correct Computation (PACC) of function $y(x)$ at accuracy τ and confidence η when, given a Lebesgue measurable discrepancy function $u\left(y(x), \hat{y}(x)\right) \in \mathbb{R}$, we have that*

$$\Pr\left(u\left(y(x), \hat{y}(x)\right) \le \tau\right) \ge \eta, \quad \forall x \in X. \tag{7.3}$$

In other terms, we are requesting that the two functions are close enough according to function $u(x)$; closeness must be intended in probabilistic terms within accuracy τ on the discrepancy satisfied with probability $\eta, \forall x \in X$. The computation provided by function $\hat{y}(x)$ is approximately correct in the sense that it approximates $y(x)$ according to $u(\cdot)$ at level τ; such a statement holds at least with probability η.

From the definition, we derive several interesting cases. Consider at first $u(x) = |y(x) - \hat{y}(x)|$. Given this loss function (7.3) can be expressed as

$$\Pr\left(|y(x) - \hat{y}(x)| \le \tau\right) \ge \eta, \quad \forall x \in X \tag{7.4}$$

Since

$$\Pr\left(-\tau \le y(x) - \hat{y}(x) \le \tau\right) \ge \eta, \quad \forall x \in X$$

if we assume that τ is small then (7.4) can be cast in a more immediate and intuitive, yet less formal, form

$$\Pr\left(y(x) \simeq \hat{y}(x)\right) \ge \eta, \quad \forall x \in X.$$

The computation provided by our algorithm is approximately correct, i.e., it provides a value which nicely approximates the true one with high probability.

Example: Scalar product

As a first example of a PACC consider the error free set-up where

$$y(x) = x^T\theta^{\text{o}} \tag{7.5}$$

with $X = [-1, 1]^d$, f_x is uniform and x and $\theta^{\text{o}} \subset \mathbb{R}^d$ represent the column vectors of inputs and coefficients, respectively. The scalar product computed in (7.5) can be a linear filter of coefficients θ^o, as those used in the waivelets, e.g., see [27, 30, 31], in a Cordic computer [28, 29] or a filter bank [32]. We request function $y(x)$ to be either implemented in a digital hardware or executed on a microcontroller. Assume for simplicity that no overflow/underflow occurs in the computation and that the truncation operator has been envisaged to reduce the number of bits needed to represent the coefficients. Finite precision representation can be modeled as the perturbation vector $\delta\theta$ affecting coefficients θ^o and leading to approximated computation

$$\hat{y}(x) = x^T(\theta^o - \delta\theta).$$

Assume that $d >> 1$ and that inputs are mutually independent. If we consider the

$$u(x) = u(y(x), \hat{y}(x)) = y(x) - \hat{y}(x)$$

loss function, then the point-wise discrepancy is the variable

$$u(x) = x^T\delta\theta$$

which, from the central limit theorem, is asymptotically Gaussian with mean $E_x[u(x)] = 0$ and variance $E_x[u^2(x)] = \sigma^2 = \frac{\delta\theta^T\delta\theta}{3}$. Finally,

$$\Pr\left(|u(x) - E_x[u(x)]| \leq \lambda\frac{\sigma}{\sqrt{d}}\right) = \text{erf}\left(\frac{\lambda}{\sqrt{2}}\right)$$

i.e.,

$$\Pr\left(|y(x) - \hat{y}(x)| \leq \lambda\frac{\sigma}{\sqrt{d}}\right) = \text{erf}\left(\frac{\lambda}{\sqrt{2}}\right).$$

The PACC computation is characterized $\tau = \lambda\frac{\sigma}{\sqrt{d}}$ and $\eta = erf\left(\frac{\lambda}{\sqrt{2}}\right)$.

Example: linear regression

Consider function $y(x) = x^T\theta^o + \zeta$ where x and θ^o are the d-dimensional column vectors of inputs and given—but unknown—parameters, respectively. ζ is a zero mean white noise of variance σ_ζ^2 ruled by Gaussian pdf f_ζ. Inputs are zero centered and independent and identically distributed, extracted according to probability density function f_x of diagonal covariance $\sigma_x^2 I_d$. Consider the n samples training set $Z_n = \{(x_1, y(x_1)), \cdots (x_1, y(x_n))\}$.

Define $\chi = [x_1, \ldots, x_n]$ to be the (n, d) dimensional matrix containing the input vectors and $Y = [y(x_1), \ldots, y(x_n)]$ the $(n, 1)$ vector of associated outputs.

If we define

$$u(y(x), \hat{y}(x)) = \hat{y}(x) - y(x)$$

the least mean squared error estimate

$$\hat{y}(x) = x^T \hat{\theta}$$

can be obtained by minimizing the

$$u_{MSE} = \hat{E}_n(u(x)) = \frac{1}{n} \sum_{i=1}^{n} \left(y(x_i) - \hat{y}(x_i)\right)^2$$

and provides the parameter estimate

$$\hat{\theta} = \left(\chi^T \chi\right)^{-1} \chi^T Y.$$

We know from Sect. 3.4.4 that the distribution of $\hat{\theta}$ is centered in θ^o. $\hat{\theta}$ can then be seen as a perturbed value of θ^o so that $\hat{\theta} + \delta\theta = \theta^o$. We can repeat the derivation carried out in the previous experiment and the point-wise error

$$u(x) = x^T \delta\theta + \zeta$$

converges to a Gaussian distribution of zero mean and variance $\sigma^2 = \sigma_x^2 \delta\theta^T \delta\theta + \sigma_\zeta^2$ provided that d is large enough.

As before, we can then write

$$\Pr\left(|\hat{y}(x) - y(x)| \leq \lambda \frac{\sigma}{\sqrt{n}}\right) = erf\left(\frac{\lambda}{\sqrt{2}}\right) \tag{7.6}$$

With the choice $\tau = \lambda \frac{\sigma}{\sqrt{n}}$ and $\eta = erf\left(\frac{\lambda}{\sqrt{2}}\right)$, $\hat{y}(x)$ represents a PACC computation of $y(x)$ at level τ and probability η.

Example: Maximum value estimate

Another interesting case emerging from applications is the one where we aggregate punctual discrepancies with the maximum operator. Define

$$u_{\max} = \max_{x \in X} u(x)$$

with $\hat{u}_{\max}$ being the estimate of such a maximum as provided by a suitable algorithm. We have a good estimate $\hat{u}_{\max}$ when

$$\Pr\left(u_{\max} - \hat{u}_{\max} \leq \tau\right) \geq \eta \tag{7.7}$$

In other terms, the (7.7) states that estimate $\hat{u}_{\max}$ is a good estimate when the probability of having the true value within distance τ is high. In a way, (7.7) is closely related to the weak law of empirical maximum.

While the (7.7) is a well-posed formulation from the mathematical point of view it is of scarce use in the practice apart from very simple cases. In fact, for a generic $u(x)$ loss function, we cannot guarantee, even in probability, that given $\hat{u}_{\max}$ belongs to a neighborhood of $u_{\max}$ within distance below τ.

As we have seen in Chap. 4, this is a well-known problem whose solution within a PACC framework requires introduction of an additional level of probability.

Comments

The probabilistic theory behind the PACC well describes the operational modality of those applications for which an exact computation suffice. If an embedded system is considered, and given the comments raised in Chap. 3, we discover that very few applications require a high accuracy in the computation, e.g., those involving financial data, all the others being affected by uncertainty that affects the correctness of the computation output.

However, the use of the PACC theory appears to be of limited use in practical cases for generic $y(x)$ and $\hat{y}(x)$ functions due to the difficulty in providing (or determining) the η and τ values requested by (7.4) or identifying a $\hat{u}_{\max}$ that satisfies the (7.7).

Fortunately, thanks to the procedures based on randomized algorithms we introduce in the sequel, we will be able to provide estimates for both η and τ and $\hat{u}_{\max}$, hence making effective and operational the PACC framework. The main outcome is that any computation, under the very weak hypothesis of Lebesgue measurability, can be effectively cast into the PACC framework that makes available a set of algorithms able to identify η, τ, and $\hat{u}_{\max}$.

7.3 The Performance Verification Problem

The performance verification problem aims at verifying to which degree a performance level is attained, computing the probability that an inequality on performance is satisfied or estimating the maximum value associated with a performance discrepancy loss function.

In the following, the key actors will be the given functions $y(x)$ and $\hat{y}(x)$ and the Lebesgue measurable figure of merit $u\left(y(x), \hat{y}(x)\right)$. It is recalled that a probability density function f_x is induced by the measure of probability space X. We already pointed out in Chap. 4 that if f_x is unknown the user should consider a uniform distribution which, under mild hypotheses about the regularity of functions $y(x)$ and $\hat{y}(x)$, acts as a worst case scenario (e.g., derived bounds tend to be overestimated).

The computationally difficult problem of evaluating the different performance associated with a given figure of merit over the whole input space X can be tamed by resorting to probability and accepting a PACC computation.

We will address in the sequel the following problems

- *The performance satisfaction problem.* Given a tolerated performance loss τ for an application, compute the probability that the discrepancy $u\left(y(x), \hat{y}(x)\right) \leq \tau$, $\forall x \in X$;

- *Figure of merit expectation problem.* Provide an estimate of the expected value $E\left[u\left(y(x), \hat{y}(x)\right)\right]$ at arbitrary accuracy and confidence levels;
- *The maximum performance problem.* The goal is to provide an estimate $\hat{u}_{\max}$ for the maximum value $u_{\max} = \max_{x \in X} u\left(y(x), \hat{y}(x)\right)$;
- *The PACC problem.* Compute those η and τ characterizing the PACC degree for the given application;
- *The minimum(maximum)-perturbed expectation problem.* Estimate the minimum/maximum value performance function $u(x)$ assumes when perturbations affect it.

Clearly, $u\left(y(x), \hat{y}(x)\right)$ can degenerate to $\hat{y}(x)$. When this happens, the considered problems must be intended as applied to function $\hat{y}(x)$.

7.3.1 The Performance Satisfaction Problem

The performance satisfaction problem aims at assessing the level of performance satisfaction of a given $\hat{y}(x)$ function in approximating $y(x)$ according to figure of merit $u\left(y(x), \hat{y}(x)\right)$ and a tolerated performance loss τ.

Examples of applications

- We designed application $\hat{y}(x)$ solving my problem. Is that satisfying the accuracy constraint set at level τ requested by the application?
- We designed solution $y(x)$ which is working well in a high performance machine where it satisfies the requested real time performance. We port it on our embedded system where it becomes solution $\hat{y}(x)$. Is porting granting a loss in execution time between the two platforms below τ?
- We designed solution to our problem $y(x)$ within a high resolution platform (e.g., Matlab, Mathematica, Simulink). Then we need to port the solution to an embedded system characterized by a given finite precision representation (limited word-length for data and inner variables, truncation mechanisms and look up tables). Is the performance loss in accuracy we are introducing tolerable if τ is what we are willingly to loose?
- Our application solution $\hat{y}(x)$ has an accuracy scalable with complexity in the sense that the solution performance (accuracy, execution performances, power consumption) scales with the solution complexity (the higher the complexity the better the performance). We would like to minimize complexity provided that performance loss τ is attained.

The above problems can be formalized as follows: Given a tolerated performance loss τ we wish to estimate the satisfaction level

$$u\left(y(x), \hat{y}(x)\right) \leq \tau, \quad \forall x \in X$$

that is to say, determine the percentage of points of X satisfying the inequality. The problem can be immediately related to the algorithm verification problem presented in Sect. 4.4.1 by simply substituting x to each occurrence of ψ. We recall the main operational steps.

The percentage of points $x \in X$ satisfying $u(x) \leq \tau$ is simply the ratio

$$n_{u(x)\leq\tau} = \frac{\int_X I(x)\mathrm{d}x}{\int_X \mathrm{d}x}$$

where

$$I(x) = \begin{cases} 1 \text{ if } & u(x) \leq \tau \\ 0 \text{ otherwise} \end{cases}$$

Since determination of $n_{u(x)\leq\tau}$ is a computationally hard problem for a generic function we move to a probabilistic problem.

In order to do that, consider the probability density function f_x defined over X and request to evaluate the probability

$$p_\tau = \Pr\left(u(x) \leq \tau\right) = \frac{\int_X I(x) f_x \mathrm{d}x}{\int_X f_x \mathrm{d}x} = \int_X I(x) f_x \mathrm{d}x$$

since $\int_X f_x \mathrm{d}x = 1$. We have seen in Chap. 4 that p_τ can be evaluated through randomization and that, given τ, we can compute the estimate $\hat{p}_n$ of p_τ by drawing n samples $x_1, \ldots x_n$ according to f_x and evaluate the indicator function

$$I\left(u(x) \leq \tau\right) = \begin{cases} 1 \text{ if } u(x) \leq \tau \\ 0 \text{ if } u(x) > \tau \end{cases}$$

and the estimate

$$\hat{p}_n = \frac{1}{n} \sum_{i=1}^{n} I\left(u(x_i) \leq \tau\right).$$

By selecting a number of samples according to the Chernoff bound

$$n \geq \frac{1}{2\varepsilon^2} \ln \frac{2}{\delta}$$

we have that

$$\Pr\left(|\hat{p}_n - p_\tau| \leq \varepsilon\right) \geq 1 - \delta$$

holds for any accuracy level $\varepsilon \in (0, 1)$ and confidence $\delta \in (0, 1)$. Value $\hat{p}_n$ is the probabilistic outcome of the performance satisfaction problem provided that small ε and confidence δ values are set. The randomized algorithm solving the problem is Algorithm 6.

7.3.2 The Figure of Merit Expectation Problem

The performance satisfaction problem returns the percentage of points satisfying a given bound on the tolerated performance associated with my solution. The expectation problem aims at evaluating the expected value of the loss function.

The problem solves the cases where

- We are interested in quantifying the expected performance loss having moved from solution $y(x)$ to $\hat{y}(x)$.
- We are interested in quantifying the expected performance of our application $\hat{y}(x)$ for a given problem, e.g., in execution on our embedded system. In such a case $u\left(y(x), \hat{y}(x)\right) = u\left(\hat{y}(x)\right)$.

In the following to ease the derivation assume that $u(x)$ is defined in the $[0, 1]$ interval. This normalization operation is simply done with a rescaling of function $u(x)$.

Define $E[u(x)]$ to be the expectation of function $u(x)$ according to f_x and follow the derivation given in Sect. 4.4.3 for the evaluation of the expectation problem with randomization. Briefly, set accuracy level $\varepsilon \in (0, 1)$, confidence $\delta \in (0, 1)$ and draw

$$n \geq \frac{1}{2\varepsilon^2} \ln \frac{2}{\delta}$$

i.i.d. samples $x_1, \ldots, x_n$ from random variable x defined over X according to f_x and generate the estimate

$$\hat{E}_n(u(x)) = \frac{1}{n} \sum_{i=1}^{n} u(x_i)$$

then,

$$\Pr\left(|\hat{E}_n(u(x)) - E[u(x)]| \leq \varepsilon\right) \geq 1 - \delta.$$

Value $\hat{E}_n(u(x))$ is the probabilistic outcome of the algorithm. By selecting small ε and δ, we obtain a good approximation. The randomized algorithm solving the problem is given in Algorithm 9.

The interesting reader should refer to Sect. 4.4.3 for the interesting relationships between the needed samples requested by the Chernoff bound and those needed from the central limit theorem.

7.3.3 The Maximum Performance Problem

The maximum performance problem aims at estimating the maximum value function $u(x)$ can assume. Its relevance in many application cases is immediate. As we have seen in Sect. 4.4.2 (were all details are given) the problem requires evaluation of

$$u_{\max} = \max_{x \in X} u(x)$$

and its analytical determination is impossible for most of functions. We saw in Sect. 4.4.2 that a manageable probabilistic version of it requires two levels of probability. From the operational point of view, once fixed accuracy ε and confidence δ, we have to draw

$$n \geq \frac{\ln \delta}{\ln(1-\varepsilon)}$$

i.i.d. samples $x_1, \ldots, x_n$ and generate the estimate $\hat{u}_{\max}$

$$\hat{u}_{\max} = \max_{i=1,\ldots,n} u(x_i)$$

then,

$$\Pr\left(\Pr\left(u(x) \geq \hat{u}_{\max}\right) \leq \varepsilon\right) \geq 1 - \delta.$$

The estimate of the maximum performance level is $\hat{u}_{\max}$. All comments raised in Sect. 4.4.2 hold. The reference randomized algorithm solving the maximum performance problem is given in Algorithm 8.

7.3.4 The PACC Problem

We have seen that the PACC problem requires the evaluation of τ and η so that

$$\Pr\left(|y(x) - \hat{y}(x)| \leq \tau\right) \geq \eta \tag{7.8}$$

holds. Then, if τ is small $y(x) \simeq \hat{y}(x)$ with probability η. The problem can be solved by considering the figure of merit $u(x) = |y(x) - \hat{y}(x)|$, although a more general discrepancy function could be considered.

Solution of the PACC problem for a given function requires the evaluation of τ for which (7.8) holds with high probability. This problem can be addressed by using the method proposed in Sect. 4.4.1 and here reproposed in its main steps. Define $p(\gamma)$ as

$$p(\gamma) = \Pr\left(u(x) \leq \gamma\right) = \Pr\left(|y(x) - \hat{y}(x)| \leq \gamma\right)$$

for an arbitrary but given γ. Estimate $p(\gamma)$ according to method given in Algorithm 6 which returns $\hat{p}_n(\gamma)$. Explore the γs by selecting arbitrary incremental points and generate set $\Gamma = \{\gamma_1, \ldots, \gamma_k\}$ s.t. $\gamma_i < \gamma_j$ $\forall i < j$. For all $\gamma \in \Gamma$ evaluate $\hat{p}_n(\gamma)$ so as to build

$$\hat{p}_\Gamma = \{\hat{p}_n(\gamma_1), \ldots, \hat{p}_n(\gamma_k)\}$$

as done in Algorithm 7.

Define $\bar{\gamma}$ to be the smallest value for the finite sequence $\{\gamma_1, \ldots, \gamma_k\}$ for which $\hat{p}_n(\gamma_i) = 1, \forall \gamma_i \geq \bar{\gamma}, \gamma_i \in \Gamma$. Having selected k according to the Chernoff bound discrepancy

$$|p(\gamma) - \hat{p}_n(\gamma)| \leq \varepsilon$$

holds with probability $1 - \delta$ and is satisfied for all γ. As such, it also holds for $\hat{p}(\bar{\gamma})$. We have that

$$\hat{p}_n(\bar{\gamma}) - \varepsilon \leq p(\bar{\gamma}) \leq \hat{p}_n(\bar{\gamma}) + \varepsilon,$$

i.e.,

$$1 - \varepsilon \leq p(\bar{\gamma}) \leq 1$$

from which $p(\bar{\gamma}) \geq 1 - \varepsilon$. The PACC problem is solved and provides $\eta = 1 - \varepsilon$ and $\tau = \bar{\gamma}$.

7.3.5 The Minimum(Maximum)-Perturbed Expectation Problem

The minimum (maximum) expectation problem aims at estimating the minimum (maximum) value of the expectation of $u(x)$ when perturbations Δ affects $u(x)$, thus providing a perturbed version of the performance function $u(x, \Delta)$. Examples of applications are the following

- Part of my embedded application has been designed with an analog technology which is subject to electronic noise. I would like to know the minimum(maximum) performance of my device subject to the fact perturbations affect it. I am happy to know the minimum(maximum) performance by taking expectation of the perturbation space.
- There is a source of uncertainty affecting my embedded system and I would like to obtain an estimate of the performance loss by taking the average over the perturbation space.

The formalization of the problem is as follows:

Consider $u(x, \Delta) \in [0, 1]$, $x \in X \subset \mathbb{R}^d$ and $\Delta \in D \subset \mathbb{R}^k$ (being X and Δ probability spaces) and focus, e.g., on the minimum problem

$$u_{\min} = \min_{x \in X} E_{\Delta}[u(x, \Delta)]$$

which is equivalent to

$$\begin{cases} \phi(x) = E_{\Delta}[u(x, \Delta)] \\ u_{\min} = \min_{x \in X} \phi(x) \end{cases}$$

solution to this computationally demanding problem is given in Sect. 4.4.4 and addressed by randomized algorithm given in Algorithm 10.

7.4 Accuracy Estimation: A Given Dataset Case

In some of previously discussed performance verification problems we were able to assess the quality of an estimator in probabilistic terms. For instance, in the figure of merit expectation problem discussed in Sect. 7.3.2, by extracting n data

$$x_1, \ldots, x_n$$

from random variable x according to f_x over X, we where able to provide an estimate $\hat{E}_n(u(x))$ for the true unknown expectation $E[u(x)]$. At the same time, we provided a quality assessment of the estimator since we wrote that

$$\Pr\left(|\hat{E}_n(u(x)) - E[u(x)]| \le \varepsilon\right) \ge 1 - \delta. \tag{7.9}$$

The designed framework implicitly permits that we can extract sequences of arbitrary length n from X so that we can meet both accuracy ε and confidence δ according to (7.9).

However, in the real life we generally encounter situations where n is given (e.g., we have n finite samples from an industrial process, n models built on the sensor data stream, etc). Whenever that is the case, (7.9) clearly must hold but n is now given and δ and ε cannot assume any arbitrary values any more. If we assume that δ must be fixed since we want our results to hold at some confidence level, e.g., $\delta = 0.95$, then, accuracy is no more arbitrary and is set as

$$\varepsilon = \sqrt{\frac{1}{2n} \ln \frac{2}{\delta}}$$

by simply inverting the Chernoff bound. Unfortunately, if n is not large enough the accuracy we should expect might be poor.

Two comments need to be made at this point. The first is that results derived by invoking bounds such as the Chernoff one are pdf free and, as such, the bound might be rather conservative. As a consequence, the derived ε might be hardly usable if the number of samples is fixed and not large enough. Second, we should remind that, given a finite number of samples, we cannot pretend to have an arbitrary accuracy since the amount of information carried by the data set is finite and depends on n. As an example, we recall that the estimated standard deviation of mean $\hat{x}$ based on n i.i.d scalar data $x_1, \ldots, x_n$ extracted from a distribution of unknown mean μ and known variance σ^2, is $\sqrt{\frac{\sigma^2}{n}}$. The quality of the estimator $\hat{x}$ scales with $n^{\frac{1}{2}}$ and cannot be improved unless we increase the number of samples n. In following subsections, we wish to assess the quality of an estimate in the case n is given.

7.4.1 Problem Formalization

Consider a data set Z_n obtained by extracting n i.i.d. samples $x_1, \ldots, x_n$ from random variable x defined over X, i.e., $Z_n = \{x_1 \ldots, x_n\}$ and construct the estimator $\Phi_n = \Phi(Z_n)$. We are interested in providing an indication of the quality ζ of Φ_n, e.g., we wish to provide a confidence interval for Φ_n.

The problem is of extreme relevance in many applications and, in particular, in embedded systems where we wish to carry out the performance evaluation of a figure of merit (estimator) having a finite and given data set to be used to compute ζ. The performance verification problem is the same introduced in this chapter with the unique difference that, here, n is fixed, e.g., we have only n i.i.d. data from sensors, n i.i.d. parameter models or extracted features, n time measurements related to the execution of a thread.

Clearly, the ideal framework would recommend to carry out the following procedure

1. Extract m independent data sets of cardinality n from X so as to generate datasets $Z_n^1, \ldots, Z_n^m$;
2. Evaluate, in correspondence of the generic i-th data set Z_n^i the estimator $\Phi_n^i = \Phi(Z_n^i)$. Repeat this procedure for all $i = 1, \ldots, m$.;
3. Estimate the quality $\zeta\left(\Phi_n^1, \ldots, \Phi_n^m\right)$ of the estimator Φ_n based on the m realizations $\Phi_n^i = \Phi(Z_n^i), i = 1, \ldots, m$.

Unfortunately, the above framework is mostly theoretical: if we have m independent datasets Z_n we should use all nm data to provide a better estimate. This means that in practical applications we have only a dataset but, at the same time, we are interested in evaluating the quality ζ of the estimator Φ_n.

The literature introduces interesting approaches for providing an assessment ζ of the quality of an estimator Φ given a limited data set n.

7.4.2 The Bootstrap Method

In the bootstrap method [236] the needed m data sets $Z_n^i, \ i = 1, \ldots, m$ are extracted from Z_n with replacement. It means that, once a sample x_j has been extracted and inserted in the generic Z_n^i set, x_j is also placed back in Z_n that keeps all its original n data. Once all estimates Φ_n^i, $i = 1, \ldots, m$ have been generated they are used to compute $\zeta\left(\Phi_n^1, \ldots, \Phi_n^m\right)$.

The Bootstrap algorithm is given in Algorithm 15. Efron and Tibshirani [236] proves that the distinct number of samples we should expect in Z_n^i is $\approx 0.632n$. This comment allows us for reducing the computational load associated with the execution of the algorithm. In fact, if we are expecting to receive approximatively $0.632n$ independent samples and the estimator requires to compute point wise terms, e.g., $u(x_i)$ for sample x_i given a generic $u(\cdot)$ function, then, there is no need to compute all n values and a weighting approach can be consider.

Algorithm 15: The bootstrap algorithm

i = 0;
while $i < m$ **do**
  Extract n samples with replacement from Z_n and insert them in Z_n^i;
  Compute $\Phi_n^i = \Phi(Z_n^i)$;
  i = i+1;
end
Evaluate the assessment $\zeta\left(\Phi_n^1, \ldots, \Phi_n^m\right)$ of the quality of the estimator Φ_n.

Example: The Bootstrap Method

Consider, as a straight example, the figure of merit expectation problem discussed at the beginning of the section. Differently from the derivation based on the Chernoff bound, we do not need to require here that the Lebesgue measurable function $u(x)$ is defined in interval $[0, 1]$ but it is sufficient that $u(x)$ is bounded for some value (not to be necessarily known).

Select $\Phi = E[u(x)]$ and $\Phi_n = \hat{E}_n(u(x))$ evaluated on Z_n. The estimate of $E[u(x)]$ we have is $\hat{E}_n(u(x))$. We are now interested in evaluating the quality of the estimate ζ, e.g., by evaluating the variance of the estimates generated with the bootstrap method. This is done by invoking Algorithm 15.

For instance, the quality of the estimator, here considered to be the variance of the estimator evaluated according to the bootstrap method Var_B, can be estimatedas

$$Var_B = \zeta\left(\Phi_n^1, \ldots, \Phi_n^m\right) = \frac{1}{m-1} \sum_{i=1}^{m} \left(\Phi_n^i - \Phi_n\right)^2.$$

The bootstrap algorithm shows to be accurate for a wide range of estimators but is also characterized by a significant computational load. Variants of the bootstrap method have been suggested in the literature, e.g., the m out of n bootstrap [237] to solve the consistency issue in applications where the bootstrap fails. The computational issue has been addressed, among the others, in the m out of n bootstrap where $Z_{n'}^i$ sets are chosen to have a smaller cardinality than Z_n^i, i.e., $n' < n$, and in [238] where a method aiming at using a small m and extrapolation techniques has been proposed and in the bag of little bootstrap method [239], which will be detailed in the next subsection.

7.4.3 The Bag of Little Bootstraps Method

The Bag of Little Bootstraps (BLB) method is a bootstrap-inspired method proposed in [239] to mitigate the problem posed by Bootstrap and associated with the required computational complexity. BLB shows to be accurate and appears to over-perform all other bootstrap methods in terms of computational complexity, hence becoming a very appealing method for Big Data.

Algorithm 16: The Bag of Little Bootstraps algorithm

```
n' = n^γ;
i = 0;
while i < m do
    Extract n' samples without replacement from Z_n and insert them in Z^i_{n'};
    j = 1;
    while j < r do
        Extract n samples with replacement from Z^i_{n'} and insert them in Z^i_{n,j};
        Compute Φ^i_{n,j} = Φ(Z^i_{n,j});
        j=j+1;
    end
    Evaluate ζ_i = ζ(Φ^i_{n,1}, ···, Φ^i_{n,r});
    i = i+1;
end
Evaluate the assessment ζ for of the quality of the estimator Φ_n as ζ = (Σ_{i=1}^m ζ_i)/m.
```

The BLB algorithm is given in Algorithm 16. The starting point of BLB is to select a smaller data set of cardinality $n' < n$. Authors select $n' = n^{\gamma}$, e.g., with $\gamma = 0.6$. Then, m subsets $Z^i_{n'}, i = 1, \ldots, n'$ are extracted from Z_n without replacement. For each subset $Z^i_{n'}$ r subsets $Z^i_{n,j}$, $j = 1, \ldots, r$ are generated with replacement (little bootstraps) and the corresponding estimators $\Phi^i_{n,j} = \Phi(Z^i_{n,j})$ evaluated. Finally, the quality of the estimator is evaluated for each little bootstrap and yields $\zeta_i = \zeta\left(\Phi^i_{n,1}, \ldots, \Phi^i_{n,r}\right)$. At the end of the procedure, we end with m "bags", the i-th one associated with quality assessment ζ_i. The final assessment ζ of the quality of the estimator Φ_n is considered to be the ensemble of the little bootstrap one as $\zeta = \frac{\sum_{i=1}^{m} \zeta_i}{m}$. r is generally chosen so that ζ_i ceases to fluctuate and, in general, this happens in correspondence with small values of r, e.g., refer to [239].

As mentioned, the BLB shares the consistency properties of the bootstrap and higher order correctness under the same hypotheses assumed by the bootstrap.

As claimed in [239], the BLB shares the fast convergence rate of the bootstrap where the procedure's output scales as $O(\frac{1}{n})$ instead of the $O(\frac{1}{\sqrt{n}})$ rate achieved by asymptotic approximations. This fast convergence rate assumes that $n' = \Omega(\sqrt{n})$, i.e, $\lim_{n\to\infty} sup|\frac{n'}{\sqrt{n}}| > 0$, and m is large enough compared with the variability observed in data samples. Satisfaction of such assumption grants that n' is significantly smaller than n (but larger than $\sqrt{n}$). Moreover, as shown in [239], the BLB is faster than the bootstrap even in a serial execution. We comment that the algorithm can be easily made parallel since each little bootstrap procedure can be parallelized.

7.5 Cognitive Processes and PACC

All mechanisms involving cognitive processing, e.g., those carried out by VM-PFC—ACC are real and, although complex and mostly unknown to us in detail, Lebesgue measurable. The PACC theory can hence be applied to cognitive processing. It is extremely natural to assume that those (sub)systems operate in probability given the highly uncertainty associated with the processing. Our actions are, in fact, mostly correct and the output to an emotion or an action mostly correct, with high probability: exactly what the PACC theory is about.

7.6 Example: Accuracy Assessment in Embedded Systems

The section aims at showing how the accuracy assessment framework can be utilized in embedded systems to assess the quality of a proposed solution and drive the designer toward the identification of the most appropriate one within a given set. The reference application is based on a neural network that, once designed in a high precision framework, needs to be ported onto an embedded system. The relevance of the example is in the diversified computation carried out by the neural computation that requires evaluation of scalar products as well as nonlinear functions.

The porting operation introduces several sources of structural perturbations whose impact on accuracy can be effectively evaluated with the framework presented in the chapter.

In particular, the chosen application refers to a neural network designed to provide a virtual sensor for the chemical process described in the `Mathworks Matlab` neural network toolbox (chemical_dataset).

The virtual sensor is inferred from the readings of other eight sensors, whose values provide the inputs of the network. The chosen network topology is feedforward, with 8 inputs ($x \in X \subset \mathbb{R}^8$), 10 neurons in the hidden layer characterized by a hyperbolic tangent activation function, and a single linear output neuron. Once the neural network was successfully trained, neural function $f(\theta, x)$ was requested to be implemented on the embedded systems.

To illustrate the effect of structural perturbations induced by the embedded system architecture on function $f(\hat{\theta}, x)$, we consider two different digital embedded implementations for the neural network that, once ported, become the approximated $\hat{f}(\theta, x)$. The first implementation is based on a 16 bits word-length solution, the second on a 8 bits one.

In this way, we investigate the performance in accuracy of the two architectural solutions, an operation which also allow us for selecting the final target platform depending on the requested accuracy, the power consumption as well as the requested area (e.g., think of an ASIC or a FPGA implementation).

Results were emulated on a 32 bits ARM Cortex M3 microcontroller. Since the ARM Cortex M3 microcontroller is not equipped with a Floating Point Unit (FPU),

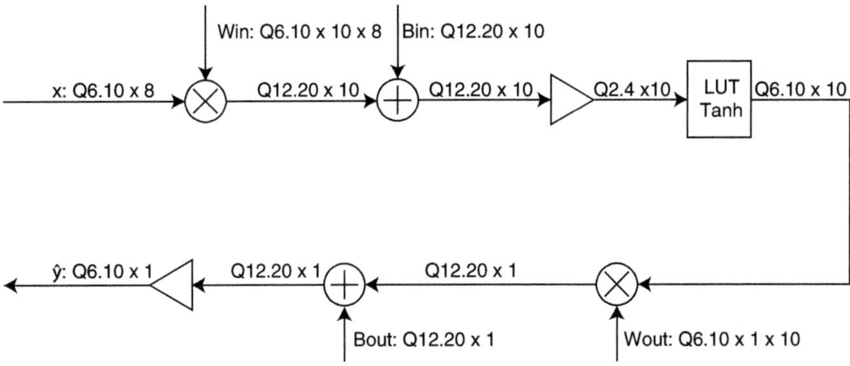

Fig. 7.1 The neural network data flow ported on a 16 bits architecture. The activation value u feeding the hidden units can be obtained with a matrix product between the 10×8 matrix Win containing the weights between the input and the hidden layer and the input column vector x; the bias value Bin is then added so that $u = Win\, x + Bin$. Inputs composing x are represented with a $Q6.10$ notation, weights Win with a $Q6.10$ notation, the outcome of their product on a $Q12.20$ notation. The bias term coded as $Q12.20$ is added and the outcome u on $Q12.20$ is reduced to $Q2.4$ to feed the LUT. The $Th(u)$ value coming from the LUT is defined on $Q6.10$ and multiplied by the weights connecting the hidden layer with the output neuron. The bias is added and the output defined on a $Q12.20$ notation reduced to a 16 bits output in the form $Q6.10$

all computations must be performed with a fixed point 2cp representation as presented in Sect. 3.2.4.

In the following, a fixed point number is represented with the $Qx.y$ notation, which implies a 2cp representation on $x + y$ bits with x bits to the left of the fixed point (integer part, sign bit included) and y bits after the point (fractional part).

We recall that a sum between two numbers defined on $Qx.y$ does not modify the position of the radix point, but overflow might occur. Instead, a multiplication between numbers $Qx.y$ and $Qk.z$ generates a value represented as $Q(x+k).(y+z)$ whose radix point is shifted to the left of y positions compared to z. This effect must be taken into account if the final value needs to be brought back to a $Qx.y$ notation. Usually, multiplications take place with operands characterized by the same Q representation.

The data flows associated with the two 16 bits and 8 bits implementations are shown in Figs. 7.1 and 7.2, respectively. The complete description of the architectural operations is detailed in the caption of Fig. 7.1.

The 16 bit implementation uses the same $Q6.10$ representation for inputs, output, and weights. With such a choice, we are safe from any possible occurrence of overflows. The sum following the product of the generic input by the corresponding weight is still performed at the full resolution of $Q12.20$ bits. Differently, in the 8 bits implementation, we adopt different resolutions for inputs, outputs, and weights (input and output values are represented using a $Q2.6$ coding, while weights are represented with a $Q5.3$ one). The neuron biases are still represented at the full resolution of $Q7.9$. After the operations shifts are introduced to bring back the obtained

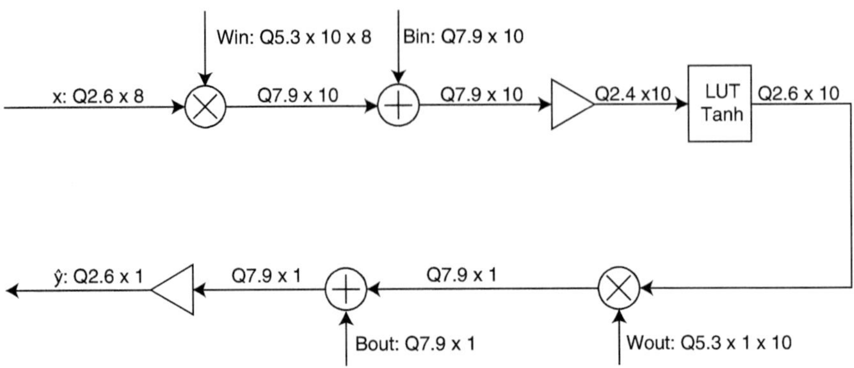

Fig. 7.2 The neural network data flow ported on a 8 bits architecture. The description of the flow is similar to that given in the caption of Fig. 7.1 with the suitable change of word length for the involved entities

numbers to the envisaged word-length. The embedded code for implementing the neural activation value ready to feed the nonlinear activation function of the hidden layer is given in Listing 7.1.

The evaluation of the nonlinear hyperbolic tangent function is particularly costly from the computational point of view. To deal with the issue, the best solution is to rely on a Look Up Table (LUT) memory enumerating the input–output relationship in correspondence of some points. The input of the memory is the neural activation value (i.e., the scalar product between the neuron inputs and the associated weights), the output is the content stored in the memory cells which represents the value the activation function assumes in correspondence of the input value. We aimed at keeping the size of the LUT as small as possible. As such, the input values where coded as unsigned $Q2.4$ values for a total of a 64 cells LUT memory; the output values follow the encoding used for the inputs and, as such, depend on the chosen architecture ($Q6.10$ and $Q2.6$ for the 16 and 8 bits architectures, respectively). We comment that the input of the LUT is unsigned. The reason is that the hyperbolic tangent (Th) is an odd function for which $Th(-u) = -Th(u)$. As such, we can represent the inputs by uniformly subdividing the interval [0, 4] (values above input 4 provide a saturated output at 1): there is no need to represent the full [−4, 4] interval with an immediate memory saving.

The approximation ability of the solution is given in Fig. 7.3 for the $Q6.10$ coding of the output.

The embedded code implementing the LUT is shown in Listing 7.2. As it can be noted in Fig. 7.3, the quantization error does not introduce a bias in the approximated function, since we opted for a rounding of the reduced argument instead of a simple truncation of the value. The cost of an extra shift and the sum needed to implement the rounding operator is well compensated by the disappearing of the bias term in the approximating the hyperbolic tangent function.

Listing 7.1: The C embedded code providing the activation value "ured" to be inputted to the LUT in the 16 bits architecture. NEURONS=10, INPUTS=8 represent the number of hidden and input units, respectively. "u" is the full precision activation value. The code returns the number of encountered overflows

```
int sumProductsIn(int16_t x[],int16_t ured[]){

  int i, j, overflow=0;
  int32_t u;

  for (i=0;i<NEURONS;i++){
    u=0;
    for (j=0;j<INPUTS;j++){
        u=u+Win[i][j]*(x[j]);
    }/*Win is a global array
    containing the input-hidden layer weights*/

    u=u+b1[i];
    overflow=overflow+reduxSums(u,&ured[i]);

  }
}

int reduxSums(int32_t arg, int16_t *ured){

/* reduces a Q12.20 value to a Q6.10 value*/

  int16_t temp;
  int32_t redux=(arg>>10);/*realign radix point*/

/*overflow management*/

  if (redux>32767){
        *ured=32767;
        return -1;
  }
  if (redux<-32768){
        *ured=-32768;
        return -1;
  }
  temp=redux&0xFFFF;
  *ured=temp;
  return 0;

}
```

Figures 7.4 and 7.5 compare the output of the virtual sensor in its two implementations with that evaluated on a high precision platform. As expected, the higher the number of bits made available the better the reconstruction performance. However, we comment that the 8 bits architecture provides a good reconstruction ability which lowers in correspondence with high peeks.

To quantify the accuracy level of the porting of the algorithm on the embedded systems, we considered the figure of merit

$$\hat{E}_N(u(x)) = \frac{1}{N}\sum_{i=1}^{N} |f(x_i, \theta) - \hat{f}(x_i, \theta)|.$$

Listing 7.2: Evaluation of the activation function value with a LUT. 16 bits architecture case. At first the activation value is transformed so as to be always positive. The rounding operator is introduced to remove the error bias otherwise present if a truncation operator is considered instead.

```
int16_t tanhQ16(int16_t u){
        int16_t out;
        uint16_t reduced;
        int16_t uc;
        /* leverage the antisymmetry of the
        hyperbolic tangent:
        Th(-u)=-Th(u) */
        if (u<0)
                uc=-u;
        else
                uc=u;
        reduced = ((uc+((uc&0x7F)>>1))>>6)&0xFF;
        /* reduce value from $Q6.10$ to $Q3.5$ with rounding
        to reduce the bias */
        if (reduced>63)
                out=(1<<10); /*saturation value*/
        else
                out=LUT[reduced];
                /* LUT is a 64 locations array
                containing the hyperbolic tangent
                values */
        if (u<0)
                return -out;
        else
                return out;
}
```

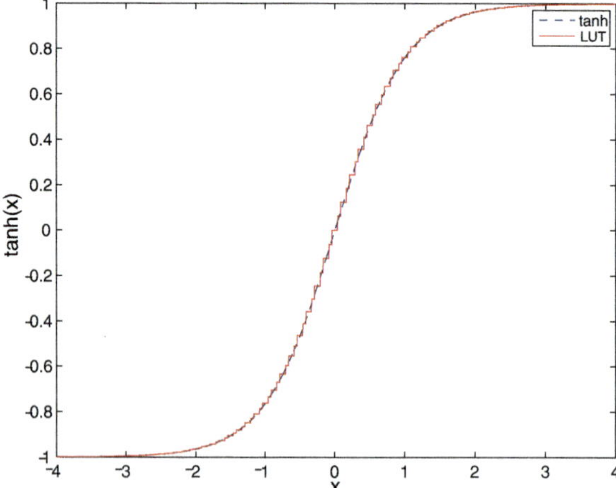

Fig. 7.3 The approximation accuracy of the hyperbolic tangent with a 64 cells LUT in the 16 bits representation for the output value

The accuracy performance assessment was carried out by following the figure of merit expectation problem delineated in Sect. 7.3.2 after rescaling the $u(\cdot)$ function. The input space was explored with a number of samples N drawn according to the Chernoff bound.

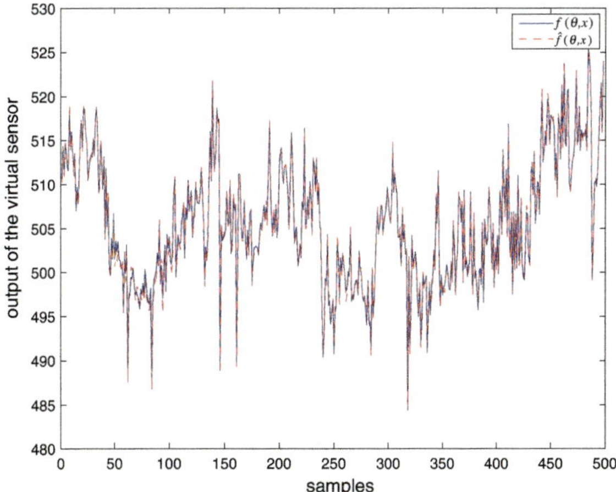

Fig. 7.4 The accuracy of the virtual sensor data stream provided by the 16 bits embedded architecture (function $\hat{f}(\theta, x)$) compared with the reference one (function $f(\theta, x)$)

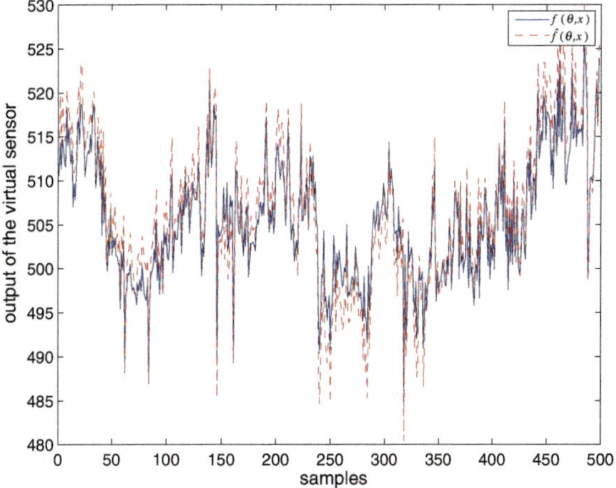

Fig. 7.5 The accuracy of the virtual sensor data stream provided by the 8 bits embedded architecture (function $\hat{f}(\theta, x)$) compared with the reference one (function $f(\theta, x)$)

The figure of merit expectation problem was solved by considering an incremental accuracy ε (the confidence value δ was fixed to 0.05). The numbers of points drawn according to Chernoff are given in Table 7.1. Since the distribution induced on the 8th dimensional input space in unknown we considered an uniform distribution. Each experiment was repeated 40 times. The expected value of the performance loss is given in Fig. 7.6 for the 16 bits architecture. Results show also the confidence interval based on one standard deviation. As expected, the larger the N the smaller the confidence interval.

Similar results are given in Fig. 7.7 for the 8 bits embedded architecture. As expected, the average error is higher for the 8 bits implementation.

Table 7.1 The number of points chosen to address the figure of merit expectation problem.

δ	ε	N
0.05	0.061	500
0.05	0.043	1000
0.05	0.035	1500
0.05	0.030	2000
0.05	0.025	3000
0.05	0.021	4000

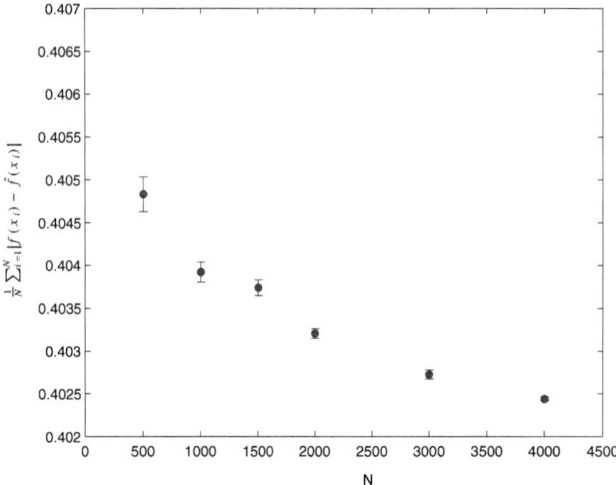

Fig. 7.6 The figure of merit expectation problem. The figure shows the performance loss associated with the porting of the virtual sensor neural network-based code on the 16 bits embedded architecture. Expected values and the confidence intervals are given

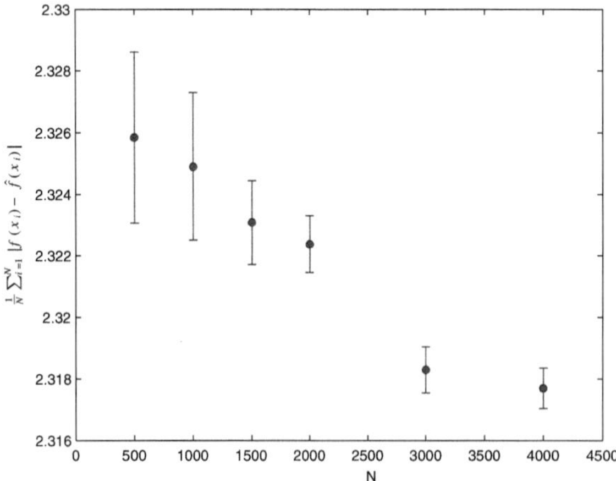

Fig. 7.7 The figure of merit expectation problem. The figure shows the performance loss associated with the porting of the virtual sensor neural network-based code on the 8 bits embedded architecture. Expected values and the confidence intervals are given

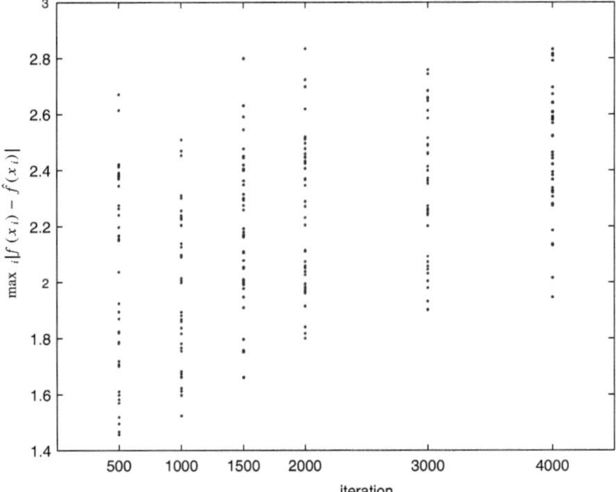

Fig. 7.8 The maximum performance problem for the 16 bits architecture. The experiment is repeated 40 times and each estimate of the maximum coming from the particular realization on N inputs is plotted

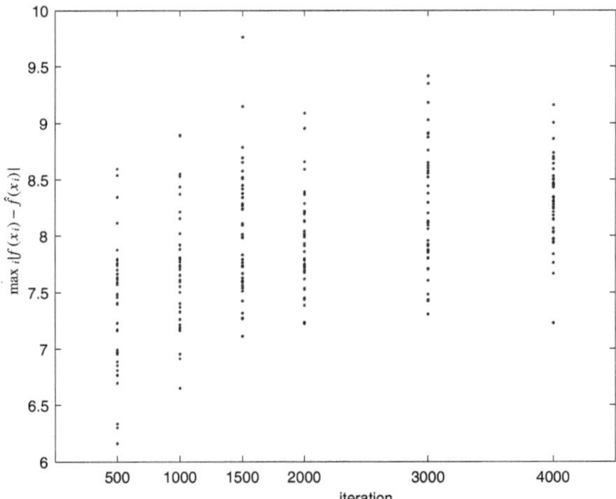

Fig. 7.9 The maximum performance problem for the 8 bits architecture. The experiment is repeated 40 times and each estimate of the maximum coming from the particular realization on N inputs is plotted

Figures 7.8 and 7.9 present the the maximum performance problem presented in Sect. 7.3.3 for the two embedded implementations. The experiment was carried out 40 times in correspondence with each chosen N ($\delta = 0.05$, $N \geq \frac{\ln \delta}{\ln(1-\varepsilon)}$); estimates of the maximum value $\hat{u}_{\max}$, $u(x) = |f(\theta, x) - \hat{f}(\theta, x)|$ are plot (one for each experiment run). No scaling to the [0, 1] interval is requested for $u(x)$. We comment that the variability of the estimates reduce as N increases and that the maximum error introduced by the 8 bits architecture is about four times that introduced by the 16 bits one. Depending on the tolerated accuracy loss and the given constraints on cost and power consumption, the designer might decide which architecture should be considered between the 16 and the 8 bits one.

Chapter 8
Intelligent Mechanisms in Embedded Systems

Many embedded systems require intelligent mechanisms to deal with those situations where either evolution or time variance requests a reaction to grant a performance level. Adaptation is the basic form of intelligence to be considered every time the embedded system has to react quickly, possibly with a very simple algorithm to minimize both time-to-reaction and energy consumption. It must be made immediately clear that simple has not to be confused with trivial. In fact, most of the times, adaptive mechanisms are the result of sophisticated techniques that provide effective actions thanks to the theory behind them. We have seen in Chap. 6 that adaptation represents that form of intelligence associated with the execution of automatic cognitive processes and it must be considered when a prompt reaction to some stimuli is more relevant than high accuracy in the answer. We will discover that adaptation plays a relevant role, e.g., in keeping under control the energy consumption of the device, maximizing the efficiency of the energy harvesting process, keeping the units clocks synchronized. Sometimes, the needed level of intelligence can scale up and a controlled "conscious" mechanism might be requested to carry out more accurate decisions. The need to reprogram the embedded device (when and how), the implementation of sophisticated mechanisms for adaptive sensing, taking advantage of group information to improve accuracy as in distributed clock synchronization are some relevant examples of "conscious" mechanisms.

The ultimate decision between which strategy must be implemented, either automatic (adaptation) or controlled (conscious) depends on the application constraints. If our embedded system has to satisfy strong real-time constraints as well as keep under control the energy consumption associated with the action to be undertaken, only very simple automatic strategies based on adaptation are expected to be the ones to be considered. Differently, if accuracy performance is more relevant than computational complexity and energy and power consumption, probably a controlled mechanism is to be preferred.

In this chapter, we focus on some fundamental mechanisms an embedded system should possess to expose basic intelligent abilities. At first, the attention will be devoted toward those low-level adaptation mechanisms that, by adapting both clock

C. Alippi, *Intelligence for Embedded Systems*, DOI: 10.1007/978-3-319-05278-6_8,

frequency and power voltage, allow the digital embedded system to reduce the consumed energy (automatic process, adaptation mechanisms). Afterward, we will investigate strategies aiming at reducing the power consumption at a higher abstraction level. In particular, we will address the adaptive sensing issue, requested by embedded systems to keep under control the power consumption in eager sensors and, at the same time, control the acquisition bandwidth. In other words, sampling is a characteristic of the application and the requested rate of data sampling depends on the application and how it evolves with time (both automatic and controlled processes, adaptation and cognitive mechanisms).

Then, we will shift the focus towards optimal energy harvesting for tiny photovoltaic cells, given the high energy density these technologies provide compared with other solutions e.g., piezoelectric, wind turbines, and Peltier cells. In energy harvesting, adaptation mechanisms are required both at the electronic and algorithm levels to maximize the harvested energy by passively tracking changes in the environment. The efficiency of the harvesting mechanism is strictly related to adaptation: no adaptive solutions have to assume that the phenomenon providing energy is weakly time variant, a situation that is hardly met in real situations, e.g., think of the intensity and direction of the wind and the intensity and orientation of solar energy (automatic process, adaptation mechanisms).

Such examples of adaptation mechanisms will naturally introduce the need for a more general framework able to deal with adaptive learning in evolving and time variant environments. Both passive and active learning methods will be introduced and detailed in Chap. 9 to cover all needs (both automatic and controlled processes, adaptation, and cognitive mechanisms).

Another level of adaptation can be considered when the embedded units are joined together to create the concept of network, e.g., a sensor network. Here, adaptation is needed at the communication level, e.g., to adapt online the parameters of the protocol or the physical communication [234, 235] or grant clock synchronization among units. Clearly, the possibility to exchange information among embedded units introduce another level of potential processing which can be either distributed or hierarchical (controlled process, cognitive mechanisms).

Self-localization of the units carried out by exploiting low cost received—signal—strength sensor data represent an advanced form of cooperation among embedded units. The algorithms can be both automatic and controlled, depending on the application setup and the required accuracy (both automatic and controlled processes, adaptation, and cognitive mechanisms).

The last section of the chapter addresses aspects related to application reprogrammability. Here, once the decision about the need to undergo reprogramming is made, suitable actions are identified and associated commands issued. Reprogrammability can be either carried out at the harware level or at the software one or both (controlled processes, cognitive mechanisms).

8.1 Adaptation at the Power Supply Voltage and Processor Frequency Levels

The power consumed by a CMOS circuit is mainly associated with the moving of charges in the CMOS gates whenever they switch. As a result, the simplified model of a CMOS circuit consisting of several gates can be viewed as a large equivalent capacitor that undergoes charge and discharge phases. Following the model provided in [159], the averaged power consumption P of a CMOS circuit can be described as:

$$P = \mathrm{ACV}_{\mathrm{cc}}^2 f + \tau \mathrm{AV}_{\mathrm{cc}} I_{\mathrm{short}} f + V_{\mathrm{cc}} I_{\mathrm{leak}}. \tag{8.1}$$

where A represents the number of active gates within the circuit (i.e., the powered transistors), C is the internal equivalent capacitance of the circuit as seen at the circuit output, V_{cc} is the supply voltage, f is the clock frequency of the device, I_{short} represents the short-circuit current flowing for time τ once a short occurs, and I_{leak} represents the leakage current accounting for the self-discharge mechanism. The first two terms of (8.1) account for a dynamic power dissipation, while the third term represents a static power contribution that accounts for the self-discharge of the capacitor induced by parasitic phenomena. (8.1) shows how the consumed power scales quadratically with the power supply and linearly with the frequency. It immediately appears that it is more advantageous to act on the power supply to reduce power consumption than acting on frequency that, in turn, would imply a smaller Million Instructions Per Second (MIPS) performance. That said, we might be interested in implementing a policy aiming at maximizing the MIPS by increasing the clock frequency and controlling the power balance accordingly by acting on the supply voltage V_{cc}.

However, there is a second relation limiting the maximum clock frequency f_{max} a circuit can support. f_{max} depends on the noise immunity, i.e., the difference between V_{cc} and the associated logic threshold V_{th} for which voltages above $V_{cc} - V_{th}$ correspond to the logic value one, and those below V_{th} to logic value zero (here we simplified the analysis by assuming a unique value for the two thresholds). The functional relationship reads

$$f_{\max} \propto \frac{(V_{\mathrm{cc}} - V_{\mathrm{th}})^2}{V_{\mathrm{cc}}} \tag{8.2}$$

Equation (8.2) states that, by reducing the voltage V_{cc} so as to reduce the power consumption, also the maximum frequency the circuit can tolerate reduces (the immunity to noise term reduces as well).

It is worth to outline that a decrease in clock frequency does not necessarily imply a loss in MIPS performance. In fact, while it is clear that computationally intensive algorithms would benefit from a higher clock frequency, heavy access to the primary RAM memory reduces such an advantage with a trade-off that is application and device specific. The presence of a cache memory improves the situation in favor of performance; execution with a code stored in a flash memory worsen it.

The above considerations have led many chip manufacturers to include voltage and frequency regulation circuitry in their microprocessors so as to support Dynamic Voltage/Frequency Scaling (DVFS) policies.

The voltage/frequency regulator can either be managed at the software level, with a code in execution on the controlled microprocessor/core, or through a dedicated hardware. In both cases the goal is to introduce adaptation mechanisms that, by acting on the voltage/frequency settings, minimize the power consumption/MIPS performance ratio.

Existing methods for identifying the most appropriate voltage/frequency settings are either classified as online or offline. In online methods, the literature suggests the use of interval and checkpoint-based solutions. Interval-based algorithms are periodically activated after the elapse of a fixed period of time. The decision about whether adapting the voltage and frequency or not is taken based on features extracted during such a time frame. Differently, checkpoint-based solutions identify offline, at compile time, the points of the application code where we should check and decide about a possible voltage and frequency adaptation. However, the scaling factors to be applied for voltage and frequency are decided online, at run time.

In offline methods, both checkpoints and scaling factors are static and decided directly at compile time before the application is executed on the embedded system.

8.1.1 Online DVFS

In online methods, the embedded system is suitably instrumented and the acquired information is used to characterize its current state. However, it should be pointed out that the considered features are not necessarily coming from physical sensors: the needed information is mostly provided by virtual sensors that, by operating at the meta-information level, provide features related to the task execution. The most used strategy to characterize the state of the embedded system is by considering performance counters for features such as the number of accesses to the cache memory, the average frequency of active processor cores and the energy delay-squared product (the term refers to the energy consumed by the processor due to the circuit input–output delay at the gate level).

By exploiting such information we can derive estimates for the embedded system performance and, subsequently, design a closed loop controller acting on the supply voltage and frequency to track a desired reference value.

8.1.1.1 Interval-Based DVFS

Interval-based DVFS algorithms are executed periodically to identify the optimal voltage/frequency setting with strategies based on the previous and the current states of the embedded system. Characterization for the system state can be obtained by inspecting the performance counters associated with the execution profile of the embedded system over the last time interval. Features or performance indexes I_ϕ need to carry information about the task/system status and must be related to voltage and/or frequency. As such, by acting on them, we directly/indirectly control the power consumption and balance it with the system performance. A first task level feature we can consider is $I_\phi = T - \tau$, where T is the execution time slot associated with the task and τ the effective time usage ($T - \tau$ represents then the available unused execution time). As a second example consider a task whose role, given input x, is to provide output $f(x)$, e.g., the neural network example of Sect. 7.6. Here I_ϕ can be associated with the number of activations of $f(x)$ per time unit. The higher the number of activations per time unit, the higher the power consumption (we might even need to increase f to satisfy constraints on real-time execution).

One of the simplest algorithms belonging to this family is the threshold method [163] where a performance index I_ϕ related to the processor performance is compared against a pair of thresholds. If I_ϕ is higher than the upper threshold or lower than the lower threshold, then the voltage/frequency setting is updated as with the threshold values. The system settings remain unchanged, otherwise.

The thresholds employed by such a method are determined heuristically. Another approach for selecting the thresholds has been proposed in [164]. There, the DVFS problem is transformed into a control problem, with I_ϕ modeled as a linear function in the voltage and the frequency. In such a model, the actual system load affecting I_ϕ is seen as a source of noise to be compensated, bringing the performance index back to the desired value. A light proportional–integral controller is then designed to track the desired performance index value.

Differently, [165] proposes a greedy method to find the optimal voltage/frequency setting by minimizing the energy over squared throughput ratio of the system. The ratio is computed for each time interval and, if different from the value associated with the previous state, a new configuration for voltage and frequency is enforced. In particular, if during the time interval the vector having voltage and frequency as components change in a given direction of the vectorial space by inducing an increment in the performance index, then, voltage and frequency are moved in the vector direction. The opposite holds.

A more sophisticated example of a whole class of methods applying machine learning to the DVFS problem is described in [166]. There, profiling information encompassing several performance counters such as number of cache hits, cache misses, number of user instructions, and loop counters are collected to constitute a train set. A classification and regression algorithm generates a decision tree to be used at run time to update voltage and frequency according to the current interval counter values.

8.1.1.2 Checkpoint-Based DVFS

Checkpoint-based methods evaluate the performance of the embedded system in correspondence with specific lines (points) of the application code where the programmer has either inserted a software interrupt or the call to a function.

This approach is particularly suitable in hard real-time systems, where the task predictability[1] is of fundamental importance. In cycle-conserving Real-Time Dynamic Voltage Scaling (RT-DVS) methods, [160] checkpoints are inserted when the task is released (e.g., when it enters a *wait* state in a time sharing based operating system) or has terminated its execution. The embedded system, by inspecting the set of active tasks, selects the voltage/frequency setting able to satisfy the tasks constraints with the least amount of energy. Once the task in execution completes or is released, the voltage/frequency is updated with the same mechanism. In other words, when a task to which is assigned time slot T ends at time τ before its allocated time, the time left $T - \tau$ allows to reconfigure both frequency and voltage to the minimum values that still grant the remaining tasks to be be executed without violating the associated constraints. In this way, all tasks guarantee constraints satisfaction but energy is saved since the voltage and the frequency are scaled down as much as possible.

Such an approach is optimized in the Look-Ahead RT-DVS algorithm [160]. In it, the needed frequencies for remaining tasks are estimated in advance thanks to scheduling actions that rearrange the tasks execution to minimize the energy consumption (still satisfying the execution constraint on each task execution). In other words, whenever possible, the scheduler anticipates the execution of tasks characterized by lower energy requirements. Power eager tasks are executed only later, possibly with a new configuration of the voltage/frequency. The hope is that the current execution of a power-eager task does not coincide with its worst-case execution that would request to set the voltage/frequency to satisfy the high performance needs.

8.1.2 Offline DVFS

Performing an offline voltage/frequency scaling for our embedded system implies that all decisions about which voltage/frequency setting should be used during the task execution are identified at compile time. Software interrupts (or calls to suitable routines) are hence invoked to change the operating values of voltage/frequency in correspondence with specific points in the program.

The simplest case is that of a hard real-time system with voltage/frequency scaling [160]. In this scenario, the voltage or frequency scaling problem is entirely determined by the satisfaction of the deadlines assigned to the tasks, given the scheduling algorithm (e.g., the earliest deadline first). In such a case, the chosen frequency is the minimum between the available ones satisfying the scheduling request. Such frequency is set at the boot, during its initial configuration. The setting is maintained during the whole task life without the possibility to evaluate at run time the system performance or intervene on it.

In case of static intra-task DVFS, the algorithm decides offline both the instant and the value of the voltage/frequency scaling on the basis of profiling data [161] or performance estimates task [162]. In both cases, the program is divided into regions

[1] A task is predictable when it is possible to determine its worst-case execution time.

of constant voltage/frequency and the scaling values are identified by solving an optimization problem that minimizes the energy consumed by taking into account the task constraints. Although attractive for their simplicity requiring no extra computation or hardware to monitor the system status, offline methods implement a simple open-loop control on the dissipated power and, as such, are subject to loose optimal performance in correspondence of perturbations (different input values, differences in the actual hardware, variations in temperature, etc.).

8.2 Adaptive Sensing and its Policies

Intelligent embedded systems can consider different strategies to acquire data from sensors. The simplest approach is based on a sequential polling where sensors are sampled one after the other, periodically, possibly within a loop. Not rarely, polling relies on a synchronous mechanism initiated by an hardware interrupt activated after the elapse of a fixed amount of time (time-sharing task scheduling modality). More in detail, a task (or thread) is initiated when new data instances need to be acquired. The due steps can be itemized as:

- The interrupt might, at first, wake up the microprocessor, possibly from a deep sleep modality where it was placed to save energy,
- afterward, the interrupt routine sends the warm up directives to the sensors (if needed) and waits (or the task is inserted in the task wait list) for available data. This operation can also be accomplished by an ad hoc routine.
- once sensors are ready, the sampling procedure is activated, acquiring data, and storing them in the memory. If a data estimation module is not available in the sensor (refer to Sect. 2.1.1), we might opt for a high frequency sampling modality where data are averaged before storage to reduce the impact of electronic noise.
- once the acquisition task is completed the routine terminates.

Previously itemized phases can be present or not depending on the application and the problem that needs to be solved. In some cases, the same routine acquires all data from the sensors, within a polling mechanism. If the interrupt routine is activated according to a given frequency, the same time stamp is generally assigned to each measurement. This might introduce significant differences among the different instants of acquisition. The consequence is that we assign the same time stamp to all data acquisitions even though there might be a significant time discrepancy between the first and the last measurement. The time error associated with a given acquisition is a random variable whose properties and features can be estimated with randomized algorithm-based methods (e.g., we simply take an average or compute the maximum expected time skew).

However, given the fact that sampling is a rather fast operation compared to the dynamics of the measured phenomenon for many embedded applications, the effect introduced by the same time stamp to all polled data within an acquisition cycle can be considered negligible. When this is not the case, the designer should pay extreme attention to the application code to grant that each measurement is associated with its accurate time stamp.

The procedure that sees the microprocessor periodically enabled to carry out a set of tasks, e.g., data sampling, and disabled afterward to save energy is called duty cycle. Duty cycling is an effective energy management strategy whenever the overhead associated with the on-off modality is negligible and justifies the energy saved in the off state.

The sampling rate is generally determined at design time and depends on the specific application, which is supposed to be time invariant. Generally, the designer sets the sampling rate based on some a priori information about the physical problem under monitoring or on his/her naïve interpretation. For instance, if we shall measure the external temperature of the environment, we might naively say that we need a sample every second, based on some personal feelings. Not rarely, the application designer claims that "sampling at 1 Hz is enough", with the statement not justified by any physics investigation or information about the real dynamics of the phenomenon. In the best case this approach shows to be very conservative, hence yielding an oversampled datastream that introduces an unnecessary extra cost in data processing, storage, and transmission. Of course, the different operations to be executed depend on the specific application.

In order to avoid/mitigate the oversampling phenomenon, we need to investigate the dynamics of the signal to be acquired, determine its finite bandwidth under the time invariance assumption for the signal and derive the Nyquist frequency f_N [52], e.g., estimated with a Fast Fourier Transform. The sampling frequency f_S to be considered for the envisaged signal is then obtained as:

$$f_S = c f_N$$

where c is a scalar value commonly set from 3 to 5 to compensate for the uncertainty associated with the estimate of the Nyquist's frequency (in general we rely on a contained series of noise affected data). In an ideal uncertainty free case where the signal has finite bandwidth, c should be set to 2. Although the method above works quite well and should be adopted when sampling a signal, it is worth recalling that we also requested the time invariance hypothesis for the signal to be sampled. This assumption in turn requires the Nyquist frequency f_N not to change. In many applications this assumption does not hold and adaptive sensing strategies should be considered to adaptively estimate f_S in order to acquire only the due samples, hence reducing the amount of data to be stored, processed and, possibly, transmitted.

Duty cycling and adaptive sensing are complementary solutions that can be adopted within an optimal energy management for embedded systems and can be combined together to double the advantage. More specifically, the operating system provides the set of commands needed to power the sensors (duty cycle activity at the sensor level) and, when the task is active according to an adaptive sensing strategy, the sensors acquire the data. The duty cycle time must be carefully designed to avoid hardly detectable side effects such as sampling when the sensor is not ready (e.g., warm up time not completed) hence yielding a not relevant information or

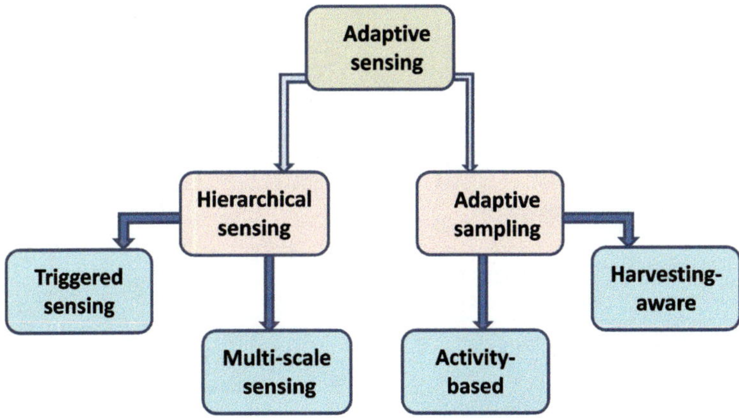

Fig. 8.1 A classification of the adaptive sensing strategies present in the literature

consuming more than the energy needed to keep the processor and the sensor always active. This latter situation arises when the sensor is kept off for a short time, surely not enough to balance the energy consumption of the sensor during its warm up [53].

We follow the classification of adaptive sensing strategies given in [54, 55] and summarized in Fig. 8.1. We assume that envisaged sensors are fault free, accurate, and compensated. These assumptions should be checked/verified before enabling each of the following techniques.

8.2.1 Hierarchical Sensing Techniques

Hierarchical sensing techniques are a form of redundant sensing where multiple sensors of the same type are available on the embedded system to monitor a specific phenomenon. Sensors differ in terms of resolution and, possibly, energy consumption, leading to measurements characterized by different precisions (sensors are assumed to be accurate). For instance, if we wish to measure a temperature, we might consider an integrated digital temperature sensor (resolution 0.5 °C) and a thermistor (resolution 0.1 °C) and a high performance thermocouple (resolution 0.03 °C). The implicit assumption behind hierarchical sensing is that resolution-poor sensors are also characterized by a lower power consumption and/or cost compared to high resolution ones. Generally, this assumption is satisfied. However, the embedded system designer has to investigate further the sensor datasheet before enabling the method.

During operational life, data are mostly acquired by low resolution sensors and decisions made out of their values. Conversely, high resolution sensors are activated only when needed to improve the accuracy of the measure or run more accurate algorithms, e.g., by following an event triggered by the algorithm designed to process the lower resolution temperature values. The situation is that depicted in Fig. 8.2.

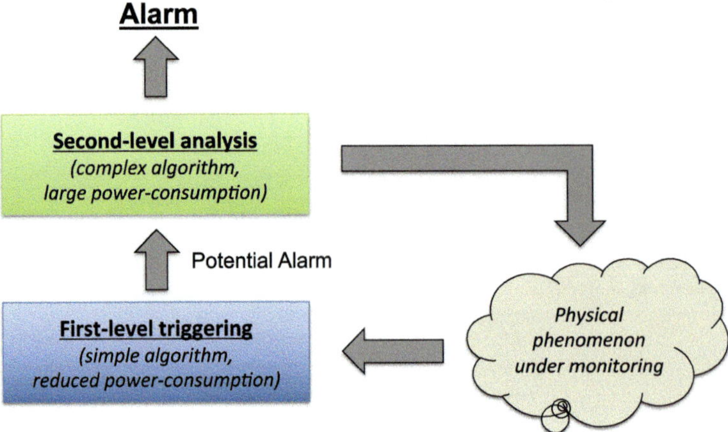

Fig. 8.2 Hierarchical sampling. Data acquired with low resolution/low power consumption sensors are processed by simple algorithms. When a potential alarm is detected (an event possibly arose), high resolution sensors generally characterized by a higher power consumption are activated. A more complex algorithm evaluates the available data to finalize the decision or take the appropriate action

In other words, the idea behind hierarchical sensing techniques is to dynamically select the sensors to be activated within the available sensor platform so as to trade-off accuracy and energy consumption. Final decisions are made by processing data coming from all sensors, or a subgroup of them depending on the sensors resolution, the expected precision and the complexity of the chosen processing algorithm. The methods belonging to hierarchical sensing are detailed in the sequel.

8.2.1.1 Triggered Sensing

In triggered sensing, instant measurements or derived features obtained by processing basic data are used to decide whether to activate an alarm because an event arose or not. For instance, by averaging low temperature sensor data we detected that a threshold has been exceeded, meaning that a constraint has probably been violated. However, given the low resolution of the sensor we cannot guarantee that the statement is true. High resolution sensors are then activated to validate the constraint violation hypothesis.

A different example taken from the rich literature is that given in [65] and referring to the case where each sensing unit has an integrated CMOS camera that can be reconfigured in terms of spatial resolution. The lower the resolution, the coarser the image, the lower the energy consumption. Low resolution scenes of the environment are quickly processed for target detection. If targets are detected, some digital cameras undergo an adaptation phase that, after reconfiguration, provides higher quality images for target detection validation. Once the decision has been taken, the cameras are reconfigured back to the low resolution modality.

Clearly, more sophisticated adaptation strategies can be taken into account to identify the right level of resolution for the image or improve the estimate of the due resolution based on a priori or historic information. However, nothing comes for free and more accurate adaptive solutions require more complex algorithms to be executed. The compromise between complexity and accuracy is application and sensor specific.

8.2.1.2 Multiscale Sensing

In multiscale sensing, we identify areas within the monitoring field that require a more accurate inspection. Different resolutions are used for different areas. In this way, we envisage a lower resolution when the receptive field is less relevant, and a higher resolution when high precision acquisitions are requested. This idea has been originally proposed in [56].

The work in [57] proposes an interesting, albeit articulated, example within a fire emergency management scenario. The field to be monitored is instrumented with static, low resolution sensors whose aim is to detect anomalies in the expected temperature profile. When an event is detected, a mobile sensor unit (mule) is sent to the area to collect additional high precision measurements as well as other available information to validate/reject the event hypothesis. Adaptation is here at the activation of the mule level.

Comments

We should comment that both triggering and multiscale sensing have their own limitations since they require sensors characterized by different resolutions and, in turn, more complex hardware and software management policies. Moreover, other key critical elements of the multiscale sensing method are the identification of the optimal placement of the sensors, the identification of the size of the field of interest for each sensor, and the assignment to each subarea of the proper type of sensor. The answers to these problems are not straightforward and either require a priori information about the application/environment or a profiling campaign that requires the acquisition of a long series of environmental data. While optimal placement, which also implies minimization of the number of sensors to be used, requires a sequential analysis and solution of an optimization problem, the influence area can be determined with a spatial Voronoi tessellation of the environment [58]. All in all, these methods assume that the environment we are looking at is time invariant. Relaxation of this hypothesis leads to an increasing level of intelligence on one hand and complexity on the other.

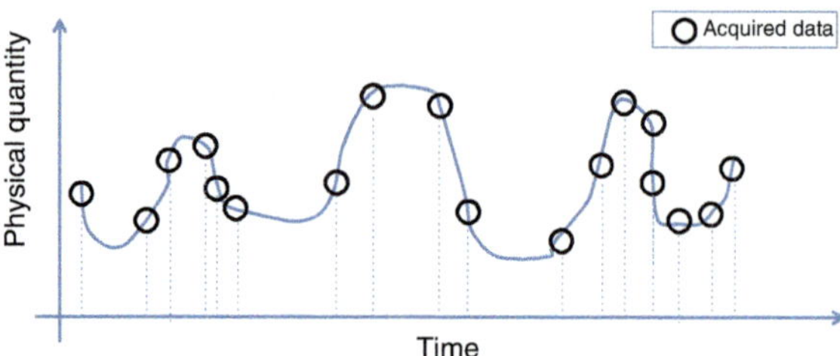

Fig. 8.3 Activity-driven adaptive sampling

8.2.2 Adaptive Sampling

Adaptive sampling methods modify the sampling rate based on the information content carried by sensed data, the available residual energy present in the batteries and the incoming or estimated harvested power. The reference situation is depicted in Fig. 8.3.

For instance, if the quantity of interest evolves slowly with time so that subsequent samples do not differ that much it does make sense to adaptively reduce the sampling rate if the energy available is an issue. We say that there exists a temporal locality when it is high the probability that the information content carried by subsequent acquisitions does not change rapidly with time. A similar spatial locality principle can be introduced by requiring that nearby spatial acquisitions provide a similar content with high probability. While temporal locality has validity at the sensor level and, hence, at the single embedded unit, spatial locality requires that units can somehow communicate in order to take advantage of the distributed information made available.

Every time we have spatial or temporal locality, we can reduce the energy consumption associated with sensing by adapting the sampling rate. In this direction, activity-driven adaptive sampling combines both locality principles to reduce the number of samples to be acquired. However, this mechanism is not necessarily always active being also function of the residual energy in the batteries and the incoming power if energy harvesting options are present.

8.2.2.1 Activity-Based Adaptive Sampling

Activity-based adaptive sampling takes advantage of situations where the application grants the validity of temporal and spatial locality.

For instance, temporal locality was used within an adaptive sampling algorithm to minimizing the energy consumption of a snow sensor [59] and extended to the general case in [60]. There, the algorithm estimates online, as sensor data come, the Nyquist frequency f_N and verifies whether f_N changes over fixed-size data windows. If a change in f_N is detected, then the new sampling frequency is determined and updated.

This detect-and-react mechanism belongs to the class of active learning algorithms and will be detailed in Chap. 9. It should be noted that a false positive in detecting the change (erroneous detection of a change in the Nyquist frequency) does only introduce an unnecessary energy consumption. In fact, the effect induced by a false positive is that a new f_S is applied although it was not needed. A false negative (no change detection when the change takes place) is negligible if the change is minimal (the magnitude of the change is small and gets confused with the stochastic fluctuation of f_N associated with a limited training dataset and noisy data). In some cases, the change might not be immediately detected, e.g., when if affects f_N with a slowly developing drift. In those cases latency induces a delay in sampling frequency adaptation.

Activity-based adaptive sampling has been used in many applications. For instance, an adaptive sampling solution for a flood warning system is presented in [61]. The system includes a flood predictor which is used to adjust the sampling rate of individual sensing units. A spatial locality approach with a backcasting scheme was proposed in [62]. The key idea is that sensing units of a sensor network should be more dense in areas where the phenomenon introduces high variance in the signal. Spatial locality for a correlation-based collaborative MAC protocol was used in [63] to selectively reduce the number of units used to send data to the base station. Spatial and temporal locality was also used in [64] applied to a robot acting as a mobile mule.

An interesting variant of the adaptive sampling method taking adavantage of temporal locality is the model-based active sensing paradigm, e.g., see [68]. The idea of the sampling technique is to build at first a model describing the signal starting from a training dataset composed of acquired data. The situation is presented in Fig. 8.4. Once the model is available, the same sampling frequency is applied until incoming data are well represented by the model, i.e., the residual between the current data and the ones provided by the model is white noise. Differently, when the residual whiteness is no more granted (simpler threshold solutions based on the standard deviation can also be used, although less accurate), the model is no more capable of explaining the data and a new sampling frequency must be estimated based on current data instances. This method naturally falls within the concept of drift detection presented in Chap. 9, and here is applied to the prediction residual sequence.

8.2.2.2 Harvesting-Aware Adaptive Sampling

The harvesting-aware adaptive sampling method optimizes energy consumption at the embedded system level by exploiting the information related to the residual energy present in the batteries as well as the predicted energy we shall expect from the energy harvester. For instance, if the voltage of the battery is low, hence implying low residual energy, and the energy expected from the harvesting module for the next hour is small, we should reduce the sampling frequency. Actions on the sensors can be differentiated depending on the information content we expect from them. If the embedded system mounts a rich sensor platform, some sensors can be disabled

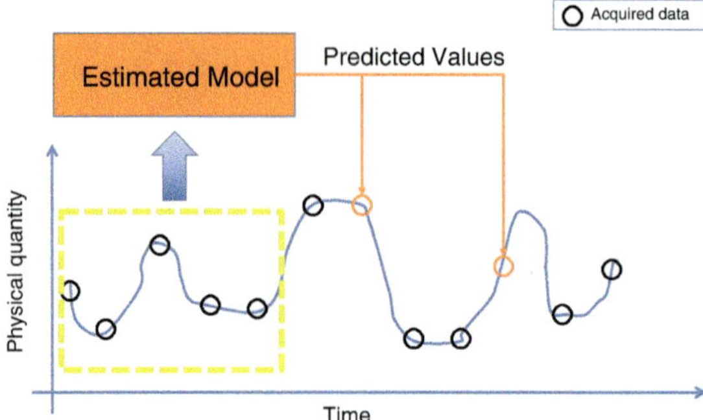

Fig. 8.4 Model-based sampling. Model-based sampling is a particular adaptive sampling mechanism where temporal and spatial locality redundancy allow us for designing a model describing, over time, incoming data. Not all incoming data are hence acquired and predicted ones used instead to generate the datastream. The consequence is that energy is saved since the effective frequency sampling reduces. When data are acquired, the discrepancy between the predicted and the real acquired value is computed and the nature of the residual investigated to decide whether it is time to generate a new model or not

depending on their relevance and power consumption. Two examples of the method have been proposed in [66] and [67] with a photovoltaic cell as reference energy harvester.

8.3 Adaptation at the Energy Harvesting Level

The problem of energy harvesting for small devices represents an extremely hot research field and a place where adaptation mechanisms do make the difference to grant efficiency in energy extraction. Energy can be extracted from photovoltaic cells, which show to be one of the most effective solutions for outdoor applications thanks to the high power density made available in sunny days. However, depending on the application scenario and the place the embedded unit is deployed, we can extract energy from other sources, e.g., from wind by using small wind turbines, from vibrations by relying on piezoelectric devices and from Peltier's cells if we have thermal gradients to exploit. For aquatic deployments, we can extract kinetic and potential energy from waves, exploit the rise and fall of the sea level caused by tides thanks to the gravitational forces exerted by the moon over the water mass or tidal currents, from temperature or salinity gradients.

All the above examples are characterized by situations where kinetic, elastic, or potential energy to be transducted into electrical power for energy storage is time invariant and depends on some uncontrollable external conditions that, if treated as a disturbance, would lead to a poor harvest efficiency. Optimal energy extraction requires the harvesting module to be able to track changes and evolutions of the environment, operation that can be done with adaptive mechanisms and ad hoc low power consuming harvesting devices, e.g., see [69–72].

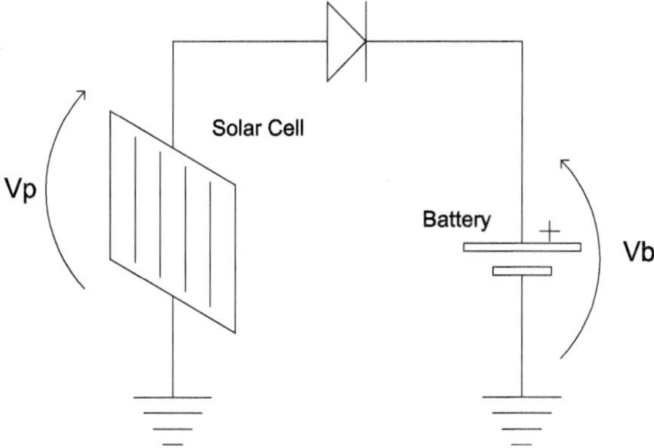

Fig. 8.5 A diode-based energy harvester

In the following, we will present details regarding the adaptation mechanism for photovoltaic cells, as it represents one of the most relevant examples of adaptation at the energy harvesting level. Time variance is caused here by several factors, e.g., changing weather conditions, aging effects and efficiency degradation in the solar cell, presence of dust and water on the cell surface, variation of the angle of incidence of the sunlight during the day time and, finally, over the year.

The simplest solution for solar energy harvesting in small embedded devices, also presented in Fig. 8.5, relies on a straight on/off charging mechanism based on a diode connecting the cell with the rechargeable battery [73].

When the tension of the cell V_p is above the voltage of the battery $V_b + 0.7$, the diode is short circuited and the current flows to the battery. Conversely, when V_b is larger than $V_p - 0.7$ the diode acts as an open switch and no power is transferred to the battery.

A diode-based solution is characterized by an extremely low cost and low power consumption (no electronic energy harvesting design is simpler than a diode) but suffers from two major limits:

- No adaptation mechanisms can be envisioned to maximize the energy harvested despite changes in the energy providing source or in the energy transduction mechanism. In fact, adaptation is prevented by the fact that the working point of the cell is set by the battery voltage and cannot be adjusted. As a direct consequence, the system cannot operate at low radiation power because the diode disconnects the solar cell from the battery.
- The size of the solar cell and its type (e.g., monocrystalline, polycrystalline, amorphous silicon) and the nominal voltage of the battery must be chosen carefully. A solution designed for a monocrystalline silicon cell dimensioned with its battery might not be appropriate for an amorphous silicon cell characterized by a much lower efficiency.

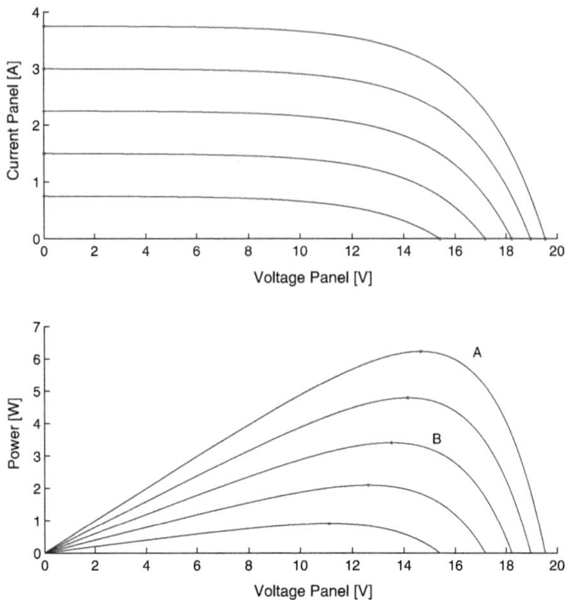

Fig. 8.6 The characteristic power delivery curve of a photovoltaic cell. The power is function of the voltage V imposed at the cell. Different solar radiation conditions generate different curves, each of which is characterized by a different maximum harvestable power. In the figure, curve A is characterized by a larger power delivery compared to curve B and is associated with a stronger solar radiation

The above problems can be solved by substituting the diode-based circuit with a Maximum Power Point Tracker (MPPT) system based on adaptive mechanisms for optimally harvesting energy.

Let us look at the power delivery curve of the photovoltaic cell given in Fig. 8.6 as function of the voltage V_p imposed at the cell. Whatever the technology of the solar cell is, the power versus the cell voltage curve $P(V_p)$ shows a convex behavior characterized by a unique maximum associated with the solar radiating conditions and the transduction mechanism [74].

Disturbances, such as aging effects and presence of dust, simply move the curve onto which the device operates into a new one. Curve A is characterized by a higher power delivery compared to curve B, e.g., it is associated with a stronger solar radiation. Clearly, if we consider a fix voltage applied to the cell (e.g., as it happens with the diode solution where the applied voltage is basically that of the battery) we might operate on a working point reasonably far from the one granting maximum energy harvesting. Consequently, given a curve and an initial value for the voltage, we shall continuously change it over time so as to converge to the optimal value maximizing the power provided to the battery. That is what adaptation within a MPPT framework does.

More specifically, the different solutions provided in the literature aim at continuously hill climbing the unknown available power curve so that the voltage to be applied to the photovoltaic cell of the solar panel tends towards the value maximizing the power delivery. Let $V_p(k)$ be the voltage currently imposed at time $t = k$ at the photovoltaic cell and $i_p(k)$ the outcoming current, so that the generated power is $P(V_p(k)) = i_p(k)V_p(k)$. Since the curve $P(V_p)$ is unknown, it is impossible to determine immediately the optimal voltage, and iterative methods must be applied to solve the problem. We assume at first the ideal case where, identified a desired voltage $V_p(k+1)$, we simply apply it to the photovoltaic cell. In that case, an adaptive gradient-based optimization can be applied directly to V_p in order to identify iteratively the voltage maximizing the delivered power, i.e.,

$$\begin{cases} V_p(k+1) = V_p(k) + \Delta V_p(k) \\ \Delta V_p(k) = \gamma \frac{dP(V_p)}{dV_p}|_{V_p(k)} \end{cases} . \tag{8.3}$$

Under the time invariance hypothesis (the solar radiation keeps the device working on a specific -yet unknown- curve), the adaptive algorithm converges to the correct value provided that the parameter gain γ amplifying the gradient function is small enough. We also comment that the curvature around the optimal voltage point is rather flat implying that even a rough estimate for the optimal point can be safely confused with the searched unknown optimal value. Moreover, this working point can be quickly reached with a relatively large γ, hence requiring few iterations to converge. γ must be selected with a trial and error approch since it depends on the particular place the embedded system is deployed (or resides) at a given time. A more sound way to identify γ for a given place is to rely on the accurate solar radiaton information made available by national energy organizations e.g., see [228], which provide the expected solar density available over the year for a given location. The optimal γ for a locality can be identified with the randomization procedure given in Chap. 7 having as a figure of merit the trade-off the energy gain reachable with high values of γ (the higher the gamma, the faster the convergence to a suboptimal value) with the loss we expect given the fact we might converge to a rough estimate.

The same holds in quasi-stationary conditions where the algorithm converges to the quasi-optimal value much faster than changes in the solar radiation. This situation is the most common one and also addresses aging effects, presence of dust or water drops on the panel and thermal excursions that change the efficiency of the photovoltaic cell. In rapidly evolving time variant situations, the algorithm has to rapidly track the change. This might require a high sampling rate, hence requesting a higher energy consumption.

Previous derivations have assumed that it was possible to directly impose voltage $V_p(k+1)$ at time $t = k + 1$ at the photovoltaic cell. However, this is not a feasible action due to electronics constraints. Thus, a Controlled Power Transfer Module (CPTM) must be introduced in the design of the harvester. The role of the module is to grant transfer of power from the input to the output by imposing that the input voltage is set to a given reference value (here the cell must be set to $V_p(k+1)$). Figure 8.7 shows the operational framework. The control loop drives V_p to the reference voltage V_s. From the electronic point of view, the CPTM can be implemented

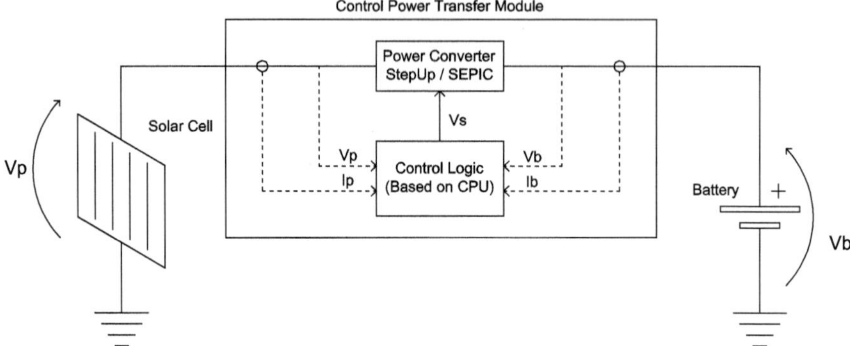

Fig. 8.7 A controlled power transfer module. The transferred power between input (cell) and output (battery) of the module is controlled by a control logic that, by taking advantage of possibly available information V_p, i_p, V_b, i_b, grants that V_p is driven to the desired reference value V_s

with several technologies, e.g., a StepUp DC/DC [76] or a Single Ended Primary Inductor Converter (SEPIC) [229].

The introduction of the controlled power transfer module allows us to transform the set of equations (8.3) in the ordered sequence of steps to be undertaken

$$\begin{cases} V_p(k) = V_p(t = \tau k) \\ V_p(k+1) = V_p(k) + \gamma \frac{\mathrm{d}P(V_p)}{\mathrm{d}V_p}|_{V_p(k)} \\ V_s = V_p(k+1) \\ V_p(t) = \mathrm{CPTM}(V_s)|V_p(t = \tau(k+1)) = V_s \end{cases} . \tag{8.4}$$

The sequence of operations given in (8.4) must be detailed in order to define the different meaning associated with time instances given the discrete–continuous representation of time. Defined τ to be the sampling period, we know that $V_p(k)$ means acquiring the value that V_p and other required information assume at time τk, or $(V_p(t = \tau k))$. By calculating $V_p(k) + \gamma \frac{\mathrm{d}P(V_p)}{\mathrm{d}V_p}|_{V_p(k)}$, we compute the value that V_p should assume at time $k + 1$. We name this as V_s. The inner control loop implemented by the CPTM activates and guarantees that at time $t = \tau(k + 1)$ all sensor information to be acquired will be characterized by having $V_p(t) = V_s$. The process iterates.

In embedded systems characterized by a limited computational ability, the adaptive algorithm can be suitably simplified to keep under control the power consumption cost of the MPPT module. This can be achieved by adopting different strategies such as the incremental conductivity approach and the perturb and observe one. These methods are detailed in the following.

8.3.1 The Incremental Conductance Approach

In the incremental conductance approach, the maximization of the harvested power is achieved by acting on the conductance $\frac{i_p}{V_p}$. By recalling that $P(V_p(k)) = V_p(k)i_p(k)$, (8.3) can be rewritten as

$$\begin{cases} V_p(k+1) = V_p(k) + \Delta V_p(k) \\ \Delta V_p(k) = \gamma \left(i_p + V_p \frac{\mathrm{d}i_p}{\mathrm{d}V_p} |_{V_p(k)} \right) \end{cases} . \tag{8.5}$$

The algorithm executed by a microcontroller acting as a control CPU follows the Eq. (8.5). More in detail, and by referring to Fig. 8.8, the embedded system acquires i_p and V_p through the ADC internal to the microcontroller and then evaluates the gradient of the conductance $\frac{\mathrm{d}i_p}{\mathrm{d}V_p}|_{V_p(k)}$ required in (8.5) by computing the incremental ratio

$$\frac{\mathrm{d}i_p}{\mathrm{d}V_p} \simeq \frac{i_p(k) - i_p(k-1)}{V_p(k) - V_p(k-1)}.$$

The approximation shows to be good depending on the sampling time and the dynamics associated with the time variance of the environment. Finally, the microcontroller executes the algorithm in (8.5) and identifies the next voltage value V_s to be assigned to the photovoltaic cell. The final ordered steps are given in (8.6)

$$\begin{cases} V_p(k) = V_p(t = \tau k) \\ V_p(k+1) = V_p(k) + \gamma \left(i_p(k) + V_p(k) \frac{i_p(k) - i_p(k-1)}{V_p(k) - V_p(k-1)} \right) \\ V_s = V_p(k+1) \\ V_p(t) = \mathrm{CPTM}(V_s) | V_p(t = \tau(k+1)) = V_s \end{cases} . \tag{8.6}$$

If the sum of the times requested by the ADC, the algorithm and the step up DC/DC module is negligible compared with the dynamics of the evolution of the solar radiation, then the set value $V_p = V_p(k+1)$ will still be a good estimate of the step to be taken along the gradient direction since the power curve did not change that much in the meantime.

8.3.2 The Perturb and Observe Approach

The perturb and observe method can be derived from (8.3) with the assumption that $P(V_p) = i_p V_p \simeq \eta i_b V_b$, where η accounts for the efficiency of the power conversion module, here assumed to be constant. Under the assumption that the voltage of the battery V_b does not change that much with instant changes in V_p (a reasonable hypothesis), we have that

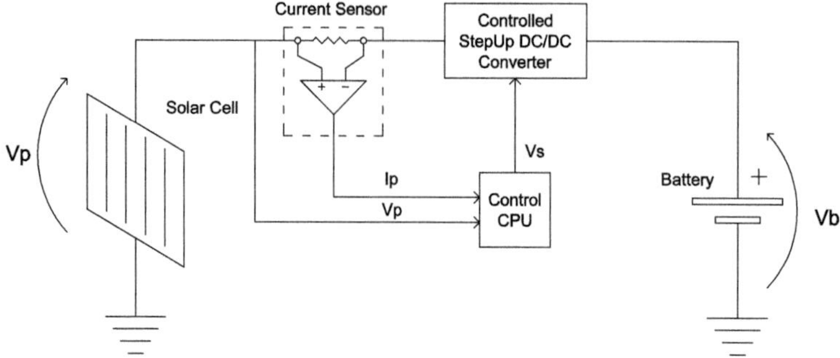

Fig. 8.8 An incremental conductance-based MPPT energy harvester with a StepUp DC/DC control power transfer module

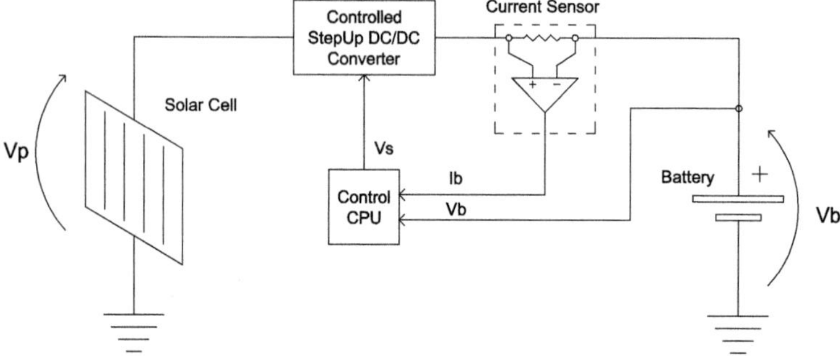

Fig. 8.9 A perturb and observe-based MPPT energy harvester with a StepUp DC/DC control power transfer module

$$\frac{\mathrm{d}P(V_p)}{\mathrm{d}V_p} = \eta V_b \frac{\mathrm{d}i_b}{\mathrm{d}V_p}$$

which, substituted into (8.3), provides

$$\begin{cases} V_p(k+1) = V_p(k) + \Delta V_p(k) \\ \Delta V_p(k) = \gamma \frac{\mathrm{d}i_b}{\mathrm{d}V_p}|_{V_p(k)} \end{cases} . \tag{8.7}$$

Clearly, γ accounts for all constant or quasi-constant terms, e.g., η and V_b, respectively. The role of the MPPT can be presented by commenting the functional block of Fig. 8.9. There, current i_b going into the battery as well as its tension V_b are measured by the introduced current and voltage sensors. Such information is provided to the microprocessor executing the adaptation algorithm whose goal is to identify the optimal voltage V_s the photovoltaic cell should be forced to. Once V_s has been identified the StepUp DC/DC converter introduces an analog control action driving V_p to the optimal value V_s by modifying the DC/DC converter duty cycle.

In line with the incremental conductance algorithm, it is assumed that the solar radiation process evolves much slower than the time required by the device to impose $V_p = V_s$. The final equations ruling the algorithm are:

$$\begin{cases} V_s(k+1) = V_s(k) + \Delta V_s(k) \\ \Delta V_s(k) = \gamma \frac{\mathrm{d}i_b}{\mathrm{d}V_p}|_{V_p(k)} \end{cases}$$

and the derivative $\frac{\mathrm{d}i_b}{\mathrm{d}V_p}$ is approximated with the incremental ratio

$$\frac{\mathrm{d}i_b}{\mathrm{d}V_p} \simeq \frac{i_b(k) - i_b(k-1)}{V_p(k) - V_p(k-1)}.$$

The final set of algorithmic steps are:

$$\begin{cases} V_p(k) = V_p(t = \tau k) \\ V_p(k+1) = V_p(k) + \gamma \left(\frac{i_b(k) - i_b(k-1)}{V_p(k) - V_p(k-1)} \right) \\ V_s = V_p(k+1) \\ V_p(t) = \text{CPTM}(V_s) | V_p(t = \tau(k+1)) = V_s \end{cases} . \tag{8.8}$$

We comment that very simple embedded systems do not have division operations that need to be emulated in software. This as an extra cost in terms of power consumption reducing the efficiency of the harvesting module. As such, in order to further simplify the algorithm to reduce the power consumption of the harvester, the designer can follow the approach suggested in [75] where only the sign of the gradient $\frac{\mathrm{d}i_b}{\mathrm{d}V_p}$ is considered. Despite of the crude approximation, it is shown in [76] that the algorithm converges almost always to the optimal value and when it does not the situation can be detected and the MPPT restarts. The algorithm evolves by introducing a perturbation in the panel operating voltage and measuring the induced current i_b; as mentioned, only the sign of the ratio is considered with an undoubted advantage in harvesting efficiency.

8.4 Intelligent Algorithms for Clock Synchronization

Real clocks are far from being ideal and are subject to clock drifts which increase, over time, the discrepancy between the time of a reference clock and the available real one. In the absence of any form of synchronization, embedded systems would loose the ability to cooperate, being not able to establish any precedence between the events happening in the system [167] or provide information hardly usable if the application is requesting distributed data acquired within the same time frame.

In fact, since sensors acquire streams of data over time, a temporal label can be assigned to each data instance. Such a label represents a relative measure of time as provided by the internal clock of the acquisition device (a timer) and refers to the time elapsed since the last relevant event arose, e.g., the boot of the embedded system, the last synchronization with respect to an external—highly accurate—clock, the start of an interrupt routine.

Depending on the nature of the acquired signal and the particular problem we want to tackle within an application, the time information can be precious or not relevant at all. In general, if the signal is composed by time-dependent instances, as it happens in most of cases, then the time information is important to carry out any subsequent processing. Differently, if data are independent, e.g., we sample instances that can be modeled as a random variable after some mean centering, probably the time information is less relevant. Within a sophisticated sensor containing a data estimation module (refer to Sect. 2.1.1) or a sensor followed by some software to improve the accuracy of the readout value, we see both mechanisms. In fact, at the higher level we have the request for a datum taken from a sensor inspecting a physical phenomenon: here, the time information is precious. Then, once the sampling procedure has been activated, we acquire a burst of data with a very high frequency compared with that of the signal dynamics and average the measurements to provide a better estimate of the real value. Within the burst, the temporal information is not relevant at all.

Unfortunately, any real clock introduces a time drift w.r.t. the ideal absolute one, even in the case we are considering some atomic clocks (for which the error on time is infinitesimal: standards agencies maintain clocks with an accuracy of 10^{-9} seconds per day). The time error can be relevant in embedded systems where a clock drift largely depends on its production quality and the external temperature. The latter term implies that the same clock can have different drifts over time depending on external conditions.

Example: Clock Drift in Embedded Systems

We inspect the performance of a given commercial clock crystal that generates a nominal 32KHz frequency clock. The datasheet of the component claims that the maximum aging expected for the first year of life is:

$$\frac{\Delta f}{f} = \pm 3\,\text{ppm} \tag{8.9}$$

where f is the clock frequency and Δf the maximum expected variation; ppm stands for parts per million. The given tolerance is valid at a reference temperature $T_0 = 25\,°\text{C}$. After the first year the clock drift tends to stabilize to a fixed rate [227]. Therefore, the maximum variation in frequency we shall expect after one year by keeping the clock at temperature T_0 is $|\Delta f| = \pm 3 \cdot 10^{-6} \cdot 32 \cdot 10^3 < 1\text{Hz}$. Let's assume, only in order to provide some quantitative information, that the aging effect stops after this first year. The time error introduced on the temporal label after any other year would be 3 s per year. This is a not negligible contribution and we should remark we are assuming that the clock is not drifting anymore, an hypothesis clashing with reality.

Moreover, we should investigate the effect of temperature on the clock. From the component datasheet we have that

$$\frac{\Delta f}{f_0} = -0.035(T - T_0)^2 \pm 10\,\% \tag{8.10}$$

in parts per million. T is the current temperature and f_0 the frequency of the clock at temperature T_0. To give a quantitative idea about the impact of temperature on Δf, assume that our device is operating at a constant $T = 35\,^\circ\mathrm{C}$ due to the board warming and that the $\pm 10\,\%$ contribution due to uncertainty is null. From (8.10) we have that the ratio $\frac{\Delta f}{f_0}$ is -3.5 ppm, in the same order of the contribution of (8.9) (that was 3 ppm).

Finally, we relax the assumption that the clock is operating at its nominal frequency. The datasheet states that the frequency tolerance $\frac{\Delta f}{f}$ is ± 30 ppm. This term is the principal error contributing to the discrepancy between the "absolute" time and the current one. However, being a structural error, it can be compensated. The easiest way to compensate it is to assume that the clock is operating at the nominal frequency and enable the timer accordingly. After a fixed known amount of time, the content of the timer is read and the temporal discrepancy used to determine the effective clock frequency. Obviously, another possibility would require to measure the exact clock frequency of the available quartz with an oscilloscope.

The first two types of clock error depend on exogenous variables, i.e., the operating temperature and aging. Since such variables evolve with time, either dedicated hardware or intelligence is required to mitigate such phenomena.

From the above it emerges that for any application requiring an accurate time stamp, the embedded system designer should pay attention to the clock synchronization issue for a given embedded system by intervening at the hardware level, at the software one, or both. If time synchronism is a main issue and we can access the GPS signal, then, by equipping the embedded system with a GPS receiver, we can get a time accuracy within 40 ns [226]. Another—less accurate—option is to synchronize the clock, e.g., by accessing the web with some application and download the current time periodically; by repeating the operation we expect to be able to mitigate the effects of randomness in the operation which, however, keeps a bias due to the time associated with the information conveyance and the information processing and update in the embedded timer. This scenario is particularly relevant also in distributed sensor networks where a set of units are deployed in a given environment with their own sensor platforms and clocks. Clock information sharing will allow these units to provide a more accurate common concept of time, possibly, with an unit (synchronization master) being able to access a GPS service to provide an improved accuracy over all other clocks. Despite the fact that many technological solutions can be envisaged to improve the overall accuracy, we focus in the following on those methods that provide a quasi-synchronous time concept among a set of distributed units.

The methods discussed in the following should be executed periodically in order to synchronize the units. Since "periodically" is a qualitative term and clock drift represents a change in stationarity due to the combination of temperature and aging effects, we should adopt methods presented in Chap. 9. In particular, given the fact that the following methods require the exchange of synchronization messages (that, in turn, have an impact on energy consumption and performances), we anticipate that active learning strategies for clock synchronization should be preferred than passive ones.

8.4.1 Clock Synchronization: The Framework

Consider a distributed sensor network composed of at least a computing element. Denote the absolute time as t and let $C_i(t)$ be the time estimate as provided by the generic ith embedded system.

Let us define the error in time e_C of the clock at the generic unit $e_C = C_i(t) - t$ and the clock drift as its derivative w.r.t. time $\frac{\mathrm{d}e_C}{\mathrm{d}t} = \frac{\mathrm{d}C_i(t)}{\mathrm{d}t} - 1$. Since the time drift introduced by the clock is very small for a well designed clock (generally not larger than few ppm of the nominal clock frequency), we can determine a positive real value $k << 1$ so that the absolute value $|\frac{\mathrm{d}e_C}{\mathrm{d}t}|$ is confined by k, i.e., $|\frac{\mathrm{d}C_i(t)}{\mathrm{d}t} - 1| < k$. At every clock cycle the temporal reference system diverges with a drift bounded by k.

The simplest model to describe the discrepancy between t and $C_i(t)$ is linear in gain $f_i = \frac{\mathrm{d}C_i(t)}{\mathrm{d}t}$ and offset $\theta_i = C_i(t_0) - t_0$, where t_0 is the initial reference time for the clock [173]

$$C_i(t) = tf_i + \theta_i.$$

f_i is called clock (time) drift and the offset θ_i is also named clock (time) skew.

If a second jth clock device is available (either on the same board or on a different unit), by recalling that the absolute time is the same for all units, we can write that

$$C_j(t) = tf_j + \theta_j = \tag{8.11}$$
$$= f_{i\to j}C_i(t) + \theta_{i\to j} \tag{8.12}$$

having defined $f_{i\to j}$ and $\theta_{i\to j}$

$$f_{i\to j} = \frac{f_j}{f_i} \qquad \theta_{i\to j} = \frac{f_i\theta_j - f_j\theta_i}{f_i}$$

as the relative drift and relative offset of clock j with respect to clock i, respectively. It is worth commenting that (8.12) holds for any couple of clocks independently from the technology being used. As such, one clock might be the one present in our mobile device, the other being associated with a time-providing service we access on the web to synchronize the local clock.

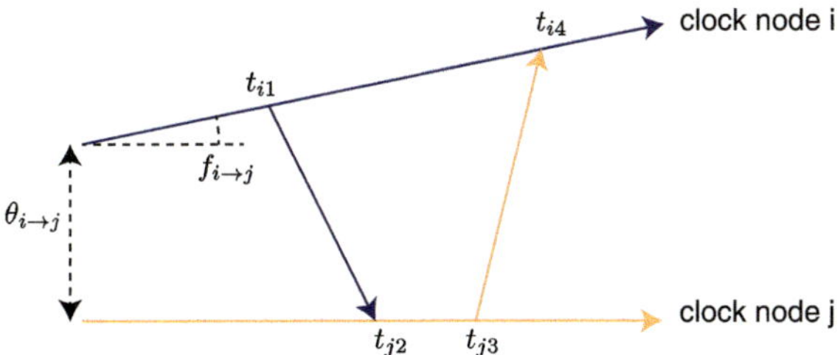

Fig. 8.10 Two-way message exchange mechanism

It follows from the equations above that, in order to synchronize clock i with clock j, the ith unit has to compute terms $f_{i\to j}$ and $\theta_{i\to j}$ and introduce corrections so that the discrepancy between the two readings becomes null or minimal.

Both hardware and software methods have been devised to solve such a problem. Hardware methods provide a very tight synchronization by employing an hardware infrastructure exclusively dedicated to synchronize clocks through electronic clock synchronization signals [168, 174]. There, the availability of a separated synchronization network/bus allows to fully characterize the system delays, e.g., by computing the expected transmission delays among the different units composing the network.

Differently, algorithmic methods exploit the information provided by synchronization messages to synchronize the local clock with the rest of the connected system. Algorithm-based methods are classified in statistic, adaptive and predictive.

8.4.2 Statistic Methods for Clock Synchronization

Statistic methods for clock synchronization, also known as probabilistic methods, update the local clock after a sufficient number of synchronization messages have been collected.

Depending on the amount of synchronization messages and the role of the envisaged entities, we divide statistic methods in "two-way message exchange methods" also known as sender-receiver methods, "one-way message dissemination methods" and "receiver-receiver methods."

8.4.2.1 Two-Way Message Exchange Methods

In two-way message exchange methods, such as the one in [246], clock synchronization affects both terminal points of the communication. Consider the generic ith unit taking unit j as a reference. At each iteration of the synchronization process, unit i sends a message with its time stamp t_{i1} to unit j.

After having received the first time stamp at time t_{j2}, node j sends a message containing time stamp t_{j3} and recorded time t_{j2} back to unit i, that receives it at time t_{i4}. This situation is depicted in Fig. 8.10.

With respect to Fig. 8.10, the following relationships hold

$$t_{j2} = f_{i\to j}(t_{i1} + \tau + \zeta) + \theta_{i\to j} \tag{8.13}$$
$$t_{j3} = f_{i\to j}(t_{i4} - \tau - \xi) + \theta_{i\to j} \tag{8.14}$$

where ζ and ξ are i.i.d random variables modeling the noise affecting the message transmission delay from unit i to unit j and vice versa, respectively. τ denotes the fixed delay introduced by the communication channel, here assumed to be symmetrical (in a symmetrical communication channel transmission in both directions has the same noise characterization, i.e., ζ and ξ have the same pdf).

We start the analysis by considering the simplifying situation where we have only clock skew (i.e., $f_{i\to j} = 1$). When this is the case, Eqs. (8.13) and (8.14) become

$$U = \tau + \theta_{i\to j} + \zeta \tag{8.15}$$
$$V = \tau - \theta_{i\to j} + \xi \tag{8.16}$$

having defined $U = t_{j2} - t_{i1}$ and $V = t_{i4} - t_{j3}$. We comment that U and V do not depend on τ but on the difference between the arrival times only.

By iterating N times the complete bidirectional message exchange mechanism of Fig. 8.10, we generate the set Z_N composed of the N measured quadruples $Z_N = \{t_{i1,k}, t_{j2,k}, t_{j3,k}, t_{i4,k}\}, k = 1, \ldots, N$.

By defining $U_k = t_{j2,k} - t_{i1,k}$ and $V_k = t_{i4,k} - t_{j3,k}$ (8.15) and (8.16) become

$$U_k = \tau + \theta_{i\to j} + \zeta_k \tag{8.17}$$
$$V_k = \tau - \theta_{i\to j} + \xi_k \tag{8.18}$$

Under the assumption that ζ and ξ are Gaussian distributed with zero mean and variance σ^2, [175] demonstrates that the Maximum Likelihood Estimation (MLE) provides the optimal estimate $\hat{\theta}_{i\to j}$ of $\theta_{i\to j}$ as

$$\hat{\theta}_{i\to j} = \frac{1}{N}\sum_{k=1}^{N}(U_k - V_k). \tag{8.19}$$

The variance of the estimator $\mathrm{Var}\left(\hat{\theta}_{i\to j}\right)$ is set by the Cramer-Rao bound

$$\mathrm{Var}\left(\hat{\theta}_{i\to j}\right) \geq \frac{1}{I(\theta_{i\to j})} = \frac{\sigma^2}{2N} \tag{8.20}$$

where $I(\theta_{i\to j}) = \frac{2N}{\sigma^2}$ is the Fisher information. We recall that the Cramer-Rao bound sets a lower bound to the quality an estimate, in this case the clock skew, can have by providing the lower bound to its variance. In the ideal case, the bound degenerates to the equality: the higher the number of complete message exchange N, the lower the variance of the estimate we should expect.

Differently, if the communication channel is such that the noise is ruled by an exponential distribution Exp(λ) of mean λ, the MLE estimate becomes

$$\hat{\theta}_{i\to j} = \frac{\min_{k=1,\dots,N} U_k - \min_{k=1,N} V_k}{2} \tag{8.21}$$

and

$$\mathrm{Var}\left(\hat{\theta}_{i\to j}\right) \geq \frac{1}{I(\theta_{i\to j})} = \frac{\lambda^2}{4N^2} \tag{8.22}$$

with the Fisher information being $I(\theta_{i\to j}) = \frac{4N^2}{\lambda^2}$.

We are ready to extend the above derivations to cover the case where the time discrepancy between the two clocks is experiencing a drift in addition to a clock skew. To make the mathematical amenable, we follow the derivation delineated in [175] and assume at first that a gaussian noise affects the communication channel. The MLE estimates of $\hat{\theta}_{i\to j}$ and $\hat{f}_{i\to j}$ are

$$\hat{\theta}_{i\to j} = \frac{\sum_{k=1}^{N}(t_{i1,k}+t_{i4,k})\sum_{k=1}^{N}(t_{j2,k}^2+t_{j3,k}^2) - Q\sum_{k=1}^{N}(t_{j2,k}+t_{j3,k})}{\sum_{k=1}^{N}(t_{j2,k}+t_{j3,k})\sum_{k=1}^{N}(t_{i1,k}+t_{i4,k}) - 2NQ} \tag{8.23}$$

$$\begin{aligned}\hat{f}_{i\to j} = &\frac{-2N[\sum_{k=1}^{N}(t_{i1,k}+t_{i4,k})\sum_{k=1}^{N}(t_{j2,k}^2+t_{j3,k}^2) - Q\sum_{k=1}^{N}(t_{j2,k}+t_{j3,k})]}{\sum_{k=1}^{N}(t_{i1,k}+t_{i4,k})[\sum_{k=1}^{N}(t_{j2,k}+t_{j3,k})\sum_{k=1}^{N}(t_{i1,k}+t_{i4,k}) - 2NQ]}+\\ &+\frac{\sum_{k=1}^{N}(t_{j2,k}+t_{j3,k})}{\sum_{k=1}^{N}(t_{i1,k}+t_{i4,k})} - 1\end{aligned} \tag{8.24}$$

where

$$Q = \sum_{k=1}^{N} t_{i1,k}t_{j2,k} + t_{j3,k}t_{i4,k} + \tau(t_{j2,k} - t_{j3,k}) \tag{8.25}$$

$$V = \sum_{k=1}^{N}(t_{i1,k}+\tau)^2 + (t_{i4,k}-\tau)^2 + 2\sigma^2. \tag{8.26}$$

The Cramer-Rao bounds associated with estimates (8.23) and (8.24) is

$$\mathrm{Var}\left(\hat{\theta}_{i\to j}\right) \geq \frac{\sigma^2(1+f_{i\to j})^2 V}{N[2V - N(\bar{t}_{i1}+\bar{t}_{i4})^2]} \tag{8.27}$$

$$\mathrm{Var}\left(\hat{f}_{i\to j}\right) \geq \frac{2\sigma^2(1+f_{i\to j})^2}{2V - N(\bar{t}_{i1}+\bar{t}_{i4})^2} \tag{8.28}$$

where term $\bar{t}_{\mathrm{im}} = \frac{1}{N}\sum_{k=1}^{N} t_{\mathrm{im},k}, m = 1,\dots,4$ represents the average of the time instants.

Although (8.23) and (8.24) provide the MLE estimates, they are impractical unless τ is available. Moreover, computation of (8.23) and (8.24) might be computationally prohibitive for typical embedded system. A less precise but computationally much simpler maximum likelihood-like estimator that takes advantage solely on the first and last message of Z_N can be considered. Define

$$
\begin{aligned}
D_1 &= t_{i1,N} - t_{i1,1} \\
D_2 &= t_{j2,N} - t_{j2,1} \\
D_3 &= t_{j3,N} - t_{j3,1} \\
D_4 &= t_{i4,N} - t_{i4,1}.
\end{aligned}
$$

In the case the channel is subject to Gaussian noise, the estimate and the quality of the estimator for the clock drift are

$$
\hat{f}_{i\to j} = \frac{D_2^2 + D_3^2}{D_1 D_2 + D_3 D_4} - 1 \tag{8.29}
$$

$$
\mathrm{Var}\left(\hat{f}_{i\to j}\right) \geq \frac{2\sigma^2(1 + f_{i\to j})^2}{D_1^2 + D_4^2 + 4\sigma^2} \tag{8.30}
$$

Likewise, in the case of channel noise modellable as an exponential pdf function, we have that

$$
\hat{f}_{i\to j} = \frac{2D_2 D_3}{D_1 D_3 + D_2 D_4} - 1 \tag{8.31}
$$

$$
\mathrm{Var}\left(\hat{f}_{i\to j}\right) \geq \frac{\lambda^2(1 + f_{i\to j})^2}{D_1^2 + D_4^2 + 4\lambda^2}. \tag{8.32}
$$

Once the drift has been estimated, we can estimate the skew. Define

$$
U_k' = U_k - \hat{f}_{i\to j} t_{i1,k}
$$

$$
V_k' = V_k + \hat{f}_{i\to j} t_{i4,k}.
$$

Then, if the noise is Gaussian

$$
\hat{\theta}_{i\to j} = \frac{\sum_{k=1}^{N} U_k' - V_k'}{2N} \tag{8.33}
$$

As far as the variance of the estimator is concerned, after the correction of the clock drift has been implemented, we are in the situation of a sole clock offset. As a consequence, the estimator has to satisfy the Cramer-Rao bound given in (8.20). Likewise, when the noise is subject to an exponential distribution we have that

$$\hat{\theta}_{i \to j} = \frac{\min_k U'_k - \min_k V'_k}{2} \tag{8.34}$$

with the variance of the estimator set by the Cramer-Rao bound given in (8.22). Although less precise, these last estimates do not require the knowledge of τ, which is mostly unknown in many situations. Their contained computational cost make such estimates suitable also for mobile embedded devices.

The pairwise synchronization can then be propagated to the units composing a network e.g., by using a spanning tree exploration method. It should be noted that, although the root of the tree should represent the reference clock to be used by all other units for synchronization, in case of a master unit failure any other unit can be promoted to the role of synchronization master.

8.4.2.2 One-Way Message Dissemination Methods

In one-way message dissemination methods, the unit i, selected to be the master, broadcasts its time information to the units composing the network. The time information received by the generic jth receiver unit is expressed by Eq. (8.13), where t_{i1} corresponds to the time stamp present in the received synchronization message and t_{j2} is the time of its arrival at node j. If we assume that τ is either known (by knowing the positions of the units we can derive it and correct the clock accordingly) or negligible with respect to the clock offset and that $f_{i \to j} \simeq 1$ (as mostly it is), Eq. (8.13) can be approximated as

$$t_{j2} \simeq f_{i \to j} t_{i1} + \theta_{i \to j} + \zeta \tag{8.35}$$

where, again, ζ accounts for the equivalent noise affecting the channel. In the Flooding Time Synchronization Protocol (FTSP) described in [172], each unit collects an increasing number of timing messages. $\hat{f}_{i \to j}$ and $\hat{\theta}_{i \to j}$ are estimated based on the dataset Z_N composed of the N measured couples $Z_N = \{t_{i1,k}, t_{j2,k}\}, k = 1, \ldots, N$ by referring to the system model in (8.35) and a least squares procedure. The clocks of the units can be periodically adjusted (and the absolute error reset) by compensating their drift with the estimates obtained during the initial message dissemination phase [242]. In this case, the assumption is that the operating temperature of the units has remained constant over time. If the temperature is time variant, the estimation procedure has to be repeated, unless enough estimates of the clock drift have already been collected at different temperatures and the values included in a look-up table.

8.4.2.3 Receiver-Receiver Methods

The receiver–receiver synchronization approach derives from the consideration that, given a time stamped message broadcasted by the master unit i, the receiving units j and k can exchange messages to each other to synchronize based on common time of the master. If unit i sends its synchronization message with time stamp t_{i1}, we can write from (8.13)

$$t_{j2} = f_{i \to j}(t_{i1} + \tau_j + \zeta_j) + \theta_{i \to j} \tag{8.36}$$

$$t_{k2} = f_{i \to k}(t_{i1} + \tau_k + \zeta_k) + \theta_{i \to k} \tag{8.37}$$

where fixed delays τ_j and τ_k account for the different paths between unit i and j and unit i and k, respectively. ζ_j and ζ_k are two random variables modeling the noise affecting the two channels as in the previous subsections. By subtracting (8.37) from (8.36) we have

$$\begin{aligned} t_{j2} - t_{k2} &= f_{jk} t_{i1} + \theta_{jk} + (f_{i \to j}\tau_j - f_{i \to k}\tau_k) + (f_{i \to j}\zeta_j - f_{i \to k}\zeta_k) \\ &= f_{jk} t_{i1} + \theta_{jk} + \tau' + \zeta' \end{aligned} \tag{8.38}$$

with $f_{jk} = f_{i \to j} - f_{i \to k}$, $\theta_{jk} = \theta_{i \to j} - \theta_{i \to k}$ and ζ' representing an equivalent noise. The term τ' can be neglected because it represents the difference between two terms that are generally very small and, then

$$t_{j2} - t_{k2} = f_{jk} t_{i1} + \theta_{jk} + \zeta'. \tag{8.39}$$

After a data acquisition campaign leading to the dataset $Z_N = \{t_{j2,l}, t_{k2,l}, t_{i1,l}\}, l = 1, \ldots, N$, we can write the linear system

$$\begin{bmatrix} t_{j2,1} - t_{k2,1} \\ \vdots \\ t_{j2,N} - t_{k2,N} \end{bmatrix} = \begin{bmatrix} t_{i1,1} & 1 \\ \vdots & \vdots \\ t_{i1,N} & 1 \end{bmatrix} \begin{bmatrix} f_{jk} \\ \theta_{jk} \end{bmatrix} \tag{8.40}$$

from which we derive the estimates $\hat{f}_{jk}$ and $\hat{\theta}_{jk}$ with a least squares method.

Should we assume that no drift is affecting our clocks, i.e., $f_{jk} = 0$ (and that τ' is negligible as before), from (8.39) we derive the Reference Broadcast Synchronization (RBS) algorithm described in [171]. There, the least square solution associated with (8.40) reduces to the estimation of the relative offset θ_{jk} which simplifies as

$$\hat{\theta}_{jk} = \frac{1}{N} \sum_{l=1}^{N} (t_{j2,l} - t_{k2,l}) \tag{8.41}$$

In the simplified scenario set by the RBS method, the need for a master unit broadcasting the time disappears (no reference to time t_{j1} is present in (8.41)) and the estimate of the clock skew reduces to a distributed negotiation of the time by the different units composing the network.

8.4.3 Adaptive Methods for Clock Synchronization

Adaptive methods, also known as direct methods, require just a single message exchange between units to estimate the clock offset. Most of the existing approaches rely on the two-way message exchange scheme described in Sect. 8.4.2.1. Under such a framework, defined

$$U = t_{j2} - t_{i1} \tag{8.42}$$

$$V = t_{i4} - t_{j3} \tag{8.43}$$

the estimate $\hat{\theta}_{i\to j}$ is simply

$$\hat{\theta}_{i\to j} = U - V. \tag{8.44}$$

The method, extremely lightweight from the computational point of view, becomes the the Lightweight Tree-based Synchronization protocol (LTS) provided in [170]. The same approach is used also in the Network Time Protocol (NTP) [176] used to synchronize clocks in computer networks. Unlike LTS, in NTP, the synchronization messages are continuously sent.

8.4.4 Predictive Methods for Clock Synchronization

Predictive methods are based on the ability to predict values of timing information by exploiting the relationship between the reference clock and the clock to be synchronized. Examples of this approach can be found in [177–179]. All these approaches rely on the fact that, thanks to Eq. (8.12), the relationship

$$\hat{f}_{i\to j} = \frac{C(t_{j,n}) - C(t_{j,n-1})}{C(t_{i,n}) - C(t_{i,n-1})} \tag{8.45}$$

holds.

8.5 Localization and Tracking

The localization and tracking problem can be solved by relying on different technologies. For instance, on the hardware side, we can adopt a GPS sensor and, whenever the satellite signal is not available for a limited amount of time, reconstruct position, and trajectory with an inertial platform mounted on the system electronic board. In indoor environments where the GPS signal is unavailable, we can consider the Ultra-Wide Band (UWB) technology that, de facto, behaves as a terrestrial GPS. The robotic community provides a plethora of solutions for allowing robots to be localized and tracked. The interested reader can find in [233] a detailed description of different methods and useful references. The localization accuracy is in the order of (few tens) of centimeters, obtained at the cost of an expensive apparatus.

In the sequel, we rather prefer to investigate further the situation where the embedded system is not rich in terms of the mounted sensor platform and intelligence has to be taken into account to provide self-localization and tracking of the unit. Of particular relevance is the case where a set of wireless communication devices are available and exchange messages inside the network for localization purposes.

The problem of localization [197, 198] can be formulated as follows: Given N generic units placed at unknown locations, and M units (anchor units) deployed at known locations, determine the coordinates of the N units with appropriate pairwise measurements of their interdistances. In some cases, anchor units might not be present or missing; when this happens, it will be possible to get only the relative displacement of the network units.

The tracking of a target inside an area monitored by a set of sensor units (sensor network) can be seen as an extension of the localization problem, where a generic node is moving and the remaining units are either anchors or nodes to be localized. Most of existing tracking methods start by requiring availability of the localization information generally derived by methods based on the principles here exposed.

Depending on the technology used to compute the interdistances between units, we have ranging methods based on the Received Signal Strength (RSS), Time Of Arrival (TOA), Angle Of Arrival (AOA), and Frequency Of Arrival (FOA). The range information is then processed by localization methods to estimate the position of the units. Such methods can be subsequently classified as centralized or distributed depending on the way the localization algorithm is performed.

8.5.1 RSS-Based Localization

From basic physics, the energy of an ElectroMagnetic (EM) wave propagating in the vacuum in a space without obstacles scales in a manner inversely proportional to the square of the traveled distance [243]. Based on this principle, localization can be carried out by exploiting the Received Signal Strength (RSS) information, i.e., the power measured by the Received Signal Strength Indicator (RSSI) circuit (sensor) present in the radio device of the unit receiving the EM wave.

The RSS localization procedure usually requires the execution of three distinct phases [195]:

1. A power/frequency exploration phase. One after the other, the units of the network broadcast short packets at different levels of power, possibly by exploring the available frequency channels. In turn, each unit transmits the messages (transmission modality) that the others receive (receiving modality). Each unit of the network fills a table where the rows refer to the transmitting units and their power levels and frequencies and the columns the receiving units. To ease the presentation, we assume in the following that the power transmission level is fixed at value P_t and that only one communication frequency is considered. Under this

simplification, the generic cell $c_{i,j}$ of the RSSI matrix contains the RSSI value associated to the packets transmitted by transmitting unit i and received by unit j. The power level represents a shared information among units in the case units transmit at a fixed power. In the case, we should opt for a more complex approach requiring exploration of different power levels or frequency channels, the power transmission level is enclosed in the information field of the received message.

2. Creation of an accurate RSS-distance model. The RSSI tuples collected between pairs of anchor units are used to generate the RSS-distance model. Such a model can have a local validity, i.e., it is the same used to infer distances among the units composing a subnetwork, or have general validity and be the same to be used for the whole network.
3. Solution of the optimization problem. The localization problem is formalized as an optimization problem whose solution provides the position of each units.

In centralized solutions, the processing eager calibration and optimization phases are carried out by a high performing central unit, e.g., as suggested in [195]. The localization algorithm assumes that the elements to be localized are spread in an open and uniform environment, so that the RSS-distance model is unique for all the nodes. The function family $m(r, \theta)$ used to fit the relationship between the received power P_r and the distance r is given by

$$P_r = m(r, \theta) = P_t \left(a + \frac{b}{r^k} \right) \tag{8.46}$$

where P_t is the transmission power. The parameter vector $\theta = [a, b, k]$ must be learned after the nodes have been deployed, following the learning mechanism presented in Sect. 3.4.1. Learning is carried out, for instance, by minimizing a SE figure of merit applied to the measurements taken between all couples of anchor nodes (x is r and y is the measured power P_r if we follow the formalization given in 3.4.1). The result is a vector $\hat{\theta}$ to which is associated model $m(r, \hat{\theta})$.

Not rarely, the optimization procedure is carried out by fixing the value of k to incremental integer positive values and then by selecting the best performing model from the pool of learned models. This approach improves the effectiveness of the learning procedure (for sensitivity issues, learning an exponent value k might be critical, with uncertainty on the estimate introducing significant variation in performance). In some cases, a logarithm is applied to (8.46) so that the power is directly measured in decibels.

Here, we assume a single model for the whole network. If the network can be partitioned in subnetworks on the basis of some locality principle (geographical niche), we might consider to provide a model for each subnetwork. [244] However, given the high level of uncertainty, the gained accuracy does rarely justify the high complexity of the processing procedure. Clearly, different models can be considered if we have a priori information or we request high accuracy in localization.

On the basis of such ranging model, the estimated distance between two generic nodes i and j of the network becomes

$$r^{o}_{i,j} = \sqrt[\hat{k}]{\frac{\hat{b}}{\frac{P_r}{P_t} - \hat{a}}} \tag{8.47}$$

We anticipate that the extraction of the k-root does not represent a computational problem for embedded systems since it can be calculated by using a precompiled Look Up Table (LUT), as we did for the hyperbolic tangent function in Sect. 7.6.

The localization problem can be solved by minimizing the figure of merit

$$E = \sum_{i,j,i \neq j} k_{i,j} a_{i,j} (r_{i,j} - r^{o}_{i,j})^2 \tag{8.48}$$

with respect to all unknown distances $r_{i,j}$ not associated with couples of anchor nodes. In (8.48) $a_{i,j}$ is a binary variable; $a_{i,j} = 1$ states that the message sent by node i was received by unit j (i.e., there is a not null entry in the RSSI matrix), $a_{i,j} = 0$ implies that there was not EM visibility between units i and j at power P_t (this is the main reason for which we might need to carry out a power exploration. Exploration in frequency is also advantageous and more relevant than exploration in power for its direct impact on the EM field). $k_{i,j}$ are weights that might be introduced to differentiate different contributions on the basis of some a priori information.

Localization of unknown units can be carried out by minimizing (8.48) with a centralized least mean squared error procedure that estimates the unknown inter-distances among the nodes. In the case we should obtain several estimates for the distances (Eq. (8.48) will provide different estimates for $r_{i,j}$ and $r_{j,i}$) a straight solution would be to average them. Identification of distances will be used in subsequent methods based on the time of arrival.

However, if we are interested in determining the coordinates of units whose positions are unknown, we consider all distances between two points $P(x_i, y_i, z_i)$ and $Q(x_j, y_j, z_j)$ and (8.48) becomes

$$E = \sum_{i,j,i \neq j} k_{i,j} a_{i,j} \left(\sqrt{(x_i - x_j)^2 + (y_i - y_j)^2 + (z_i - z_j)^2} - r^{o}_{i,j} \right)^2 . \tag{8.49}$$

In (8.49) we shall avoid computing those terms for which both P and Q are anchors (their distance is known). Variables to be determined are those (x, y, z) not associated with anchor nodes. We also comment that (8.49) can be solved with any effective minimization method, e.g., based on Genetic Algorithms or Sequential Quadratic programming.

When an exploration in power (different power levels) or in frequency (different frequencies) is carried out, (8.48) and (8.49) have to take them into account.

Comments

The problem of node localization in embedded systems based on RSS is complicated because of several reasons but should be preferred to reflection methods, e.g., those based on particle filters [232], for the much lower computational load requested.

Solution to the RSS-based localization problem requires an accurate model that, based on the polynomial power attenuation law w.r.t. the traveled distance, depends on the geometry of the node deployment, the presence of obstacles, the type of the terrain, the considered transmission power, and the envisaged communication frequency. In particular, if the antenna is rather close to the soil, say 30 cm from it, then we should not be surprised to find models scaling as $\frac{1}{r^3}$ or $\frac{1}{r^4}$, with the immediate consequence that the received power assumes small values after few meters and many units are needed to cover a large area. Moreover, the presence of obstacles introduces consistent reflections whose interference, either constructive or destructive, significantly changes the EM progation field, which moves away from behaving like $\frac{1}{r^k}$. The situation is particularly critical in indoor environments, where the presence of obstacles induce a very articulated complex structure for the EM field due to EM interference mechanism [245]. As a result, the indoor EM field is time variant, a situation that negatively impacts on localization accuracy and requires ad hoc adaptive strategies, e.g., we identify at first a change in stationarity and react afterwards. Clearly, both in indoor and outdoor applications the accuracy in localization improves by increasing the number of units N and the number of anchors M. This comes at the cost of a more expensive optimization problem caused by the high number of variables to be estimated (the assumption that units are positioned on a plane, characterized by equation $z = \bar{z}$, $\bar{z}$ being a constant, somehow mitigates the problem).

As a last comment, we observe that, due to the computational load, a centralized processing approach is generally preferred. In the distributed approach, the a priori information is hardly integrated in the problem due to the limited processing capability possessed by the units. However, some authors consider a distributed approach, e.g., that proposed in [199], that treats the deployed units as the neural nodes of a Self-Organizing Map (SOM) neural network.

8.5.2 Time-of-Arrival Based Localization

In time of arrival and Time Difference Of Arrival (TDOA) methods [197, 198, 200, 201], the distance between two generic nodes i and j is derived by processing the time of arrival information of signals/messages exchanged between them.

Let us consider Fig. 8.10 for an immediate understanding of the terms used here. Under the hypothesis of clock synchronization between the units, the distance between a node i sending a message with timestamp t_{i1} and a node j receiving it at time t_{j2} is given by

$$\tau = t_{j2} - t_{i1} \tag{8.50}$$
$$r \simeq c\,\tau \tag{8.51}$$

where c is the speed of light. Such approximation holds since the difference in time between the time of arrival of an EM wave traveling through the air and the time of arrival of an EM wave traveling in the vacuum is negligible compared to the other uncertainties involved in a typical localization procedure. It must be noted that this simplified model assumes a direct line of sight between the sender and the receiver.

The localization approach shows to be accurate whenever a dedicated hardware, mostly based on a dedicated radio, has been designed and is in use. In such a case, the time required to process the incoming radio signal introduces a fixed, deterministic additional delay that can be accurately estimated with a calibration procedure. An alternative approach takes advantage of the existing radio used for unit-to-unit communication and it is worth a deeper analysis. Here, the radio transmits a series of messages and localization is carried out by processing the timing information. In particular, in the case of a one-way message exchange mechanism, the temporal information needed in (8.51) is estimated by processing the time stamps associated with the sender and the receiver. Clearly, in order to provide high quality estimates for t_{j2} and t_{i1} we must assume that clocks are synchronized and that the overheads introduced by the interaction between hardware and software (structural delays) can be estimated and compensated with high accuracy. We recall that characterization of the estimates should be done with a number of samples set by the Chernoff bound to grant that the estimate is close to the unknown expected value.

The situation improves by considering a full two-way message exchange approach for which

$$\tau_{RT} = t_{i4} - t_{i1} = 2\tau + \tau_r \tag{8.52}$$
$$\tau_r = t_{j3} - t_{j2} \tag{8.53}$$
$$r \simeq c\,\frac{\tau_{RT} - \tau_r}{2} \tag{8.54}$$

τ_{RT} is the round-trip time of the messages exchange and τ_r is the known (measurable) response time of the procedure in execution on unit j. The effectiveness of (8.54) depends on the accuracy of the time estimates. After having explored the robustness and sensitivity issues in Chaps. 4 and 7, it is clear that even small uncertainties in the estimate can have a tremendous impact in the accuracy of the distance r, a situation that worsen in correspondence of small distances.

In TDOA methods, node k deployed at an unknown location broadcasts a localization signal/message which is received at least by two synchronized receivers i and j whose locations are known. Under the assumption that the units belong to the same plane (e.g., $z = \bar{z}$), we have that

$$
\begin{aligned}
r_{ik} &= c(t_{i2} - t_{k1}) \\
r_{jk} &= c(t_{j2} - t_{k1}) \\
r_{ik} - r_{jk} &= c(t_{i2} - t_{j2})
\end{aligned}
\tag{8.55}
$$

System 8.55 describes an hyperbolic function having its foci in the positions of anchor nodes i and j. It can be proved that if we have at least three anchor nodes, then the location of the generic node k is identified by the intersection of all possible hyperbolas. Given the complexity of the intersection phase (hyperbolas do provide a feasibility area for the locations due to the presence of uncertainty), the localization problem is generally carried out at a centralized level.

8.5.3 Angle-of-Arrival Based Localization

In the Angle-Of-Arrival (AOA) localization method each unit mounts a set of antennas placed at some known angular values within a polar coordinate system centered on the units. The unit to be localized broadcasts signals/messages which are received by the antennas of neighboring units and used to estimate the internode distances as perceived by each antenna. Such distances are then used to compute the angle of arrival of the signals or, in other terms, the angle between two nodes [197, 198]. Polar coordinates, in terms of angular value and distance, are then provided. For instance, in the simple case of two antennas a and b deployed at distance l from the unit to be localized and mounted at 180 degrees, under the assumptions that units are placed on the same plane ($z = \bar{z}$), the estimates for the distance r and the angle α w.r.t. a reference axis (the line connecting the two antennas) are [202]

$$
\hat{r} = \sqrt{\left(\frac{r_a^2 - r_b^2}{4l}\right)^2 + r_a^2 - \left(\frac{r_a^2 - r_b^2}{4l} + l\right)^2}
\tag{8.56}
$$

$$
\hat{\alpha} = 90^\circ \pm \arcsin\left(\frac{\sqrt{4(l^2 + r^2)^2 - (2(l^2 + r^2) - (r_a - r_b)^2)^2}}{4lr}\right)
\tag{8.57}
$$

where r_a and r_b are the distances of the unit to be localized from antennas a and b, respectively, obtainable, e.g., with a RSS or a time-based method.

8.5.4 Frequency-of-Arrival Based Method

In the Frequency-Of-Arrival (FOA) or lighthouse method, an anchor unit rotates with angular speed ω a laser beam of angular width b. The unit receiving this signal estimates its distance from the anchor unit by measuring the duration τ of the time elapsed under the beam exposition. By knowing speed and width of the beam, the distance estimate is calculated as:

$$\hat{r} = \frac{b}{2\sin(\omega \cdot \tau/2)}. \tag{8.58}$$

8.6 Adaptation at the Application Code Level

When designing an embedded solution, we mostly assume time invariance for the application. Time invariance requires that either the algorithm behind the application does not change with time or the mechanism generating the data the algorithm relies upon does not evolve. However, the time invariant hypothesis is strong and, even though sometimes it can be met and surely holds over a short/medium period of time, it is hardly satisfied in the long run. A bug requiring the code to be updated, the insurgence of a situation not expected during the application design phase, a structural change in the interaction between a sensor and the environment are examples of time invariance. Whenever one or more of the above situations arise, the application code becomes obsolete and must be modified to match the application performance. Learning in a time invariant environment, as presented in Chap. 9, provides mechanisms for understanding when it is time to update the application as well as solutions for updating it. Here, differently, we introduce the basic mechanisms that allow the code application to undergo updates whenever needed.

Let us consider a generic program to be executed on an embedded system and its set of constraints that need to be satisfied. The maximum power consumption, the output accuracy, the execution time are some examples of performance onto which we add time constraints. We recall that the performance (constraints) verification problem was addressed in probabilistic terms in Chap. 4 and Sect. 4.7.

The program, partitioned in atomic functional segments of code, representing the smallest blocks that can undergo a change, can be modeled as a graph [210]. Each block is associated with a node of the graph that is connected by arcs that represent the feasible execution flow. Remote reprogrammability of the code can act at different levels by changing the graph with feasible actions, defined as those satisfying the set of constraints. In general, the code associated with the graph is present in the embedded system memory (but different strategies might be enabled that require the blocks on demand). The actions we can contemplate are:

- Substitute the code associated with a node. The body composing the block is substituted.
- Substitute a subgraph (or the entire graph). When this is the case, it is possible that the new subgraph has a different topology. However, the interface with the unchanged subgraph is maintained.
- Activate/remove arcs. Thanks to a parametric reprogrammability, the execution flow is modified by enabling/disabling some parts of the code. However, all the program code is kept into the memory (i.e., the topology of the node does not change).

In the following subsections, we will present the most relevant strategies for remote code reprogrammability.

Algorithm 17: A parametric program. The code is parameterized in the parameter set $\theta = \{\theta_1, \theta_2, \theta_3, \cdots, \theta_n\}$. Depending on the values assigned to the set θ the code modifies its properties.

```
i = 0;
enable-sensor();
while i < θ1 do
    data[i] = sample();
    if θ2=0 then
        data[i] =lowPass(data[i]);
    end
    i = i + 1;
end
disable-sensor();
if θ3 = 1 then
    dataF =average(data);
else
    dataF = weighted average(data, θ4, ··· , θn);
end
output(dataF);
```

8.6.1 Remote Parametric-Code Reprogrammability

In parametric reprogrammability, the application code contains a set of parameters θ that, depending on the assumed value, activate/disable/modify the code to be executed. Within a graph description of the functional code, parameters are associated with arcs which can be either present or removed (the parameter value is null). The points were parameters are inserted in the code are fixed and defined at compile time; the type and the value the parameters can assume are specific for the given application code. By remotely changing the current value of the parameters set we change the execution modality. The decision about whether changing the parameter values or not is either decided by the user or relying on some (semi)automatic tools.

The example pseudocode given in algorithm 17 describes a sophisticated data acquisition task providing an accurate filtered sensor reading for subsequent processing. At first, a procedure enables the active sensor (a duty cycle is envisaged for the energy-eager active sensor). Once the sensor is ready to operate (warm up phase completed), the sampling loop acquires a set of samples, possibly filters them and carries out a data processing either based on an average or weighted average before outputting it. Once the sampling phase ends, the sensor is switched-off.

The code is parametric in n parameters composing the set $\theta = \{\theta_1, \theta_2, \theta_3, \ldots, \theta_n\}$, n is fixed. Parameter θ_1 is integer and represents the number of samples that need to be acquired once the task is activated. Parameter θ_2 assumes a binary value enabling or disabling the low-pass filter action. Parameter θ_3 controls the final processing leading to the output value $dataF$. If $\theta_3 = 1$, a simple average is taken on the data vector $data$, otherwise the output of the processing is the scalar product between vector *data* and a corresponding vector of parameters. The maximum number of parameters is equal to θ_1 and must be less or equal than $n - 3$.

Assume that the parameter set θ is coded so that each component is assigned with a numerical value. However, different strategies can be envisaged to code the information in a smaller number of bytes, hence reducing the number of bytes to be transmitted to the remote unit. For simplicity, let's consider the first case, assume $n = 9$ and that we wish to acquire $\theta_1 = 4$ data, disable the low-pass filter and activate the weighted average procedure with parameters $\frac{1}{8}, \frac{3}{8}, \frac{3}{8}, \frac{1}{8}$. The parameter vector to be sent to the unit for updating the program is then $\theta = \{4, 1, 0, \frac{1}{8}, \frac{3}{8}, \frac{3}{8}, \frac{1}{8}, 0, 0\}$. We comment that the last two parameters, θ_8, θ_9, are set to zero since we acquired 4 samples out of a maximum number of 6.

Within a parametric reprogrammability, it is the designer who decides which constants need to be made parametric. At deployment time, the system starts with a vector θ with values (default values) assumed to be the best configuration based on some a priori information. Then, at run-time, the designer or an automatic tool might decide to modify the code. A new parameter vector is issued with the new configuration and sent to the unit. After a safe completion of the parameter update procedure, the task operates with a new modality.

The advantage of a parametric program is in the contained number of bytes that need to be updated.

8.6.2 Remote Code Reprogrammability

In remote code reprogrammability, the program or subparts of it are changed at run time. This provides a higher flexibility compared to the parametric reprogrammability, but requires the availability of a run time support for reprogrammability. Moreover, it should be clear that the ability to change the behavior of the system after its deployment introduces the risk of facing unexpected side effects that might cause the system to crash.

Algorithm 18: Remote code reprogrammability. The program is the code equivalent to the one in algorithm 17 with the configuration $\theta = \{4, 1, 0, \frac{1}{8}, \frac{3}{8}, \frac{3}{8}, \frac{1}{8}, 0, 0\}$

```
i = 0;
enable-sensor();
while i < 4 do
    data[i] = sample();
    i = i + 1;
end
disable-sensor();
dataF = weighted average(data, 1/8, 3/8, 3/8, 1/8);
output(dataF);
```

In the pseudocode example given in algorithm 18 we present the same code of algorithm 17 associated with the configuration $\theta = \{4, 1, 0, \frac{1}{8}, \frac{3}{8}, \frac{3}{8}, \frac{1}{8}, 0, 0\}$. Here, to ease the management, reprogrammability requires to substitute all the program code.

8.6.3 Decision Support System

Within the above reprogrammability framework, intelligence can be applied at two levels. At the higher level, machine intelligence can help or substitute the human operator in decision making, e.g., by identifying the most appropriate actions (modification of parts of the application code, setting a new parameter configuration) to grant quality of service and maximize the efficiency of the system. Here, the change can be either recommended and the operator makes the final decision based on his/her expertise, or automatic. When it is automatic, we enter in the realm of the Machine-to-Machine (M2M) interaction: decisions are taken by machines and affect machines. At the lower level, intelligence is used to assist the reprogramming infrastructure to perform changes at the code in execution in a safe way. We name the former case as "application update planning" and the latter as "application update validation."

8.6.3.1 Algorithms for Application Update Planning

The change manager, namely the module identifying the most appropriate reprogramming actions to be assigned over time to a self-adaptive, autonomic, or reprogrammable embedded system, can be conveniently modeled with a Monitoring, Analysis, Planning, Execution, (Knowledge) MAPE(K) framework [211, 222]. Within a MAPE(K) formalism, the change manager, i.e., the management process behind a reprogrammable embedded system, is composed of the four phases depicted in Fig. 8.11. More specifically, the change manager is composed of the

1. Monitor process. The monitor process is responsible for collecting and processing that information acquired from the embedded system that characterizes the current status of the embedded system.
2. Analysis process. The analysis process, by exploiting information associated with the current status of the system and its history, determines whether a change in the code of the embedded system is needed or not.
3. Planning process. This phase identifies which segments of the application code need to be changed and how the change should be carried out to obtain the best results.
4. Execution process. This module implements the decisions made by the planning process by translating decisions into actions and orders for the effectors. For instance, in the case of a parametric reprogrammability model, the process generates the commands that allow the new parameters to be sent to the embedded system. Similarly, in a dynamically reprogrammable framework, the process

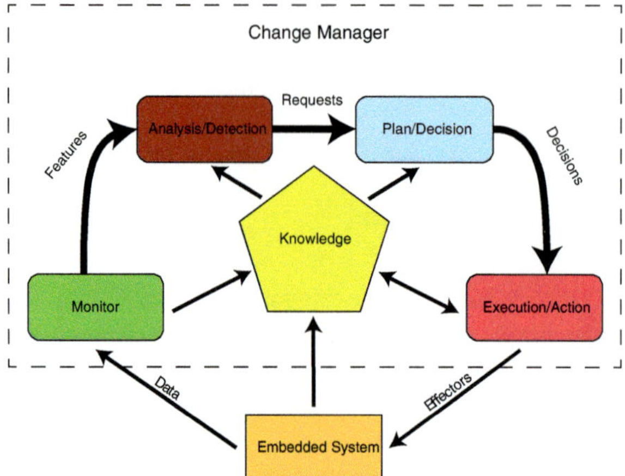

Fig. 8.11 The MAPEK cycle: Monitor -Analyze -Plan -Execute with Knowledge. The change manager identifies when it is the time to intervene and reprogram the application in execution on the embedded system. Data acquired from the embedded systems are used to characterize its state. The change manager, once identified the need to intervene with a change in the code, implements the due actions through the end effectors

leads to a semi-automatic generation of the code to be used to update or integrate the firmware in execution in the embedded system.

The functional modules composing the MAPEK take advantage of a shared knowledge base that contains, among the others, a set of action-performance instances explaining how a given action was effective in the past (or it is believed to be effective) in achieving the objectives set by the embedded system. In some other cases, the knowledge base is rich and takes the form of an explicit model that estimates how taken decisions will affect the future states of the embedded system.

The relation between the change manager and the embedded system context can be framed within a bayesian decision theory, e.g., see [212]. In such a framework, the change manager plays a game against the nature.[2] The nature defines the operational context for the embedded system $y \in \mathscr{Y}$, unknown to the change manager, and provides observations $x \in \mathscr{X}$ that can be acquired by the change manager along with the state information $s \in \mathscr{S}$ of the embedded system.

The role of the change manager is to make a decision or, which is the same, select an action a from the space of feasible actions $\mathscr{A}$. After having applied a specific action a to the embedded system a reward is obtained, represented by the value assumed by the utility function $U(y, s, a)$, which measures how the chosen action a is compatible with y.

A decision strategy can then be represented as a function $\delta : \mathscr{X} \times \mathscr{S} \to \mathscr{A}$ so that

[2] here, under the term "nature," we have all those external actors that affect the embedded system's context.

$$\delta(x, s) = \mathrm{argmax}_{a \in \mathscr{A}} E[U(y, s, a)] \tag{8.59}$$

with expectation taken w.r.t. a given policy $\pi(t, y, s, a)$ representing the probability that the action chosen at time t is a, given the status $[y, s]$.

The use of utility functions to evaluate the effectiveness of taken actions in an autonomic system has been introduced in [213] where it is assumed that the state of the context is not hidden and measurable ($\mathscr{X} = \mathscr{Y}$). In such a case (8.59) reduces to a straight optimization problem. The limits of an approach based on an optimization framework reside in the amount and quality of the knowledge needed to build the decisional system and the ability to design an appropriate utility function connecting the states of the environment with the choices operated by the change manager. To overcome this limit the literature has introduced the "tabula rasa" learning method, that builds the knowledge by starting from little or not at all built-in domain knowledge in the knowledge base. However, since some knowledge about the domain the embedded system works and how it interacts with the environment is mostly available, we can incorporate such information directly in the decision process, e.g., by considering reinforcement learning algorithms [214, 215] that learn directly from the interaction of the embedded system with its operational context.

In reinforcement learning, suitable action strategies issued at time t are updated on the basis of the utility value evaluated at time $t + 1$. Good strategies get a reward and their weights are increased, loosing strategies are penalized. In this way, the decision policy is updated online during the operational activity of the system. In a reinforcement learning setup, the utility function $U(y, s, a)$ represents the average of all rewards obtainable by starting from a state-action pair, or more formally, taking actions following a policy π

$$\begin{aligned} U(y, s, a) &= E_\pi[R_t | y_t = y, s_t = s, a_t = a] \\ &= E_\pi[\sum_{k=0}^{\infty} \gamma^k r_{t+k+1} | y_t = y, s_t = s, a_t = a] \end{aligned} \tag{8.60}$$

where the γ parameter is used to reduce (discount) the effect of rewards r obtainable in the distant future and π is the set of probabilities associated with each action-state pair.

A modified version of the R-learning algorithm provided in [215, 217] allows to account for previous knowledge as described in algorithm 19.

The learning algorithm does not require any a priori knowledge about the context and its relationship with the embedded system (in the original formulation the utility/action-value functions were all initialized to zero, hence implementing a tabula rasa approach). Clearly, any a priori available knowledge about the embedded system can provide an improved starting point.

If, during the operational life of the system, the relation between the external context and the device should change, the learning algorithm would react, hence inducing an immediate update of the utility function. On the long run, the utility function values identified by the algorithm result to be an approximation of Eq. (8.60).

Algorithm 19: R-Learning method. α and β are sufficiently small positive parameters.

Initialize $U(y, s, a)$ and the rewards by using a priori information about the available system
Initialize the average expected reward under the current policy ρ with an arbitrary value, e.g., 0
while *true* **do**
  Estimate y and update s with the current state of the device
  Choose action a given (y, s) by using a behavior policy (e.g., ϵ-greedy on U)
  Take action a and observe r,(y', s')
  $U(y, s, a) = U(y, s, a) + \alpha[r - \rho + \max_{a'} U(y', s', a') - U(y, s, a)]$
  if $U(y, s, a) = \max_a U(y, s, a)$ **then**
    $\rho = \rho + \beta[r - \rho + \max_{a'} U(y', s', a') - \max_a U(y, s, a)]$
  end
end

The method requires that the set of feasible actions, given a state, is limited. Yet, it is possible to modify the set, by taking care to update policy π by properly setting the probabilities associated with each choice. In other words, adding new actions involves a coherent update of the probability values of π.

As an example of the framework, we consider as embedded system the Electronic Control Module (ECM) of an hybrid car. We focus on the mechanism allowing the car to switch between the fuel engine and the electric propulsion to minimize the fuel consumption over a given trajectory $s = \{s_1, s_2, \ldots s_n\}$. composed of n path segments s_i.

Example of feasible paths are "slope," "flat terrain," "smooth downhill." Here, the reward function depends to the fuel needed to cover a segment of the trajectory i.e., $r_t = r(s_i, \text{en}) = \frac{1}{\text{fuel}(s_i, \text{en})}$ where en is a variable denoting the usage of internal combustion f or the electric propulsion e. In addition, the policy considers the actual speed of the car, the remaining fuel and the battery voltage. Depending on the nature of the path (e.g., slopes) and the amount of remaining fuel/battery (low battery voltage), the ECM of the car might find more convenient to use fuel than an electric propulsion. The initial values of U are derived from a mathematical model of the car. The current driving and the conditions of the terrain under different weather conditions determine both the actual reward and the state, hence providing the values to update the utility function and the average expected reward. Since the utility function embeds the behavior policy, by updating the utility function we update the choices planned by the power-track controller.

8.6.3.2 Algorithms for Application Update Validation

In the case of parameterized reprogramming, the designer has identified well in advance, before the deployment phase, the feasible changes the code might undergo. In doing this, we also grant that performances are satisfied and that the code is correct. In general, we should not expect unwished behaviors by activating a change in the code following a new parameter set. Differently, in the case of a dynamic reprogrammability, the situation is more complex and we should be sure that new actions do not introduce problems in the application. The problem is even more relevant in embedded devices due to the strict coupling between the device, the operating system (mostly real time) and the application.

The main focus of the research on code validation has been devoted to investigate the legitimacy of the code change and the reliability of the source providing the new code. In other words, it appears that it is more important that the code is given by a loyal provider than the correctness of the content. By not considering the case where the code is changed maliciously it remains the issue of verifying if the proposed change does not introduce side effects at the system level. This can be granted e.g., by introducing a mechanism that keeps a default—error free—code in the memory of the device as well as the new code in execution. The old code is used to restart the device shall the change manager discover that something went wrong with the thread in execution (e.g., by monitoring some heartbeats and verifying if functional constraints are met). When a problem is detected by the change manager the current thread in execution is aborted (a preemption mechanism must be considered) and the default thread is executed that grants the system to operate, yet at a low performance/effectiveness level, but in a safe modality. The code that created the problem must be fixed, most of times by the intervention of an operator that, by inspecting the log file, identifies and fixes the problem. The new—hopefully corrected—code is then uploaded in the embedded system. The default code is preempted (but kept in the flash memory) and the new one goes in execution.

8.6.4 Online Hardware Reprogrammability

Field-Programmable Gate Array (FPGA) are programmable devices composed of arrays of logic modules and routing channels that can be programmed to implement custom hardware functionalities. The internal complexity and richness of the FPGA depends on the particular device family which differentiate among them on the content made available, e.g., they contain basic logic operators, LUT, flip-flops, registers, ALUs and memories in the simplest implementations and microcontrollers, I/O interfaces, control modules, and CPU cores in more sophisticated versions.

FPGAs are characterized by an offline reprogrammability ability that involves parts or the whole FPGA device and is carried out offline by uploading the "bitstream" files. Such files contain information for configuring routing switches, LUT and programmable devices. It is immediate the affinity with the parametric programmability modality presented above since bitstream files contain those parameters granting the reconfiguration of the device. In general, reprogramming requires a human intervention, at least to upload the new bitstream code, and involves the whole device.

The appearance of Dynamic Partial Reconfiguration (DPR) FPGAs on the market has opened a new dimension to the hardware application reconfigurability aspect that can now be carried out online, thanks to the enhanced functionalities made available.

The main advantage of a DPR FPGA consists in the possibility of reprogramming the FPGA online, in an incremental way, by relying on a core present in the FPGA that executes the reprogramming application. If the functional code to be placed in hardware is characterized by a sequential execution and cannot be fully implemented in the device due to its complexity then, the hardware associated with the already used functionalities can be freed and reprogrammed (in such a case, the bitstream file refers only to a sub part of the whole device). Since reprogramming time and reprogrammed area are directly proportional, it becomes evident here the advantage that this technology brings to hardware reprogrammability.

As an example let us consider a simple functional flow that requires execution of three modules A, B, C characterized by two functional dependencies between modules A and B and A and C. Module A has to be completed before the other two can be executed in parallel. Assume that the FPGA cannot implement all modules together, for size limitations. Within a DPR framework, module A is implemented at first. After its completion, task A is freed and the available area is reprogrammed to host tasks B and C. During reprogrammability we might have different implementations for both B and C. The final choice depends on some application constraints and aims at finding a trade-off between performance (the more the area dedicated to the task the higher the performance) and energy consumption (the less the area dedicated to a task the lower the energy consumption). Depending on the current energy and time constraints the change manager, namely,the decisional module solving the trade-off, identifies the final solutions to be placed in the FPGA. The change manager is in execution in a processor core embedded in the FPGA and, generally, follows the MAPEK cycle of Fig. 8.11. Moreover, it monitors the execution thanks to the presence of sensors (e.g., it measures the effective energy consumption of the actual configuration) and controls that the status of the task to be completed assumes feasible values.

Reference [223] proposes a formalism that describes the functionalities of an embedded application as a Finite State Machine (FSM) where, depending on the output of some logical conditions (e.g., associated with constraints satisfaction) actions are taken. The decisions taken by the change manager are associated with a controller that transforms the adaption problem to a Discrete Controller Synthesis (DCS) one. The DCS explores the FSM graph and inspects the constraints associated with those variables that can be controlled. By acting on controllable variables the DSC optimizes a given control objective [231], e.g., it acts on the control variables such as the voltage and frequency (see the DVFS policies presented in Sect. 8.1) to minimize the consumed energy. In controlling the variables it grants that other constraints are satisfied, e.g., some temporal deadlines associated with the tasks completion.

Other authors follow a similar approach but apply heuristics to the change manager to decide which changes need to be implemented in the FPGA, e.g., see [224]. Information about the system performance is there obtained by employing an approach based on heartbeats: the system records at periodic time instants (heartbeats) the level of advancement of the task, so that it can make decisions about the efficiency of the current configuration. An hardware/software partitioning phase is then carried out that, by taking advantage of information provided by heartbeats, decide whether executing the task entirely on the FPGA, on an embedded core, or adopt a

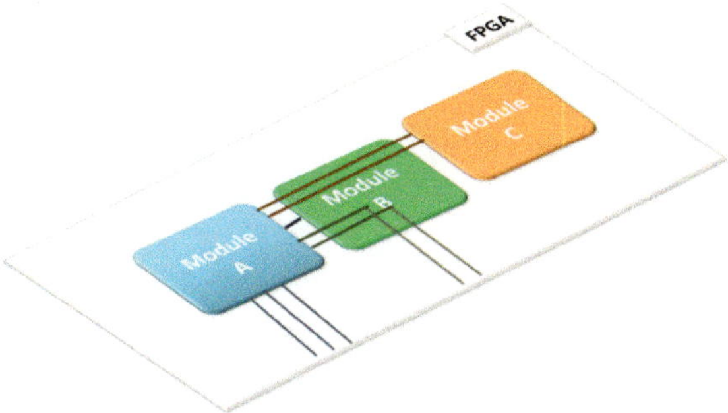

Fig. 8.12 FPGA feed-through signals: reprogramming of module *B* introduces side effects on modules *A* and *C* due to the presence of passing through signals

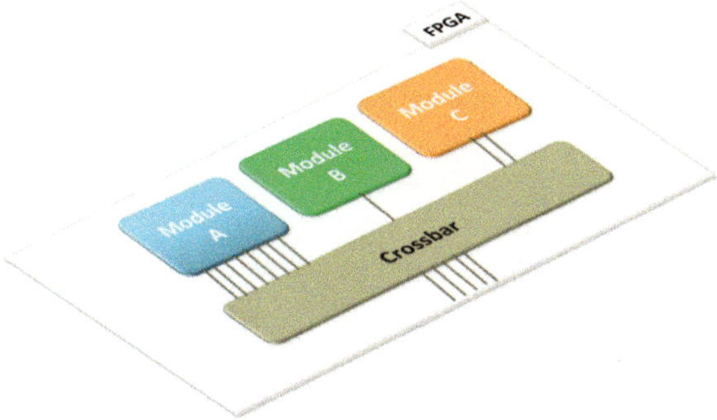

Fig. 8.13 FPGA with a crossbar module. The introduction of the crossbar module removes the unwished dependencies that might arise between modules due to signal lines connecting the modules

mixed approach where the computation is distributed between the processor and the FPGA device. Different strategies can be adopted over time, depending on existing constraints. Obviously, the change manager has to balance the gain in performance provided by a FPGA-based hardware solution with the time requested to reprogram the FPGA.

All above approaches assume that it is possible to intervene on the FPGA once decisions about which reconfiguration solution should be implemented have been made. However, at the current state of the technology, this is a strong assumption that might be difficult to meet in some cases. Consider, as an example, a generic module of the FPGA that, in order to collaborate with other modules or access to the external world with the FPGA pins, has to cross the area of a second module, with paths that act as feed-through signals. The situation is depicted in Fig. 8.12 where the lines of module *A* have to cross module *B* to reach module *C* and the external I/O. Any reconfiguration affecting module *B* would indirectly affect modules *A* and *C* due to the presence of feed-through signals.

It should also be commented that a module accessing an external line cannot be relocated anymore due to technological constraints. In fact, a change to a line accessing an external port would require a global reprogramming operation involving the whole device. This issue can be solved by adopting a crossbar solution as suggested in [225] that allows the different modules to connect to external peripherals and

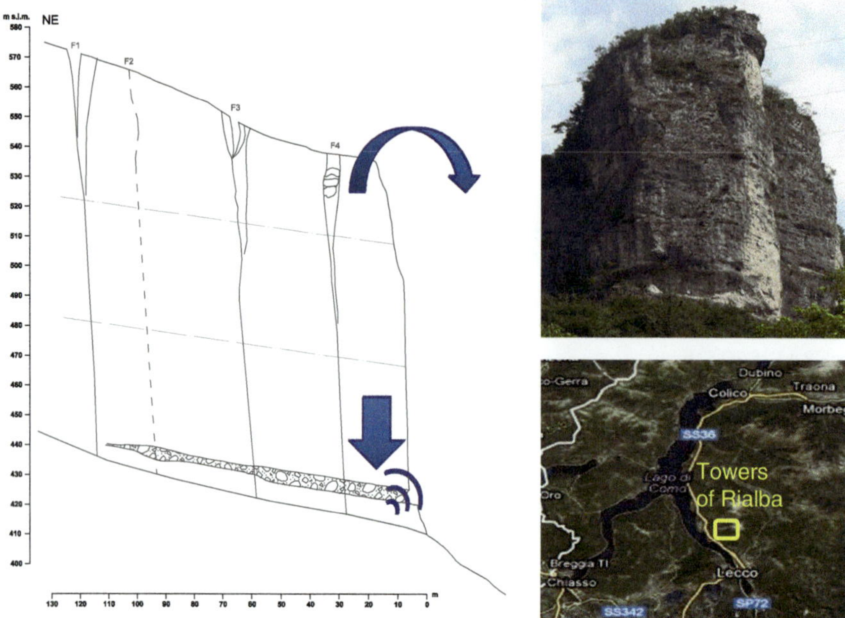

Fig. 8.14 The Rialba towers is a set of four 100m tall limestone columns insisting on a layer of clay. Gravitation is inducing a shift of the structure posing the rightmost tower at the risk of toppling. The lower parts of the structure are subject to stress causing the coalescence of fractures and the subsequent emission of microacoustic signals. The upper part, due to toppling, is subject to a rightward shift inducing enlargement of existing fractures and changes in inclination of the tower

memory with a reconfigurable multiple bus. In other words, the crossbar acts like a proxy between the internal FPGA modules and the external ones.

Figure 8.13 presents the new configuration for the three modules of the experiment associated with Fig. 8.12 with the introduction of the crossbar block. Reconfiguration of module *B* does not affect anymore the other two modules.

8.6.5 An Application: The Rialba Monitoring System

The Rialba towers are a rock tower-like limestone complex overlooking an area covered with critical infrastructures (motorway, railway, electric distribution lines) in the Lecco province, north Italy. The towers insist on a layer of clay and experience a shift that, one day, will lead to a rock toppling/collapse as depicted in Fig. 8.14.

Given the risk insisting in the area a fairly complex sensor network has been designed and deployed at the Rialba towers [218, 219, 230] to monitor the area. Since the towers are exposed to a rock toppling risk the sensor network is composed of two parts, the lower investigating insurgence of microacoustic emissions associated with the coalescencing of fractures, the upper inspecting for enlargement of the

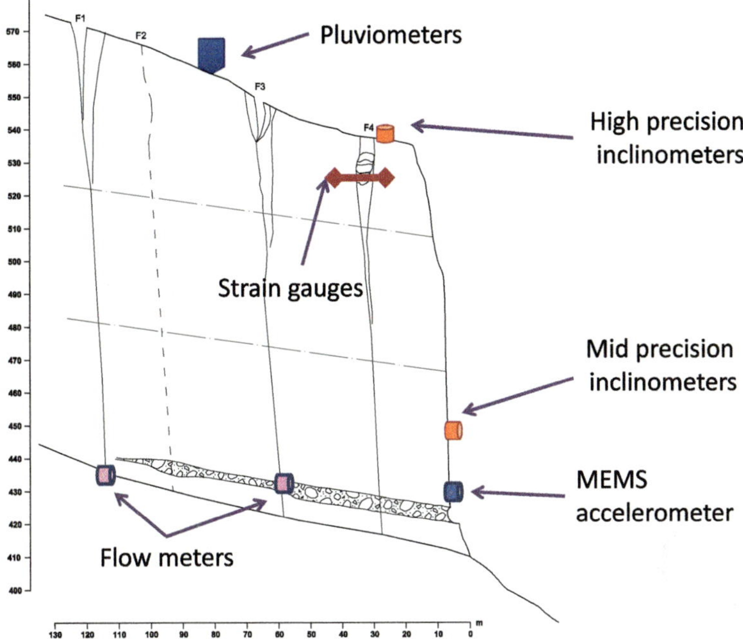

Fig. 8.15 The Rialba towers monitoring network. The upper system is associated with a wireless sensor network technology composed of three leaf units and a gateway. Leaf units mount strain gauges to evaluate possible enlargement of fractures and high resolution inclinometers. The gateway mounts a pluviometer and sends data to a remote control room for further processing. The lower system is composed of a hybrid wired-wireless technology. 5 units connected through a CAN fieldbus acquire information related to microacoustic emissions and the water flow spilling from two springs

interdistances between the towers and variations in inclination. The reference figure is Fig. 8.15 where the types of envisaged sensors are shown.

More specifically, a hybrid sensor network is the technology envisaged at the base of the towers. It is composed of a gateway system, managing energy harvesting operations and data communication both locally and remotely, and five Sensing and Processing Units (SPU), acquiring acoustic emission signals related to microfractures in the rocks as well as the flow of water coming out from two springs. A picture of the SPU is given in Fig. 8.16. SPUs continuously acquire data at 2KHz and immediately inspect incoming signals for the presence of microacoustic emission characterized by a burst nature. For each channel of the MEMS accelerometer the energy of the incoming signal is computed over two sliding windows of different size. When the ratio between the energy of the short size window and the long size one is above a threshold, i.e, the current signal is well above the background noise, then it is believed that the signal contains a microacoustic burst. The sliding window mechanism allow us for introducing adaptation to changes in the pdf of the electronic noise.

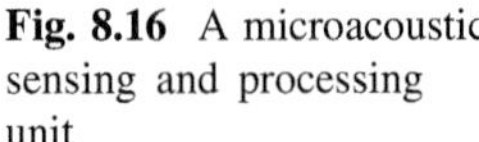
Fig. 8.16 A microacoustic sensing and processing unit

Following a positive detection the signal is flagged as valid and stored together with the content of a pre-trigger buffer designed not to loose the initial part of a burst. The dimension of the buffer is parametric.

The system controller can decide if the incoming signal must be down-sampled; this action is set by a configurable parameter. An offset and a again corrections can be applied to the signal, hence implementing compensation mechanism to improve accuracy and precision (see Chap. 2). The signal can then be filtered with parametric coefficients for a FIR filter.

Duty cycling parameters have been introduced to control the energy consumption of the system by setting the wake-up time for the gateway, the data collection frequency from SPUs and their remote transmission towards a control room. Low frequency data, e.g., the flowmeters or temperature information are acquired with lower frequency and averaged to improve the quality of the data as presented in Chap. 2.

The upper part of the deployment is composed of a fully wireless sensor network (see Figs. 8.17 and 8.18) that collects data from high precision inclinometers and strain guages, together with pluvial data and temperature information. The gateway structure is similar to the one present in the lower subsystem. Most mechanisms for managing data acquisition, even though different from the technological point of view, show a behavior similar with that presented for the lower subsystem.

Energy harvesting has been carried out by implementing the MPPT solutions presented in Sect. 8.3. More specifically, the lower system has a centralized harvesting system at the gateway level mounting a perturb and observe technology. Differently, the upper one has a distributed energy harvesting with a photovoltaic cell for each unit.

Fig. 8.17 The gateway of the upper subsystem (*left*) and a leaf unit (*right*)

Fig. 8.18 A high precision clinometer (*left sensor*) and the strain guage sensor (*right sensor*)

A parametric reprogrammability paradigm (see Sect. 8.6) has been envisaged for the monitoring system. As such, code adaptation for the system was implemented by controlling the value of some structural parameters. Parameters control aspects such as energy management, data acquisition, data quality enhancement, even triggering. An example of some parameters adopted for controlling a SPU are listed in Table 8.1.

By following the MAPEK framework (see Sect. 8.6.3.1) the human operator validates the quality of incoming data and, if not satisfied, intervenes by remotely changing the parameters of the system. Since the acquisition of microacoustic bursts

Table 8.1 SPU software parameters

Module affected	Parameter	Functionality	Values
Filtering	FE	Enable parametric filters	{True, false}
Filtering	FT	FIR coefficients (per axis)	128-elements integer vectors
Filtering	TLTMA	Enable moving average, tilt signals	{True, false}
Filtering	TMPMA	Enable moving average, temperature signal	{True, false}
Event detection	STA	Short-term average window size	{16, 32, 64, 128}
Event detection	LTA	Long-term average window size	{64, 128, 256}
Event detection	A_SET	Recorded axes set	{x, y, z, xy, xz, yz, xyz}
Event detection	EVTWS	Event window size (per axis)	{16, 32, 64,128, 256, 512, 1024, 2048}
Event detection	PRTWS	Pre-trigger size (per axis)	{0, 8, 16, 32}
Event detection	DS	Downsampling	{1, 2, 4, 8, 16, 32}
Event detection	EVTTH	Event thresholds (per axis)	(0-255)
Event detection	DSIZE	Data window size	{256, 512, 1024, 2048}
Event detection	PSIZE	Pre-trigger window size	{256, 512, 1024, 2048}
General HW	MDG	Digital amplifier gains (3 channels)	(0-255)
General HW	MAG	Analog gain switch (adds $20dB$ to MDG)	{True, false}

was the mostly unknown phase (few information is known about the expected nature of the bursts) rewards were assigned to the setting of a given parameter configuration (parameters are those controlling the MEMS accelerometers, e.g., offset and gains, pre-trigger buffer dimension, downsampling) by experts. If not satisfied by the quality of received bursts (also other events can be acquired by changing the sensitivity and the filter parameters) the operator introduced changes in their configuration so as to reduce the false positive ratio.

Chapter 9
Learning in Nonstationary and Evolving Environments

Previous chapters have developed methods and methodologies for solving specific aspects involving intelligent processing on embedded systems and presented techniques for their performance assessment. However, if we look carefully at those methods, we can observe that we have commonly assumed that the process generating the data acquired by our sensors was not changing with time (stationarity or time invariance assumption).

We say that a data-generating process is *stationary* when generated data are i.i.d. realizations of a unique random variable whose distribution does not change with time. Thus, stationarity applies to stochastic processes. We say that a process is *time invariant* when its outputs do not explicitly depend on time. Less formally, in the former case the parameters characterizing the pdf do not change over time, in the latter, the transfer function of the-possibly dynamic-system does not have an explicit time dependency.

In some cases, nonstationarity and time invariance are related concepts. For instance, we will see in the chapter that inspection for time variance in some cases can be achieved by extracting features from the transfer function and verifying a potential change in stationarity. Stationarity was requested either directly by requiring i.i.d data streams or features, or indirectly, by requesting that the application or the model learned from the data was fixed before being implemented in the embedded system. Stationarity/time invariance is requested by performance assessment methods, e.g., the PACC framework, where the given Lebesgue measurable function, albeit affected by uncertainty, is fixed. The same holds for a robustness analysis that initially operates on a stationary/time invariant processing flow.

All in all, we mainly assume stationarity/time-invariance in our applications but are aware that such an assumption represents a first order, yet in many cases, a reasonable hypothesis for the data-generating process.

However, real-world processes are often affected by *concept drift*, i.e., changes in their inherent structure, which results in having the process departing from its initial, stationary (or time invariant) conditions. Concept drift could be due, for example, to a natural evolution of the environment, changes in the operational schema of a system,

C. Alippi, *Intelligence for Embedded Systems*, DOI: 10.1007/978-3-319-05278-6_9,

aging effects (e.g., structural changes in the transduction mechanism of a sensor), as well as faults affecting a cyber-physical system (e.g., abrupt or slowly developing drift). Nonstationarity and time variance can then be modeled as instances of concept drift. We have *gradual concept drift* when concept drift continuously evolves with time (e.g., a drift type of evolution). Conversely, we have an *abrupt concept drift* when concept drift is characterized by a sudden abrupt type of change (e.g., an abrupt type of evolution).

As an example, assume that the data-generating process admits a specific parametric expression $f(\theta, x) \in Y \subset \mathbb{R}$, $\theta \in \Theta \subset \mathbb{R}^d$, $x \in X \subset \mathbb{R}^l$, which can be either the transfer function of the system or the pdf in case of random variables. Let us consider the situation where the parameter vector is affected by a slowly developing concept drift, that shifts a given θ_0 toward a perturbed state, characterized by parameter vector $\theta_0 + \delta\theta$ belonging to the neighborhood of θ_0. By expanding with Taylor differentiable $f(\theta, x)$ in θ_0 we have that

$$f(\theta_0 + \delta\theta, x) = f(\theta_0, x) + \frac{\partial f}{\partial \theta}^T |_{\theta_0} \delta\theta + O(\delta^2\theta) \tag{9.1}$$

with $\frac{\partial f}{\partial \theta}$ and θ column vectors. The stationarity/time invariance hypothesis assumes that no perturbations affect the system or that the perturbation is negligible (the gradient term is negligible compared with the driving term $f(\theta_0, x)$). Expansion given in (9.1) has been written for a slowly developing concept drift continuously affecting the parameters. However, nothing changes if concept drift introduces an abrupt type of perturbation of small magnitude on θ_0.

Clearly, if we want to design effective intelligent embedded systems they must be able to deal with time invariant/nonstationary situations to guarantee good performance also in situations where the system or the environment where they operate in evolves with time. We name *learning in a nonstationary or evolving environments* all those aspects involving learning mechanisms for evolving environments.

The literature addressing learning in nonstationary or evolving environments classifies existing approaches as passive or active depending on the learning mechanism adopted to deal with the process evolution, e.g., [77]. We say that the approach is *passive* when the application undergoes a continuous training without explicitly knowing whether concept drift has occurred or not. Differently, within an *active* approach a triggering mechanism, e.g., a Change Detection Test (CDT), is considered to detect a change in the process generating the data and the application evolves and adapts only when the change has been detected.

This chapter presents passive and active methods for learning in an evolving environment. At first we introduce the learning approaches, detailing afterwards the key elements needed for a successful adaptation.

9.1 Passive and Active Learning

Learning in an evolving environment is specialized in the literature according to the chosen learning method, the way available data instances are used in the training process, and the type of envisioned application. Most learning methods follow either the active or the passive approach, depending on the way the available data instances are used both in the training phase and in the operational life. In nonstationary environments, both passive and active methods can be pursued to cope with concept drift: the most suitable method typically depends on the type of envisioned application.

9.1.1 Passive Learning

Since in passive approaches neither a priori nor derived information is available about the potential concept drift, we are completely blind about the fact a concept drift had, has, or will happen. Adaptation strategies must then be compulsive and carried out passively, without taking advantage of the information provided by incoming data. In fact, as new data come the application is reconfigured, adapted, or relearned depending on its nature and constraints. For instance, if a model M_1 is initially built from data, then a sequence of models $M_2, \ldots M_t$ is generated over time as data come during the operational life.

Passive methods can now be classified depending on the way incoming data are processed:

- *Online learning*. In online learning the new model M_t is obtained by updating the previous model M_{t-1} with data acquired at time t. To derive the online learning procedure we consider at first the traditional, offline, training mechanism given in Sect. 3.4.1. There the model parameters in θ are estimated over a training set Z_N by minimizing the empirical risk

$$V_N(\theta, Z_N) = \frac{1}{N} \sum_{i=1}^{N} L\left(y_i, f(\theta, x_i)\right)$$

leading to the estimate

$$\hat{\theta} = \arg\min_{\theta \in \Theta} V_N(\theta, Z_N).$$

Without any loss in generality, assume that function minimization is carried out by considering the straight backpropagation procedure implementing a simple gradient descent algorithm. Parameter θ at iteration $i+1$, i.e., θ_{i+1} can be expressed as

$$\theta_{i+1} = \theta_i - \eta \frac{\partial V_N(\theta, Z_N)}{\partial \theta}|_{\theta_i}$$

where η is the learning rate. Training stops when some terminal conditions are met.

In online learning, the application-model is continuously trained by exploiting new instances and $V_N(\theta, Z_N)$ simplifies as

$$V_N(\theta, \{(x_i, y_i)\}) = L(y_i, f(\theta, x_i))$$

where (x_i, y_i) is the running supervised sample provided at time i. Parameters update becomes

$$\theta_{i+1} = \theta_i - \eta \frac{\partial L(y_i, f(\theta, x_i))}{\partial \theta}|_{\theta_i} \tag{9.2}$$

and η is a sufficiently small positive scalar in the simplest version of backpropagation. Many variants exist; the interested reader can refer to [100] for further details. Not rarely, the loss function is the squared function $L(y_i, f(\theta, x_i)) = (y_i - f(\theta, x_i))^2$. It should be noted that, at each time instant $i + 1$, a new couple (x_{i+1}, y_{i+1}) is given and the procedure is reiterated. The training algorithm is shown to converge to the optimal value of parameters minimizing the empirical risk, provided that the loss function is quadratic and a sufficiently small η is given. Since a sample-by-sample training method can be a time-consuming operation, a batch modality can be introduced to mitigate such an issue and stabilize the learning procedure. Thus, model parameters are updated asynchronously, at specific *time events* when a batch of n data is made available. All data instances in the batch are considered to have the same relevance and could be gathered from—possibly overlapping—n-dimensional data windows, even if the overlapping mechanism does not find any justifiable reason. If we denote by $Z_{n,i} = \{(x_i, y_i), (x_{i-1}, y_{i-1}), \ldots (x_{i-n+1}, y_{i-n+1})\}$ the batch of n data at time event i, then the learning procedure following the (9.2) becomes

$$\theta_{i+1} = \theta_i - \eta \frac{\partial V_N(\theta, Z_{n,i})}{\partial \theta}|_{\theta_i}. \tag{9.3}$$

- *Ensemble learning*. In ensemble learning, several individual models are activated at the same time instant, and the output of the ensemble is obtained by aggregating the outputs of each individual model. There are no specific restrictions about the individual models, which could be trained and updated during the operational life according to an online learning scheme. We comment that models to be selected neither need to belong to the same model family (the dimension of the parameter vector θ may vary) nor the same model hierarchy (models are generic entities not constrained to belong to the same model class).

  Most often, the aggregation consists in averaging the models' outputs, namely the ensemble provides a weighted average of the outputs of the individual models. However, several different aggregation mechanisms have been considered in the rich literature, which does not necessarily address the issue of learning in evolving environments. In fact, the use of ensemble of models is shown to be beneficial in many circumstances and this claim has theoretical justifications, e.g., [240], where it is proven that an ensemble of models behaves better than the single generic model, even though not necessarily better than the best performing one (which is however hard to identify in a noisy environment).

Let us denote by $M = \{M_i(\cdot), i = 1, \dots, k\}$ the set of individual models composing the ensemble. Since the simplest aggregation mechanism in ensemble learning consists in a weighted average, the output of the ensemble in correspondence with instance x is

$$y(x) = \sum_{i=1}^{k} w_i M_i(x)$$

where $\{w_i, i = 1, \dots, k\}$ are suitably chosen weights typically yielding a linear convex combination of the individual models' outputs.

A viable option to cope with concept drift is to assign larger weights to the individual models that have been more recently trained or updated. This method is effective in developing gradual concept drift where the time locality property surely holds. Differently, if we have an abrupt type of concept drift, then we introduce a spurious effect due to the "step" introduced by the abrupt change. The spurious effect vanishes after k time events if each time event is associated with a new model (after k time events models will be associated with data affected by the concept drift only and, therefore are coherent). In the specific case where each individual model M_i is trained at time event i in an online learning manner, the ensemble may act as a windowing over the k most recent models, mitigating or discarding those models older than k events.

Weights aggregation can be set in different ways, depending on the a priori information we have about the application or the developing class of concept drift. Within an instance selection framework, it is possible to select at first the individual models to be aggregated. For instance, we might select only $l < k$ models which are better suited for describing the current observation. The ensemble would provide output

$$y(x) = \sum_{i \in \mathscr{A}} w_i M_i(x),$$

where $\mathscr{A}$ is a set of cardinality l containing the indexes of the individual models selected. In this case we are able to better deal with an abrupt type of concept drift, still dealing with developing ones.

In other situations, we might set the weights depending on the accuracies of the individual models evaluated on a common validation or test set. For instance, if model $M_i(x)$ is characterized by a mean squared error in validation or test σ_i^2 then the weight w_i can be chosen as

$$w_i = \frac{\frac{1}{\sigma_i^2}}{\sum_{j=1}^{k} \frac{1}{\sigma_j^2}}.$$

When all the individual models need to be equally treated or when there is no a priori information about the effectiveness of each individual model, all weights are naturally set to $\frac{1}{k}$.
Weights can undergo adaptation following the evolution of the environment.

9.1.2 Active Learning

Active learning is a learning paradigm that assumes interaction between the learner and an oracle or some other information source. In the case of learning in nonstationary environments the oracle can be an automatic triggering mechanism able to detect concept drift. Such a triggering mechanism typically operates on features extracted from the data, that are assumed to be stationary when the data-generating process is stationary or time invariant, but are expected to propagate the effects of concept drift once a change arises. Typically, such triggering methods are change-detection tests (CDT) or Change-Point Methods (CPM), which will be described in detail in the rest of the chapter.

Once concept drift is detected the application/model/service must be retrained. We consider, as an illustrative example, the embedded system setup of Fig. 9.1 where the data stream provided by sensors feeds the application/service and is inspected by either a CDT or a CPM. When the trigger detects a change in the data stream, the application/service is reconfigured/retrained by means of cognitive mechanisms, possibly exploiting additional data coming from nearby sensors, when these are inserted in a network. We refer to this paradigm as *detect and react*.

For instance, if an agent detects that the temperature sensor of my mobile is no more accurate (as a consequence of concept drift) an information exchange modality can be activated: the App will inspect nearby weather stations or other smartphones composing the local network to provide an estimate of the correct temperature. Calibration and compensation mechanisms are introduced as a reaction aspect on our unit. Also, the triggering mechanism might need to be retrained since its configuration might be obsolete having been configured in an old state. If no change is detected no reconfiguration is requested at the triggering mechanism and the application level.

When a network of distributed embedded systems is available, the situation is that depicted in Fig. 9.2. In such a case, intelligence can be present both at the embedded system and at the upper management level of the distributed system, where the information is collected and processed for decision making. If this is the case, communication among the components of the intelligent system must be present and activated to deliver relevant information.

The most simple triggering mechanisms are deterministic and based on fixed thresholds. When the features exceed the thresholds, a change is detected.

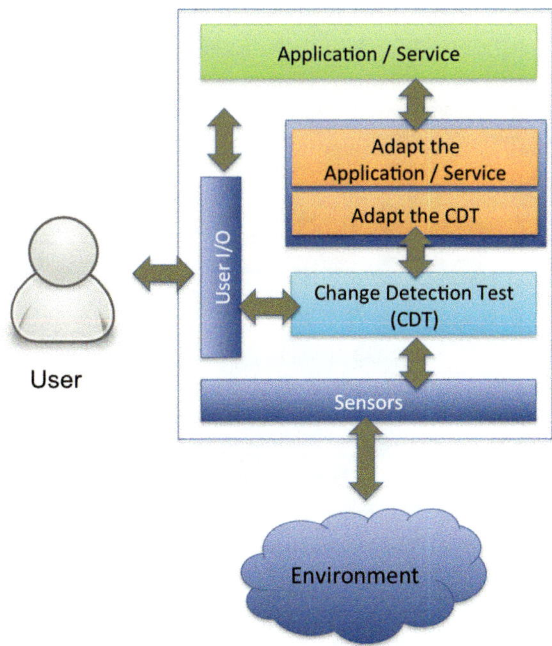

Fig. 9.1 The overall active detect and react methodology for an intelligent embedded system. A triggering mechanism inspects for concept drift (e.g., *through a CDT*). When concept drift is detected the application in execution must be adapted to track the change and, consequently, the CDT is reconfigured on the new operational state

Example: A Fixed Threshold

Consider a scalar feature x which is distributed as an i.i.d. random variable having expectation μ_x and variance σ_x^2. The feature could be the average classification error computed over time in a classification scenario, a measurement, or the result of an uncertainty-affected computation. Changes could be detected in a straightforward manner by setting a threshold $T = \lambda\sigma_x$, which, according to the Tchebychev inequality, implies that the probability of x exceeding T in stationary conditions is less than $\frac{1}{\lambda^2}$. All comments and derivations given in Chap. 3 also apply here.

That said, when $|x - \mu_x| > \lambda\sigma_x$ the threshold is violated and the trigger provides a positive response by detecting a change. The situation is that given in Fig. 9.3 where a threshold (the dashed line) is set. Once instances (samples) are above the threshold a change is detected.

However, since the realizations of x are independent, the probability of having false positives after n observations is $1 - (1 - \frac{1}{\lambda^2})^n$, which rapidly tends to 1 as n grows. By referring to Fig. 9.4 false positives are those data instances above the threshold line not associated with a true change.

We comment that a false positive might be an unpleasant—but not dramatic event—within a "detect and react" mechanism where the detection of concept drift is followed by a reaction that forces the application to undergo an adaptation phase. In fact, when false positives arise the activated reaction will introduce an unnecessary and undue computation with the effect that a new model/application/service will be

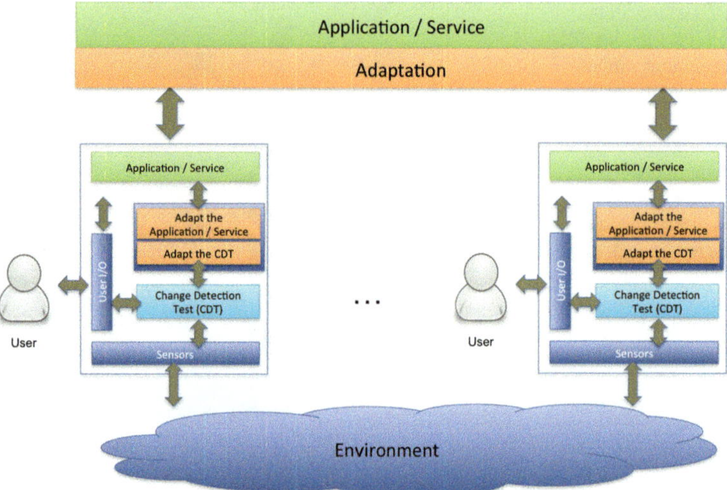

Fig. 9.2 The overall active detect and react methodology for a distributed intelligent system. The application/service is distributed and takes advantage of the information provided by the units composing the distributed platform. Adaptation can operate, with a simple strategy, at the embedded systems level and, at the same time, at the distributed network layer where more sophisticated algorithms can be executed. The outcome of the algorithm introduces adaptation at the distributed application layer and, thanks to remote reprogrammability, to the embedded distributed units

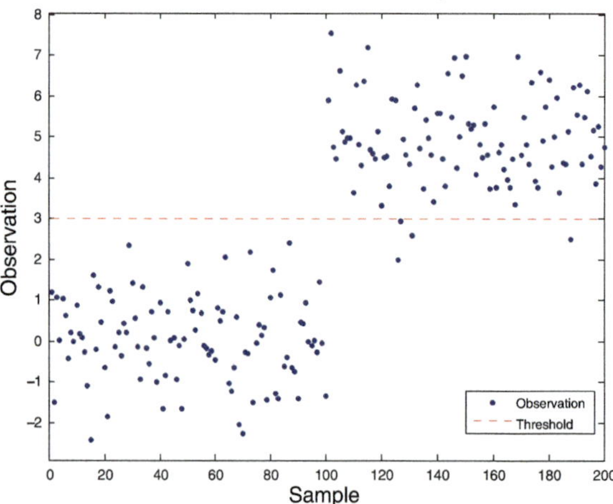

Fig. 9.3 Triggering with a fixed threshold: a change is detected when observation x is above a threshold value, here set to 3

configured even though it was not necessary. Running such unnecessary computation might be not appealing in embedded systems driven by strict real-time execution constraints and/or whenever energy consumption is an issue. As such, we should try to keep the false positive rate as small as possible.

In some other applications, a false positive might be a strongly unpleasant event. For instance, think of the case where a vision system detected in an airport a face

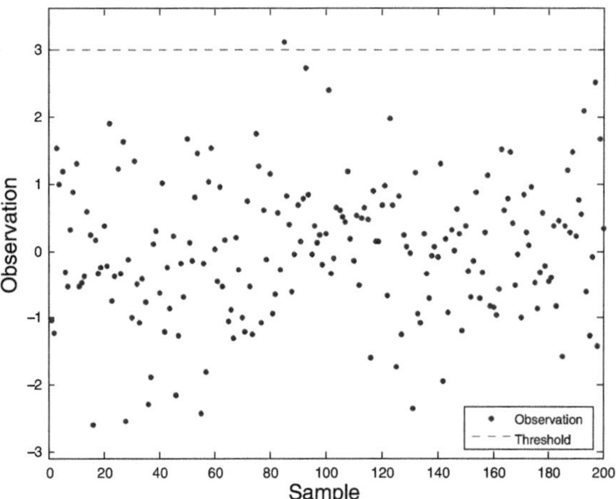

Fig. 9.4 Triggering with a fixed threshold: a false positive arises in correspondence to the sample over the threshold

of a "dead or alive wanted" person that was you! Another situation where false positives are not welcome is associated with fault detection. Here, false positives will erroneously claim that concept drift is associated with a fault in the plant, in a sensor, in a system module when the device is working properly. For the sensitivity and relevance of the issue we will address the fault diagnosis aspect in Chap. 10 where intelligence will play a main role.

To mitigate the above problem we might decide to introduce filters, e.g., a median, at the outputs of the triggering mechanism, hence implicitly assuming that if concept drift arises it will be of permanent and not of transient type. Although this solution is simple and might be effective in some applications and under specific assumptions about the nature of the fault, for others we might need a more accurate statistic-based triggering mechanism reducing false positives and negatives.

More sophisticated stochastic triggering mechanisms of CPM and CDT type will be presented in Sect. 9.2. We anticipate that the main difference between the two triggering methods resides in the way data are processed for decision making. CPMs operate on a fixed set of data to take a decision about the presence of concept drift, although some extensions have been proposed to mitigate the problem so as to address a sequential analysis. As such they are mostly inadequate to process streams of data. Moreover, the CPM computational cost might become prohibitive, making it hardly usable in embedded systems. On the other hand, CPMs are very effective in detecting concept drift, the false positive rate can be controlled at design time and latency in detection is low (CPMs show to be very responsive). Conversely, CDTs are able to operate at the data stream level, their computational cost is contained and, hence, suitable for intelligent embedded systems. The cost we have to pay is associated with an increased latency and the difficulty to guarantee a fixed false positive rate.

Tables 9.1 and 9.2 list the main stochastic triggering methods suggested in the related literature by classifying them according to the parametric/non parametric feature, respectively. Parametric tests require knowledge of the probability density

Table 9.1 Parametric triggering mechanisms for concept drift detection

Name	Test family	Change (A/D)	Entity under test	Type	Notes
Z-test	Statistical hypothesis test	Abrupt	Mean	1D	Assumes normality and known variance [89]
t-test	Statistical hypothesis test	Abrupt	Mean	1D	Assumes normality [89]
F-test	Statistical hypothesis test	Abrupt	Variance	1D	Assume normality [89]
Hotellings T-square statistic	Statistical hypothesis test	Abrupt	Mean	ND	Assumes normality [92]
SPRT	Sequential hypothesis Test	Abrupt	Pdf	1D	Minimizes the stopping time, nonparametric extensions are available [88]
CUSUM	Sequential change-point detection	Abrupt	Pdf	ND	Minimizes the worst detection latency [87]
Parametric CPM	Sequential change-point detection	Abrupt	Depends on the statistics used	1D/ND	Sequential version of a change-point method [93]

function and/or prior information about the concept drift, whereas nonparametric tests are more flexible and require little—mostly reasonable from the application point of view—hypotheses.

In detail, the tables present the family to which the method belongs, either a statistical hypothesis test designed on a given data set or sequential and, as such, suitable to address data stream-based applications. The "change" column shows which type of concept drift the method has been designed for while the "entity under test" column presents the key features used by the test to operate. "Type" refers to the nature of numerical data the test can receive, which is either scalar (univariate test, 1D) or multidimensional (multivariate test, ND). Finally, key references to the test as well as comments are given in the last column to complete the overview.

9.2 Change Point Methods

Change point methods inspect a given data sequence to check its stationary, i.e., whether the samples composing the sequence are independent realizations of a unique random variable or not. The problem is solved by checking if the sequence contains a change point, i.e., a specific time location beyond which the data distribution has changed.

Table 9.2 Non parametric triggering mechanisms for concept drift detection

Name	Test family	Change (A/D)	Entity under test	Type	Notes
Mann-Whitney U test	Statistical Hypothesis test	Abrupt	Median	1D	Rank Test Error based [186]
Kolmogorov-Smirnov test	Statistical Hypothesis test	Abrupt	Pdf	1D	Also goodness of fit test [90]
Mann Whitney Wilcoxon test	Statistical Hypothesis test	Abrupt	Pdf	1D	Rank-based [186]
Kruskal-Wallis test	Statistical Hypothesis test	Abrupt	Median	1D	Mann-Whitney based [91]
Pearson's chi-squared test	Statistical Hypothesis test	Abrupt	Pdf	1D	Goodness of fit and test of independence [80]
Distribution-Free CUSUM	Sequential change-point detection	Abrupt	Median	1D	Nonparametric extension of the CUSUM test [86]
Mann Kendall	Sequential change-point detection	Abrupt	Mean	1D	Designed to analyze climate change [79]
Multi-chart detection algorithm	Sequential change-point detection	Abrupt	Median	1D / ND	Detection of intrusion systems [85]
CI-CUSUM	Sequential change-point detection	Abrupt, Drift	PDF, sample moments	1D/ND	Computational intelligence based [84]
ICI change detection test	Sequential change-point detection	Abrupt, Drift	Mean and variance	1D	Exploits the Intersection of Confidence Interval (ICI) rule [83, 94]
Hierarchical change detection test	Sequential change-point detection	Abrupt, Drift	Mean and variance	1D	Based on a hierarchy of change detection tests [82]
Shiryaev-Robert Extension	Sequential change-point detection	Abrupt	Median	1D	Nonparametric extension of the Shiryaev-Robert test [81]
Mood	Statistical Hypothesis test	Abrupt	Dispersion	1D	Based on ranks [93]
Lepage	Statistical Hypothesis test	Abrupt	Location and dispersion	1D	sum of Mann-Whitney and Mood statistic [93]
Nonparametric CPM	Sequential change-point detection	Abrupt	Depends on the statistic used	1D	Sequential version of a change-point method [93]

9.2.1 Change Points

We say that given data sequence

$$\mathcal{X} = \{x(t), t = 1, \ldots, n\},$$

contains a change point at time/sample $\tau < n$ if subsequences

$$\begin{aligned}\mathcal{A}_\tau &= \{x(t),\ t = 1, \ldots, \tau\}, \\ \mathcal{B}_\tau &= \{x(t),\ t = \tau + 1, \ldots, n\},\end{aligned} \tag{9.4}$$

are distinct i.i.d. realizations of two different unknown random variables distributed as $\mathcal{F}_0$ and $\mathcal{F}_1$. The problem detection can be rewritten as

$$\tau \text{ is a change point if } x(t) \sim \begin{cases} \mathcal{F}_0, & \text{for } t < \tau \\ \mathcal{F}_1, & \text{for } t \geq \tau \end{cases}, \tag{9.5}$$

The change point problem is then converted into an equivalent problem asking if $\mathcal{A}_\tau$ and $\mathcal{B}_\tau$ are sets generated from the same or different distributions.

9.2.2 Set Dissimilarity

A straightforward solution to determine whether a given τ is a change point or not consists in formulating a two-sample hypothesis test on the subsequences $\mathcal{A}_\tau$ and $\mathcal{B}_\tau$. In the hypothesis test, the null (H_0) and the alternative (H_1) hypotheses are composed as

$$H_0 : x(t) \sim \mathcal{F}_0 \ \forall t \tag{9.6}$$

$$H_1 : x(t) \sim \begin{cases} \mathcal{F}_0, & \text{if } t < \tau \\ \mathcal{F}_1, & \text{if } t \geq \tau \end{cases}. \tag{9.7}$$

To test the above hypothesis we need a statistic $\mathcal{T}$, assessing the dissimilarity between $\mathcal{A}_\tau$ and $\mathcal{B}_\tau$ defined in (9.4). Denote by $\mathcal{T}_\tau$ the value of such statistic $\mathcal{T}$

$$\mathcal{T}_\tau = \mathcal{T}(\mathcal{A}_\tau, \mathcal{B}_\tau), \tag{9.8}$$

in comparing $\mathcal{A}_\tau$ and $\mathcal{B}_\tau$. Following a standard hypothesis testing procedure, H_0 can be rejected when the value of $\mathcal{T}_\tau$ exceeds a suitable threshold $h_{n,\alpha}$, corresponding to a given confidence level α and depending on n. In this case, it is possible to claim that $\mathscr{A}_\tau$ and $\mathscr{B}_\tau$ are generated from different distributions (and $\mathscr{X}$ is, hence, not stationary), taking into account the percentage α of false positives.

Example: Evaluating the Dissimilarity of Two Sets

Consider, as an example, the case where data in $\mathscr{A}_\tau$ and $\mathscr{B}_\tau$ are Gaussian distributed with the same variance and we aim at investigating if they share the same expected value. We choose as test statistic D the standardized difference between the two sample means, which leads to a *two-sample t test*. The test statistic is

$$D_\tau = \sqrt{\frac{\tau(n-\tau)}{n}} \frac{\bar{\mathscr{A}_\tau} - \bar{\mathscr{B}_\tau}}{S_\tau} \tag{9.9}$$

where $\bar{\mathscr{A}_\tau}$ and $\bar{\mathscr{B}_\tau}$ denote the sample means evaluated on $\mathscr{A}_\tau$ and $\mathscr{B}_\tau$ respectively and S_τ is the pooled sample variance evaluated on $\mathscr{A}_\tau$ and $\mathscr{B}_\tau$. The threshold $h_{n,\alpha}$ for the statistic D is provided by the Student t distribution with $n-2$ degrees of freedom.

9.2.3 The Change Point Formulation

When the test statistic corresponding to a specific partitioning of $\mathscr{X}$ does not provide enough statistical evidence to reject H_0 we can only claim that the particular τ is not considered as a change point at the given confidence level, hence implying that no change in stationarity happened at sample τ. All other points composing the sequence need to be checked for being potential change points by considering all possible partitions of $\mathscr{X}$. The change point formulation provides a rigorous framework for testing the presence of a change point in a sequence $\mathscr{X}$. Within the CPM framework, e.g., [183], the null and alternative hypotheses for change point method are formulated as

$$H_0 : \forall t,\ x(t) \sim \mathscr{F}_0 \tag{9.10}$$

$$H_1 : \exists\, \tau\ x(t) \sim \begin{cases} \mathscr{F}_0, & \text{if } t < \tau \\ \mathscr{F}_1, & \text{if } t \geq \tau \end{cases}. \tag{9.11}$$

Each feasible time location in $\mathscr{X}$ has to be considered as a candidate change point. More in detail, for each candidate change point $s \in \{2, \dots, n-1\}$,[1] the sequence $\mathscr{X}$ is partitioned into two nonoverlapping sets $\mathscr{A}_s = \{x(t),\ t = 1, \dots, s\}$

[1] The actual range of s depends on the minimum number of samples needed to compute $\mathscr{T}$ from $\mathscr{A}_s$ and $\mathscr{B}_s$.

and $\mathscr{B}_s = \{x(t),\ t = s+1, \dots, n\}$. Set dissimilarity is measured as recommended in Sect. 9.2.2 by means of a suitable test statistic $\mathscr{T}$, which is evaluated for each change point candidate, yielding $\{\mathscr{T}_s, s = 2, \dots, n-1\}$. The most likely change point for sequence $\mathscr{X}$ is finally the one maximizing the statistic

$$M = \underset{s=2,\dots,n-1}{\operatorname{argmax}} \left(\mathscr{T}_s\right). \tag{9.12}$$

corresponding to the value $\mathscr{T}_M$ of $\mathscr{T}$

$$\mathscr{T}_M = \max_{s=2,\dots,n-1} \left(\mathscr{T}_s\right). \tag{9.13}$$

To finalize the test, $\mathscr{T}_M$ has to be compared with a predefined threshold $h_{n,\alpha}$, which guarantees a controlled rate α of false positives. Besides α, the threshold depends on the statistic $\mathscr{T}$ and the cardinality n of $\mathscr{X}$.

When $\mathscr{T}_M$ exceeds $h_{n,\alpha}$, the CPM rejects the null hypothesis, and $\mathscr{X}$ is claimed to contain a change point at M, the location maximizing (9.13). In these circumstances, besides claiming that $\mathscr{X}$ is not stationary, the CPM also provides M, an estimate of the change point instant τ. Conversely, when $\mathscr{T}_M < h_{n,\alpha}$, there is not enough statistical evidence to reject the null hypothesis, and no change point is identified within $\mathscr{X}$. The above can be formalized in the final outcome of the CPM test

$$\begin{cases} \text{The estimated change point in } \mathscr{X} \text{ is } M & \text{if } \mathscr{T}_M \geq h_{n,\alpha} \\ \text{No change point identified in } \mathscr{X}, & \text{if } \mathscr{T}_M < h_{n,\alpha} \end{cases}. \tag{9.14}$$

It is important to comment that, often, the major issue for a CPM is the definition of the thresholds $\{h_{n,\alpha}\}$. In fact, even when the distribution of the statistic $\mathscr{T}$ is known for any partitioning of $\mathscr{X}$, the distribution of its maximum $\mathscr{T}_M$ is hard to be computed. Asymptotic results are available for some test statistics which, however, are often inaccurate at low sample size. We comment that also when it is possible to provide an approximation for the distribution of the maximum, the outcome might not be appropriate. For instance, as discussed in [183], the Bonferroni approximation tends to be over-conservative as n grows. For these reasons, thresholds are often computed with the Monte Carlo method or, even better, with randomized algorithms.

Example: The CPM

Figure 9.5 illustrates the operating principle of a CPM relying on the Student t statistic D (9.9). Figure 9.5a presents a sequence $\mathscr{X}$ composed of 500 data. A change point is injected at $\tau = 350$, so that

$$x(t) \sim \begin{cases} \mathscr{N}(0,1), & \text{if } t < 350 \\ \mathscr{N}(-1,1), & \text{if } t \geq 350 \end{cases}. \tag{9.15}$$

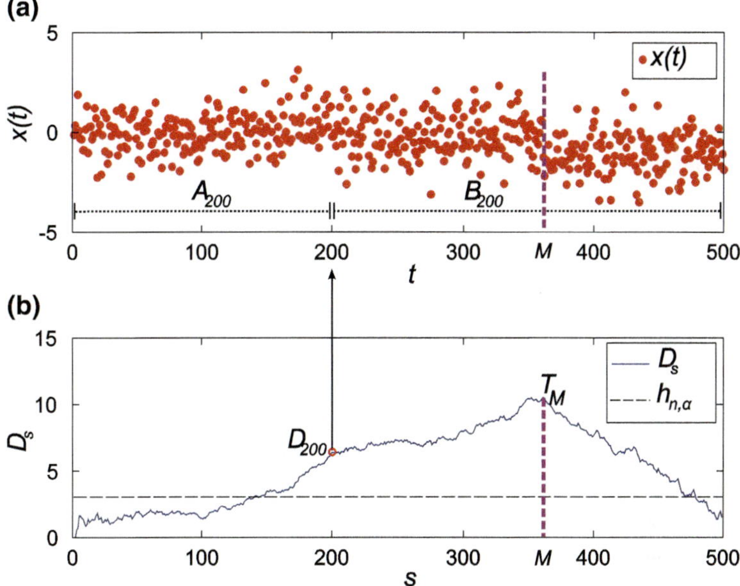

Fig. 9.5 An example of a CPM based on a Student t statistic. Data in (**a**) are distributed according to (9.15). The values assumed by the corresponding test statistic $\{D_s, s = 2, \ldots, 499\}$ are reported in (**b**). The estimated change point M, and the corresponding value of the test statistic $\mathcal{T}_M$ are also reported. For illustrative purpose the figure also shows the partition of $\mathcal{X}$ in $\mathcal{A}_s$ and $\mathcal{B}_s$ when $s = 200$ together with the corresponding value of the statistic $D_{s=200}$

Figure 9.5b illustrates the values of the statistics D_s as a function of $s = 2, \ldots, 499$.

The threshold corresponding to $\alpha = 0.05$, i.e., $h_{500,0.05} = 3.225$ was provided by the CPM package [184], implemented in the statistical R language. Other CPMs can be designed to detect shifts in the mean of a Gaussian random variable, e.g., [183].

9.2.4 Test Statistics Used in CPMs

Very often the test statistic $\mathcal{T}$ measures the dissimilarity between two sets by comparing the estimates for both expected value (sample mean) and variance (sample variance). This choice is motivated by the fact that, in practice, a change in the distribution as per (9.5) would also affect its first moments [185]. It is also preferable to employ nonparametric test statistic since, often, the distribution (even before the change point) is unknown.

Several nonparametric statistics are based on the rank computation, such as the Mann–Whitney [186] (to assess changes in the location), the Mood [187] (to assess changes in the scale), and the Lepage ones [188] (to assess both changes affecting the location and the scale). A CPM based on the Mann–Whitney statistic was introduced in [189] together with a CPM for Bernoulli random variables.

A different approach consists in locating change points by comparing the empirical distributions over two sets of data, as in the CPMs [190] that are based on the Kolmogorov–Smirnov and the Cramer Von Mises [191] statistics. So far, we mentioned only test statistics for scalars, however, the change point formulation can be used to analyze multivariate data, such as the CPM in [192], which relies on the Hotelling T^2 statistic.

9.2.5 Extensions Over the Basic Scheme

The change point formulation was originally presented as an offline processing tool. However, the methods have recently gained a lot of attention and CPM solutions for online and data streams have been provided. Such extensions basically consist in iterating the CPM at each new sample arrival [205]. It comes out that the computational complexity of such CPMs would endlessly grow, hence pushing the research toward the proposal of variants keeping in mind the computational complexity and the memory requirement of the methods [185,190]. In particular, a streaming adaptation is required when the test statistic $\mathscr{T}$ is computationally demanding (such as in test statistics based on the rank computation).

Another relevant issue is how to set the thresholds for online CPMs. First of all, it does not make any sense to control the probability of a false positive as in a hypothesis test. In fact, the test has to be iterated as a new sample arrives and the control of false positives has to be intended within a sequential scenario. Therefore, the thresholds have to be set to guarantee a fixed Average Run Length (ARL) of the test [87], namely the expected number of samples before the test yields a false positive during the operational life. Second, the probability that $\mathscr{T}_M$ exceeds $h_{n,\alpha}$ at the n-th sample has to be here conditioned to the fact that $\mathscr{T}$ never exceeded the threshold in previous $n-1$ samples. For these reasons, thresholds $\{h_{n,\alpha}, n > 0\}$ have to be computed numerically, through simulations, as in [183]. The `CPM` package [184] implements several CPMs based on different test statistics and provides also the thresholds $\{h_{n,\alpha}\}$ for both offline (traditional) CPMs and their online versions. Such thresholds could be loaded in a LUT as we move the CPM to an embedded system.

Any CPM (9.10) requires that data are either i.i.d. in the whole sequence (stationarity) or before and after the change point (nonstationarity). This may seem a restrictive assumption, since in real applications data are often characterized by different forms of dependence. When this happens, and we wish to operate with a CPM, it is necessary to move in a feature space where the i.i.d. assumption is met. A possibility is to operate in the model space by designing suitable models, e.g., of regression or predictive type, to fit the observations, and then analyze the residuals, see [193]. However, the residuals may not be i.i.d. since, quite often, the obtained model has a model bias component: in this case it can be convenient to aggregate several CPMs in an ensemble, as described in [204].

In what follows Change Detection Tests (CDTs) will be presented as statistical techniques designed having in mind online and sequential monitoring.

9.3 Change Detection Tests

There exists a large literature for concept drift detection mostly based on statistical hypothesis tests which, generally, require knowledge of the probability density function of the process generating the data and/or priors about the structure of the concept drift, e.g., a fault or a change in the environment. Again, the reference is that of Tables 9.1 and 9.2. In the parametric class of CDTs we find classic textbook tests such as the Student t-test and the Fisher f-test, addressing changes affecting the mean and the variance of the extracted features, respectively.

Nonparametric tests are more flexible tools, which require weaker assumptions, mostly tolerable at the application level. For instance, the Mann–Whitney U-test and the Wilcoxon test are nonparametric tests designed to detect a single change point and cannot support a sequential use, as sensing datastreams require. Differently, MannKendall and CUSUM are widely used tests adequate for a sequential analysis as the recently introduced ICI-based and hierarchical tests. In the section we present and detail three CDTs representing effective sequential nonparametric solutions to be implemented in embedded systems.

9.3.1 The CUSUM CDT Family

Complex and effective nonparametric tests generally require a configuration phase to fix test parameters at design time. The traditional CUmulative SUM control chart (CUSUM) is a sequential analysis technique designed for change detection that guarantees an appreciable change detection accuracy when a priori information about concept drift and the process generating the data are available. We present in the sequel two CDT methods that extend the traditional CUSUM by relaxing some of its restrictive assumptions. The first test extends the CUSUM by allowing the designer to automatically identify the configuration of the test parameters (adaptive CUSUM). The change detection ability of the adaptive CUSUM is based on the analysis of the evolution over time of the mean and the variance of some features extracted from the data-generating process. The second test, named Computational Intelligence CUSUM (CI-CUSUM), extends the first one by considering a richer set of features to improve efficiency in detecting changes in stationarity.

9.3.1.1 The Adaptive CUSUM CDT

Let $X = \{x(t), t = 1, \ldots, N\}, x(t) \in \mathbb{R}$ be a sequence of instances coming from the data generating process ruled by probability density function $f_\theta(x)$, which we assume to be unknown and parameterized in the parameter vector $\theta \in \mathbb{R}^n$.

Assume that the stochastic process changes its statistical behavior at unknown time T^o. This is generally modeled by considering a transition from parameter vector θ_0 to θ_1, associated with the pdfs $f_{\theta_0}(x)$ and $f_{\theta_1}(x)$, respectively. As with CUSUM, we evaluate the discrepancy between the two pdfs at time t, by computing the log-likelihood ratio

$$s(t) = \ln \frac{f_{\theta_1}(x(t))}{f_{\theta_0}(x(t))} \text{ for each } t = 1, \ldots, N$$

and the cumulative sum

$$S(t) = \sum_{\tau=1}^{t} s(\tau).$$

CUSUM identifies a change in X at time $\hat{T}$ when $g(t) = S(t) - m(t)$, the difference between the value of the cumulative sum $S(t)$ and its current minimum value $m(t) = \min_{\tau=1,\ldots,t} S(\tau)$ exceeds a given threshold h, namely

$$\hat{T} \text{ is the earliest time when } g(t) \geq h$$

CUSUM assumes that key parameters θ_0, θ_1 and h are available at design time. The assumption is generally hard to be satisfied but parameters can be estimated with the following procedure. Generate at first the cumulative sequence $Y = \{y(1), y(2), \ldots, \}$, where each s-th instance $y(s)$ represents the value of the sample mean estimated over a sliding nonoverlapping window of width n taken from X

$$y(s) = \frac{1}{n} \sum_{t=s(n-1)+1}^{sn} x(t)$$

From the central limit theorem the distribution of Y can be approximated with a Gaussian distribution provided that n is large enough. The basic CUSUM can then be applied to sequence Y. The first K configuration instances of X constitute the configuration set that is used to generate the training set of Y, whose cardinality is K/n (K is conveniently selected among the multiples of n). The whole procedure is depicted in Fig. 9.6. The parameters θ_0 characterizing the Gaussian distribution are the mean and variance of Y, i.e., $\theta = [\mu, \sigma^2]$, estimated on the train set. The parameters θ_1 are obtained through the identification of a neighborhood confidence for θ_0.

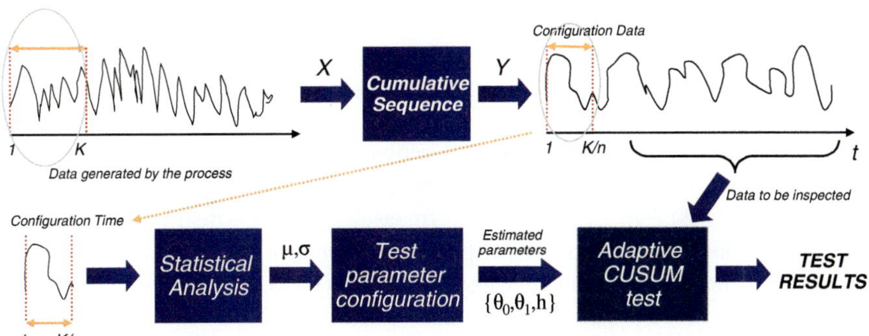

Fig. 9.6 The operational procedure for the adaptive CUSUM test. Data stream X undergoes a sequential windowing as data come in. When n samples are available, a data window is completed and ready to be averaged to generate the transformed instance $y(s)$. The distribution of $y(s)$ is approximately Gaussian, thanks to the central limit theorem, provided n is large enough. The basic CUSUM test can be applied with parameters $\theta = [\mu, \sigma^2]$. Needed parameters θ_0, θ_1 and threshold h are estimated on the training set

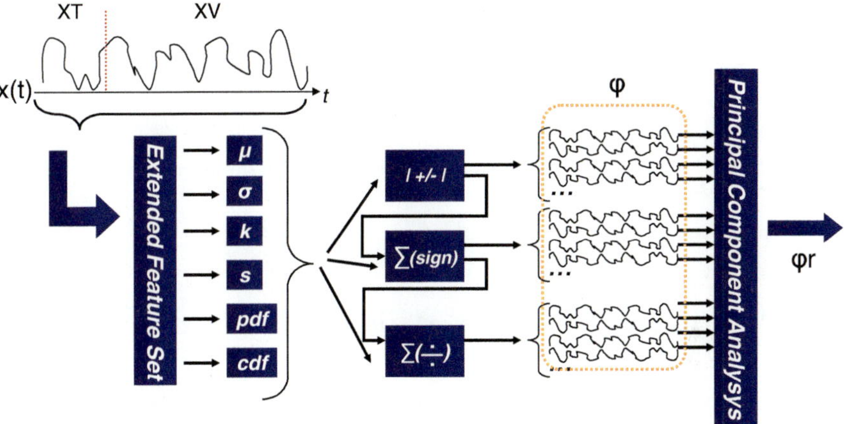

Fig. 9.7 The feature extraction and reduction phases of the CI-CUSUM. A rich set of features is extracted from the input signal to compose the feature set φ. Features extracted from the operational set XV are contrasted with those evaluated on the training configuration set XT. A PCA technique yields the reduced feature vector φ_r

9.3.1.2 The CI-CUSUM CDT

The CI-CUSUM represents an interesting extension of the adaptive CUSUM and turns to be much more powerful than the basic CUSUM and the Adaptive CUSUM since any feature can be extracted from the data stream to take advantage of different sensitivities in concept drift detection. The reference figure is Fig. 9.7.

Features φ are selected to be sensitive to concept drift. In particular, since data extracted from the training set XT are assumed to be i.i.d., the considered features are evaluated in a differential way to amplify the discrepancy between the current feature and the reference one associated with the training stationary state.

Envisaged features contain some well-known moments such as the mean μ, the variance σ^2 (to assess changes in the mean and variance of the distribution), the kurtosis $kurt$ and skewness $skew$ indexes (measuring how the distribution is peaked or flat and the lack of symmetry of a distribution, respectively), as well as information derived from the pdf and cumulative density function (cdf) of the signal. The running index is then contrasted with the corresponding one evaluated on the training set and features aim at amplifying the discrepancy between the two. For instance, feature $\varphi_1(t) = |\mu_0 - \mu_V|$ aims at amplifying discrepancies in the mean value. μ_0 is the value of the mean evaluated on the training set XT and subscript V refers to the test set, i.e., the index must be evaluated on data up to the running one (training data excluded). The basic features are

$$\varphi_1(t) = |\mu_0 - \mu_V|, \varphi_2(t) = |\sigma_0 - \sigma_V|, \varphi_3(t) = |kurt_0 - kurt_V|, \varphi_4(t) = |skew_0 - skew_V|$$

$$\varphi_5(t) = \int_x |pdf_0(x) - pdf_V(x)|dx, \varphi_6(t) = \int_x |cdf_0(x) - cdf_V(x)|dx$$

$$\varphi_{7\leq j\leq 12}(t) = \left\{\sum_{v=1}^{t-1} sgn\left(\varphi_{j-6,v+1} - \varphi_{j-6,v}\right)\right\}, \varphi_{13\leq j\leq 24}(t) = \left\{\sum_{v=1}^{t-1}\left(\frac{\varphi_{j-12,v+1}}{\varphi_{j-12,v}}\right)\right\}.$$

In particular, features $\varphi_5(t)$ and $\varphi_6(t)$ evaluate the discrepancy between the running pdf and cdf and that induced by the training set, respectively.

Features $\varphi_7(t)$ to $\varphi_{12}(t)$ investigate changes in the sequence of signs in consecutive elements and $\varphi_{13,t}$ to $\varphi_{24,t}$ the cumulative sum of the ratio of consecutive elements. To reduce the complexity of the feature space we performed a PCA on φ which provides a transformed feature φ_r. Since the pdf of φ_r is not a priori available, we operate as in the adaptive CUSUM case. In detail, we take the average of φ_r over nonoverlapping windows and invoke the central limit theorem which provides an approximated multivariate Gaussian distribution for the transformed variable φ' characterized by mean M and covariance matrix C. The mean M_0 and covariance matrix C_0 of φ_r are estimated on the training set and provide the nominal reference configuration $\theta_0 = [M_0, C_0]$. The adaptive CUSUM procedure is invoked that computes the alternative hypothesis for the change detection test $\theta_1 = [M_1, C_1]$. The CI-CUSUM is now configured and assesses over time $\varphi'(t)$ by checking whether it belongs to distribution $\mathcal{N}(M_0, C_0)$ or not by measuring the discrepancy between the two multivariate probability density functions at time t through the log-likelihood ratio mechanism

$$s(t) = \ln \frac{\mathcal{N}_{M_0,C_0}(\varphi'(l))}{\mathcal{N}_{M_1,C_1}(\varphi'(l))} \text{ for each } l = 1, \cdots, t.$$

The adaptive CUSUM test can now be applied and either returns detection of concept drift or claims that concept drift is not present.

9.3.2 The Intersection of Confidence Intervals CDT Family

The Intersection of Confidence Intervals (ICI) CDT and its evolutions [94] detect concept drift affecting a data stream by monitoring the evolution of suitable features extracted from incoming data. Features must be i.i.d. and Gaussian distributed, at least before concept drift occurs. The assumptions might appear strong and far from any engineering reality, in particular the i.i.d. one. However, this is not the case in many real applications provided that suitable transformations are invoked.

For instance, the method can be used to inspect sequences of residuals, e.g., associated with the discrepancy between a predictive model describing the data stream and the real data as they are acquired. When the test detects a change in the residual then concept drift is detected. This issue will be further addressed in the sequel. We now present the principal features of the ICI-CDT family.

9.3.2.1 The ICI-CDT

In the ICI-CDT, features are extracted by windowing the available data in disjoint subsequences composed of n instances. For each subsequence we compute the sample mean and the sample variance which are Gaussian distributed thanks to the central limit theorem for the former and and ad hoc transformation [95] the latter. More in detail, named s the s-th subsequence, the extracted features are

$$M(s) = \frac{\sum_{t=(s-1)n+1}^{ns} x(t)}{n}, \text{ and } V(s) = \left(\frac{\left(\sum_{t=(s-1)n+1}^{ns} (x(t) - M(s))^2 \right)}{n-1} \right)^{h_0}, \tag{9.16}$$

The parameter h_0 is the exponent of the power-law transform devised in [95] to generate an approximated Gaussian distribution for the sample variance. h_0 is estimated from the sample cumulants computed on training data O_{T_0}.

The ICI-CDT is configured on the two sequences of features $\{M(s), s = 1, \ldots, S_0\}$ and $\{V(s), s = 1, \ldots, S_0\}$, being $S_0 = T_0/n$ extracted from O_{T_0},

We compute the means $\hat{\mu}^M_{S_0}$, $\hat{\mu}^V_{S_0}$ and the standard deviations $\hat{\sigma}^M_{S_0}$, $\hat{\sigma}^V_{S_0}$ of the two features over the training set, i.e.,

$$\hat{\mu}^M_{S_0} = \frac{\sum_{s=1}^{S_0} M(s)}{S_0}, \quad \text{and} \quad \hat{\sigma}^M_{S_0} = \sqrt{\frac{\sum_{s=1}^{S_0} (M(s) - \hat{\mu}^M_{S_0})^2}{S_0 - 1}}. \tag{9.17}$$

and

$$\hat{\mu}^V_{S_0} = \frac{\sum_{s=1}^{S_0} V(s)}{S_0}, \quad \text{and} \quad \hat{\sigma}^V_{S_0} = \sqrt{\frac{\sum_{s=1}^{S_0} (V(s) - \hat{\mu}^V_{S_0})^2}{S_0 - 1}}. \tag{9.18}$$

These estimates define the confidence intervals for the mean and standard deviation features that, under the stationary condition, are defined as

$$\begin{aligned} \mathscr{I}^M_{S_0} &= [\hat{\mu}^M_{S_0} - \Gamma\hat{\sigma}^M_{S_0}, \hat{\mu}^M_{S_0} + \Gamma\hat{\sigma}^M_{S_0}], \\ \mathscr{I}^V_{S_0} &= [\hat{\mu}^V_{S_0} - \Gamma\hat{\sigma}^V_{S_0}, \hat{\mu}^V_{S_0} + \Gamma\hat{\sigma}^V_{S_0}], \end{aligned} \tag{9.19}$$

with $\Gamma > 0$ controlling the amplitude of the confidence interval and, then, the probability that features belong to the interval under the stationary assumption.

Once training is perfected the CDT becomes operational and can be used to assess changes in stationarity in the data stream. Every time n data are made available, a new sequence s is created and features extracted to populate the $\mathscr{I}^M_s$ and $\mathscr{I}^V_s$.

The intersection of confidence intervals rule (ICI-rule) [96] can then be applied. The ICI-rule verifies whether the new feature instance can be intended as a realization of the existing Gaussian distribution. If not, concept drift is detected in the data stream.

From the operational point of view, the sample mean of all the feature values is computed, together with the confidence interval of the corresponding estimator which is expressed as (9.19). As soon as the intersection of all the confidence intervals up to the current one results in an empty set, the basic ICI-CDT detects a change. Thus, we detect a concept drift in the subsequence $\hat{s}$ if

$$\begin{aligned} \bigcap_{s<\hat{s}} \mathscr{I}^M_s \neq \emptyset \text{ and } \bigcap_{s\leq\hat{s}} \mathscr{I}^M_s = \emptyset \text{ or} \\ \bigcap_{s<\hat{s}} \mathscr{I}^V_s \neq \emptyset \text{ and } \bigcap_{s\leq\hat{s}} \mathscr{I}^V_s = \emptyset \end{aligned} \tag{9.20}$$

and the detection time $\hat{T} = n\hat{s}$ corresponds to the rightmost term of the subsequence $\hat{s}$.

Concept drift is associated with those feature(s) that yielded the empty intersection. Figure 9.8 illustrates how the ICI-rule operates. To reduce the computational

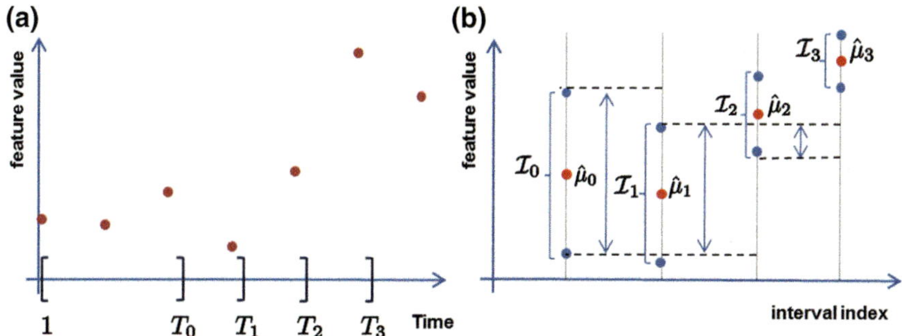

Fig. 9.8 An illustrative example of the ICI rule in the setting used for change detection: **a** feature values and the set of intervals $\{[1, T_0], [1, T_1], [1, T_2], [1, T_3]\}$, **b** the corresponding polynomial zeroth-order estimates and their confidence intervals. The ICI rule selects the interval $[1, T_2]$, since $\mathscr{I}_0 \cap \cdots \cap \mathscr{I}_2 \neq \emptyset$ and $\mathscr{I}_0 \cap \cdots \cap \mathscr{I}_3 = \emptyset$. The brackets in (**b**) represent the confidence intervals; the arrows their intersections

load, the average feature and the intersection of confidence intervals are computed incrementally, and each feature is separately processed.

The whole procedure is summarized in Algorithm 20. As pointed out in [94] the basic ICI-CDT is particularly effective but introduces a structural limitation inducing a structural false positive when time passes.

Despite the fact that this problem can be tolerated in many detect and react mechanisms, it is important to design a test that does not introduce structural false positives as time passes. This problem can be solved by considering a second test, built on the top of the basic ICI-CDT that verifies, once activated, if a false positive has been generated by the first CDT or the raised alarm should be considered a proper concept drift. For its layered structure the test is named Hierarchical CDT.

9.3.2.2 The H-CDT

The Hierarchical CDT (H-CDT) has been designed to mitigate the structural problem posed by the ICI-CDT, for which false positives are generated as time passes. The H-CDT is a hierarchical sequential change detection test structured into two processing levels. The first level is composed of the ICI-CDT test and the second is a statistical test validating/rejecting the change hypothesis. The ICI-CDT operates sequentially as presented in the previous subsection and, when it detects a change in the sequence $x(t)$ at time $\hat{T}$, it activates the upper test to validate the detection by checking if the data sets before and after the estimated $\hat{T}$ are consistent with the change hypothesis.

For such change validation purposes, we need to acquire a set of N additional data $O_{\hat{T}} = \{x(t), t = \hat{T}, \ldots, \hat{T} + N\}$ generated after $\hat{T}$, which are considered to have been potentially generated from the new state of the data-generating process, namely, after concept drift. The adjective "potentially" is appropriate since a false

Algorithm 20: The basic ICI-CDT

Compute $\{M(s), s = 1, \ldots, S_0\}$, being $S_0 = T_0/n$;
$\hat{\mu}^M_{S_0} = \sum_{s=1}^{S_0} \frac{M(s)}{S_0}$;
$\hat{\sigma}^M = \sqrt{\sum_{s=1}^{S_0} \frac{\left(M(s)-\hat{\mu}^M_{S_0}\right)^2}{S_0-1}}$, $\hat{\sigma}^M_{S_0} = \frac{\hat{\sigma}^M}{\sqrt{S_0}}$;
Define $\mathscr{I}^M_{S_0} = \left[\hat{\mu}^M_{S_0} - \Gamma\hat{\sigma}^M_{S_0}, \hat{\mu}^M_{S_0} + \Gamma\hat{\sigma}^M_{S_0}\right]$;
Compute h_0;
Compute $\{V(s), s = 1, \ldots, S_0\}$;
$\hat{\mu}^V_{S_0} = \sum_{s=1}^{S_0} \frac{V(s)}{S_0}$;
$\hat{\sigma}^V = \sqrt{\sum_{s=1}^{S_0} \frac{\left(V(s)-\hat{\mu}^V_{S_0}\right)^2}{S_0-1}}$, $\hat{\sigma}^V_{S_0} = \frac{\hat{\sigma}^V}{\sqrt{S_0}}$;
Define $\mathscr{I}^V_{S_0} = \left[\hat{\mu}^V_{S_0} - \Gamma\hat{\sigma}^V_{S_0}; \hat{\mu}^V_{S_0} + \Gamma\hat{\sigma}^V_{S_0}\right]$;
Set $s = S_0$;
while ($\mathscr{I}^M_s \neq \emptyset$ *and* $\mathscr{I}^V_s \neq \emptyset$) **do**
  Set $s = s + 1$;
  Wait for n observations, until a new subsequence is populated;
  Compute $M(s)$ and $V(s)$ from observations in the subsequence according to 9.16;
  Compute $\hat{\mu}^M_s = \frac{(s-1)\hat{\mu}^M_{s-1}+M(s)}{s}$ and $\hat{\sigma}^M_s = \frac{\hat{\sigma}^M}{\sqrt{s}}$;
  Compute $\hat{\mu}^V_s = \frac{(s-1)\hat{\mu}^V_{s-1}+V(s)}{s}$ and $\hat{\sigma}^V_s = \frac{\hat{\sigma}^M}{\sqrt{s}}$;
  $\mathscr{I}^M_s = \left[\hat{\mu}^M_s - \Gamma\hat{\sigma}^M_s; \hat{\mu}^M_s + \Gamma\hat{\sigma}^M_s\right] \cap \mathscr{I}^M_{s-1}$;
  $\mathscr{I}^V_s = \left[\hat{\mu}^V_s - \Gamma\hat{\sigma}^V_s; \hat{\mu}^V_s + \Gamma\hat{\sigma}^V_s\right] \cap \mathscr{I}^V_{s-1}$;
end
Concept drift detected at $s = \hat{s}$ within time interval $\left[(\hat{s}-1)n+1, \hat{s}n\right]$,i.e., $\hat{T} = n\hat{s}$

positive might arise and, hence the information present in the $O_{\hat{T}}$ set is compatible with that provided by the training set O_{T_0} onto which the method was configured (T_0 refers to the last time instant associated with the training set).

We comment that, if estimate $\hat{T}$ is highly accurate, then we might expect to improve the accuracy of the method by considering the whole set $\{x(t), t < \hat{T}\}$ instead of O_{T_0}. The reason behind the last statement is associated with the fact that if $\hat{T}$ is a good estimate for the concept drift time then, the presence of data associated with new state in $\{x(t), t < \hat{T}\}$ is likely to be negligible. In what follows, we rather prefer to be conservative and consider O_{T_0} instead: this choice goes also in the direction of operating with a reduced computational load, which is a relevant issue for embedded systems.

Statistical affinity between sets O_{T_0} and $O_{\hat{T}}$ should be evaluated with a proper statistical test, e.g., Kolmogorov–Smirnov test or other hypothesis tests such as those given in Table 9.2. However, the required computational load of the Kolmogorov–Smirnov is too high and unfeasible for a large class of embedded systems where high MIPS cannot be provided. The problem of comparingthe data distribution over O_{T_0}

and $O_{\hat{T}}$, can be conveniently simplified to the problem of comparing the expected values of the features (9.16) of the ICI-CDT over O_{T_0} and $O_{\hat{T}}$ by means of an Hotelling's test.

In particular, the Hotelling's test is a multivariate hypothesis test, which we apply to compare the values of the features (9.16) arranged in two-dimensional vectors $F = [M(s), V(s)]$. These feature vectors are extracted from the time interval O_{T_0} (onto which the ICI-CDT was configured) and on the interval $O_{\hat{T}}$ (which are expected to describe the new state of the process). From each of these sets, the sample means $\overline{F}(O_{T_0})$ and $\overline{F}(O_{\hat{T}})$ and the pooled sample covariance matrix are computed. The null hypothesis H_0 is formulated as

$$H_0 : \overline{F}(O_{T_0}) - \overline{F}(O_{\hat{T}}) = \underline{0} \tag{9.21}$$

where $\underline{0}$ represents the two-dimensional vector of null components. Finally, the Hotelling T^2 test [241] can be executed to reject the null hypothesis at a predefined confidence level α. If the Hotelling test rejects the equivalence hypothesis then a change is considered to be present in the feature set and the change hypothesis raised by the ICI-CDT is validated. In turn, the hierarchical test detects concept drift.

Conversely, if there is not enough statistical evidence to reject the null hypothesis, then the ICI-CDT introduced a false positive and must be retrained on the original stationary state O_{T_0}. The Hotelling's test applied to the the features (9.16) shows to be a particularly effective solution to assess changes detected by the ICI-CDT.

The H-CDT is therefore an adaptive test which reacts when false positives are introduced by the ICI-CDT and shows to be a great test to be used in embedded systems. The H-CDT algorithm describing from a high-level perspective is given in Algorithm 21.

Interestingly, if we want the hierarchical test to operate in a sequential manner, after each change validation, we are able to retrain the ICI-ICT and we also update the reference set O_{T_0} used in the Hotelling test. In fact, if the change has been validated at time $\hat{T}$, the set $O_{\hat{T}}$ contains instances associated with the new state of the data-generating process, thus the inspection for concept drift proceeds. The algorithm is summarized in Algorithm 22.

9.3.2.3 An Improved Estimate for the Concept Drift Detection Time

The H-CDT described in the previous section has the main drawback of having to wait for N observations after $\hat{T}$ before proceeding with the change-validation phase. This is of course not appealing in an online monitoring scenario, also because the detection $\hat{T}$ is typically characterized by a structural delay (correct detections provided by most of CDTs come typically after the unknown change-time instant T^o). The idea is hence to improve the estimate of T^o once the change has been detected, to recover part of the samples between T^o and $\hat{T}$ for improving change-validation efficiency and possibly CDT reconfiguration. Therefore, the improved estimate of T^o, which we denote by $\overline{t}$, is expected to satisfy $T^o \leq \overline{t} \leq \hat{T}$. Once $\overline{t}$ has been

Algorithm 21: The hierarchical change detection test H-CDT. The test is initially configured on the training set O_{T_0}. Once concept drift is detected by the ICI-CDT, Hotelling test is activated. If the Hotelling test validates the concept-drift hypothesis then an alarm is raised by the H-CDT and concept drift is validated. When the ICI-CDT detection is not validated, a false positive is found, no concept drift alarm is raised, and the ICI-CDT needs to be reconfigured on the initial training data.

Train the ICI-CDT on O_{T_0};
while *(1)* **do**
    Extract features $M(s)$ and $V(s)$ out of the data stream;
    if *(ICI-CDT detects a change in the features) AND (Hotelling test validates the change)* **then**
        Conceptdrift= true;
        retrain ICI-CDT onto $O_{T_0} = O_{\hat{T}}$;
    else
        false positive: retrain ICI-CDT onto O_{T_0}
    end
end

Algorithm 22: The H-CDT within the active learning modality. When concept drift is validated at time $\hat{T}$ the application is reconfigured and the H-CDT retrained on the new instances.

Train the ICI-CDT on O_{T_0};
while *(1)* **do**
    Extract features $M(s)$ and $V(s)$ from the data stream;
    if *(ICI-CDT detects a change in the features) AND (Hotelling test validates the change)* **then**
        Conceptdrift= true;
        React on concept drift at the application level;
        retrain ICI-CDT onto $O_{T_0} = O_{\hat{T}}$;
    else
        false positive: retrain ICI-CDT onto O_{T_0}
    end
end

computed, we could use the observations $\{x(t), t = \bar{t}, \ldots, \hat{T}\}$ to define $O_{\hat{T}}$ without delaying the validation procedure

$$O_{\hat{T}} = \{x(t), t = \bar{t}, \ldots, \hat{T}\}$$

that contains more than n samples, thus increasing the significance during validation and configuration w.r.t. the approach described in the previous section. However, when the ICI-CDT is very quick $O_{\hat{T}} = \{x(t), t = \bar{t}, \ldots, \hat{T}\}$ may not contain enough samples, and it would be preferable to wait for at least N sample before activating the change-validation and reconfiguration procedures.

Algorithm 23: The refinement procedure leading to the improved estimate for the change time instant $\bar{t}$.

1 Provide $\hat{T}$;
2 Compute $T_1 = T_0 + (\hat{T} - T_0)/\lambda$;
3 $i = 1$; continue = *true*;
4 **while** *(continue = true)* **do**
5   Apply the ICI-CDT to $[0, T_0] \cup [T_i, \hat{T}]$, detecting at $\hat{T}_i$;
6   Compute $T_{i+1} = T_i + (\hat{T} - T_i)/\lambda$;
7   Define $T_{\min} = \min\left(\hat{T}_j\right), j = 1, \ldots, i$;
8   **if** $(T_{min} < T_{i+1})$ **then**
9   continue = *false*;
    end
10     $i = i + 1$;
end
11 Define $\bar{t} = T_{\min}$.

The key point of the proposed solution is that the ICI-CDT introduces a structural detection latency that increases as time passes [94]. This undesirable behavior can be exploited to design a post-detection procedure that, starting from $\hat{T}$ yields a better estimate $\bar{t}$ as illustrated in Algorithm 23.

The algorithm operates as follows. Given concept drift detection from the ICI-CDT at time instant $\hat{T}$ split the interval $[T_0, \hat{T}]$ in two intervals $[T_0, T_1]$ and $[T_1, \hat{T}]$, with $T_1 = T_0 + (\hat{T} - T_0)/\lambda$ defined according to the user-set parameter $\lambda > 1$ (line 2). Apply the ICI-CDT to dataset $[0, T_0] \cup [T_1, \hat{T}]$ (line 5), leading to a detection at time $\hat{T}_1$. Note that $\hat{T}_1$ is a more accurate estimate of the change time T_0, since the test operates on a shorter sequence w.r.t. the one which provided the initial detection $\hat{T}$. Interval $[T_1, \hat{T}]$ is further split into two intervals $[T_1, T_2]$ and $[T_2, \hat{T}]$ where $T_2 = T_1 + (\hat{T} - T_1)/\lambda$ (line 6). If $T_2 > \hat{T}_1$, the procedure stops, and $\bar{t} = \hat{T}_1$.

Otherwise, the procedure iterates: at the i-th iteration, the ICI-CDT is executed on $[0, T_0] \cup [T_i, \hat{T}]$, providing the estimate $\hat{T}$ (line 5). The interval $[T_i, \hat{T}]$ is then split by point $T_{i+1} = T_i + \frac{\hat{T} - T_i}{\lambda}$ (line 6). The procedure ends when T_{i+1} is larger than $T_{\min}$, the earliest detection identified during the iteration of the procedure (line 7). Finally, $T_{\min}$ is the best estimate of T^o obtainable according to this procedure, The improved final estimate is hence $\bar{t} = T_{\min}$.

The refinement procedure is visualized in Fig. 9.9.

Comments

The estimate $\bar{t}$, which is provided by all the ICI-CDTs, makes it particularly appealing for active (detect and react) learning frameworks, since they provide set $O_{\hat{T}}$ that contains instances associated with the new state of the process generating the data. These instances, in the data stream domain, can now be used to reconfigure the appli-

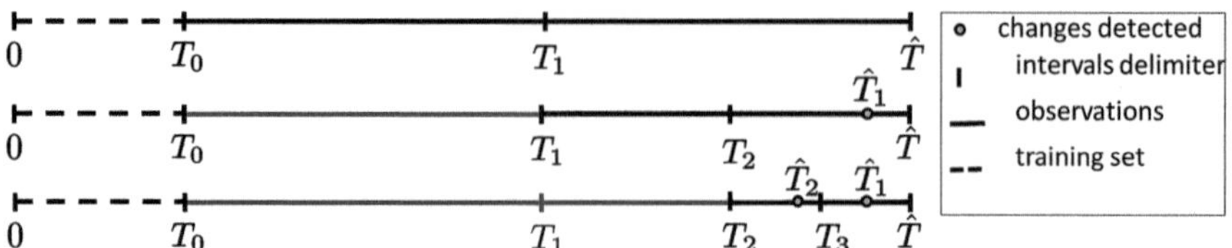

Fig. 9.9 The ICI-based time change estimate refinement procedure: an example with $\lambda = 2$. Initially, (*first line*) a change is detected by the ICI-CDT in correspondence with time $\hat{T}$ and the procedure starts by computing $T_1 = T_0 + \frac{\hat{T}-T_0}{\lambda}$. The ICI-CDT is then executed onto interval $[0, T_0] \cup [T_1, \hat{T}]$, resulting in a detection at $\hat{T}_1$ (*second line*). This procedure iterates by computing $T_2 = T_1 + \frac{\hat{T}-T_1}{\lambda}$ and the test is executed on interval $[0, T_0] \cup [T_2, \hat{T}]$. The procedure terminates when $T_3 > \hat{T}_2$, being $T_2 = \min\{\hat{T}_j\}$. The output is $\bar{t} = \hat{T}_2$, and $[\hat{T}_2, \hat{T}]$ is assumed to be generated by the process in the novel state, i.e., after concept drift

cation, besides the CDT itself. Moreover, the H-CDT shows to be computationally lighter than CI-CUSUM [101] and in most applications involving embedded systems should be preferred. For this reason the code of the hierarchical CDT has been made freely available and can be downloaded from the link given in [102].

We recall that insurgence of false positives introduces a processing load, since it leads to unnecessary reconfiguration, and that this might also reduce the performance of the application. In fact, if we mistakenly abandon the a priori rich training set O_{T_0} for the new one $O_{\hat{T}}$ following the false positive we should expect to end up with a consistent data set of lower cardinality.

At the same time, the presence of a false negative is also critical, since when no concept drift is detected, the adaptation mechanism is not activated.

As a last note we investigate the effects induced by slowly developing gradual concept drift. It is expected that this concept drift will not be detected in its early stages, but most probably later, when the influence of the concept drift on the features level grants detection. However, latency in detection is the cost we have to pay in correspondence to slowly developing gradual concept drift. Moreover, given the type of CDTs we are considering, a slowly developing concept drift results in a sequence of concept drift detections, a detection profile being symptomatic of a gradual concept drift evolution.

The literature has proposed CDTs specifically designed to manage situations with slowly developing concept drift under some assumptions about the evolution model for the concept drift, mostly following polynomial functions of a fixed order. The interested readers can refer to [97].

9.3.3 Amygdala—VM-PFC: The H-CDT

The H-CDT is a pure example of a cognitive mechanism. There, the lower processing level based on an ICI-CDT quickly processes the input stimuli like in the amygdala and provides a first reaction outcome. A threat (perceived change) is immediately detected (automatic process) and actions are promptly taken (e.g., we immediately react when we see a gun oriented toward us irrespective of other extra information). Afterwards, the emotional state is passed to the VM-PFC which perfects the taken action with a more articulated processing and comes back to the amygdala with a new control action (e.g., when we realize that the gun is indeed a fake water gun, held by a kid). The counterpart of VM-PFC corresponds to the higher level of the H-CDT, where the Hotelling test either validates or rejects the concept drift hypothesis raised by the lower level CDT. When the change hypothesis is rejected, the action invoked by the ICI-CDT is aborted, the state rolled back, and the ICI-CDT is reconfigured after the false positive detection. There is no evidence that the VM-PFC levels reconfigure the amygdala even though a negative feedback is likely to be provided.

9.4 The Just-in-Time Learning Framework

Availability of a CDT within a sequential framework allows us for designing applications characterized by an active learning modality. The chosen CDT detects concept drift (detection modality) and the application reacts accordingly (reaction modality) by adapting to the new state. This active learning modality is known in the literature as *Just in Time* (JIT) learning meaning that the application reconfiguration to track changes in the environment is activated exactly when needed, i.e., in correspondence with concept drift detection, in contrast with passive solutions where learning is always enabled.

We instance the JIT mechanism to an application to ease the presentation and to the classifier case for its relevance in applications. In JIT classifiers a CDT identifies concept drift affecting incoming data and the classifier-based application undergoes a reconfiguration phase to track the change in stationarity. A unique characteristic of JIT classifiers compared with other classifiers following the active learning modality is that, when no change is detected, the classifier continues integrating new supervised information made available to improve the classification accuracy.

A high level description of the JIT adaptive classification framework is given in Algorithm 24. The framework is very general and can host any type of CDT and classifier. Clearly, we should consider effective low complexity CDTs and classifiers having in mind, as final target, embedded systems.

The JIT framework is very general and can deal with any type of concept drift from *abrupt concept drift* to *gradual concept drift*. In the abrupt case we need to release obsolete data used for training the classifier and replace them with novel supervised instances characterizing the new operational condition, then the classifier is trained on the new training set, e.g., [94]. Differently, a frequent management activity involving both the update of the training set and retraining is needed when we encounter gradual concept drift, which are seen as a sequence of abrupt concept drifts following the detection mechanism. Extensions of the mechanism meant to deal with gradual concept drift has been suggested in [97].

Algorithm 24: The JIT adaptive classifier. New data instances are integrated in the classifier as they come until concept drift is detected. When the CDT detects a change the supervised instances associated with the $O_{\hat{T}}$ data set are used to reconfigure the classifier.

```
Configure the JIT classifier and the CDT;
while (true) do
    input receive new data;
    if (CDT detects concept drift) then
        Characterize the new process state;
        Configure the JIT classifier and the CDT on the new process state;
    else
        integrate available extra information in the JIT;
    end
    Classify the new input samples;
end
```

In the sequel we focus at first on the core JIT for abrupt concept drift and address afterwards the gradual concept drift case. We use, as reference CDT the ICI-based family although any CDT can be adopted.

9.4.1 Observation Model

Consider, for sake of simplicity, a two-class classification problem. The operational framework can be formalized as follows.

Let $x \in X \subset \mathbb{R}^d$ be an i.i.d. random variable and $y \in \{\omega_1, \omega_2\}$ the associated binary classification output. The pdf of the inputs at time t

$$p(x|t) = p(\omega_1|t)p(x|\omega_1, t) + p(\omega_2|t)p(x|\omega_2, t), \tag{9.22}$$

depends on the pdfs of the outputs $p(\omega_1|t)$ and $p(\omega_2|t) = 1 - p(\omega_1|t)$ and the conditional probability distributions $p(x|\omega_1, t)$ and $p(x|\omega_2, t)$. In general, these distributions are unknown.

Let $O_T = \{x(t), t = 1, , T\}$ be the data sequence at time T and $Z_T = \{(x(t), y(t)),\ t \in I_T\}$ the knowledge base of the classifier at time T, which contains the supervised couples $(x(t), y(t))$, i.e., $y(t)$ is the classification label associated with the observation $x(t)$, and I_T is the set containing the arrival times of supervised samples up tp time T.

We further assume that the samples acquired before T_0 have been generated in stationary conditions. The set O_{T_0} is then used to train the CDT, while $Z_0 = \{(x(t), y(t)),\ t \in I_0\}$ represents the initial knowledge base (KB) of the classifier, being I_0 the set of supervised samples in O_{T_0}. Assume that at time instant $T^{\mathrm{o}} > T_0$ a change in stationarity occurs with a subsequent change in the distribution of x: also the distribution after the change is unknown.

In the JIT framework the CDT inspects the process by operating on the data sequence O_T and, in some of its variations, by also exploiting supervised sample information.

9.4.2 The JIT Classifier

With reference to Algorithm 24, the JIT classifier undergoes an adaptation phase whenever the CDT detects concept drift. Otherwise, it integrates available new information in the training set to improve the classification accuracy over time.

9.4.2.1 React to the Change: Updating the Classifier

Retraining the adaptive JIT classifier requires learning the model of the data generating process after the change. Thus, the set of features $Z_{s|t>\bar{t}} = Z_{s|[\bar{t},\hat{T}]}$, i.e., the data observations in time interval $[\bar{t}, \hat{T}]$, represent the new state of the process generating the data following concept drift. Such instances must be used to retrain the classifier (in the time domain t) and the ICI-CDT (in the s domain).

Any consistent classifier, where consistency requires as necessary sufficient that the model family is a universal function approximator, can be considered in the JIT framework. However, if we have embedded systems in mind, then not all classifiers are equally valid. Computational complexity and memory usage must be taken into account when designing the application also for the indirect effect on power consumption in energy-aware applications. Feedforward neural networks, k-NN classifiers, Radial basis function neural networks are examples of consistent classifiers [100], SVM and regularized kernel classifiers are consistent classifiers depending on the particular choice of the loss function and the implementation algorithm [99]. However, the training phase is a costly operation for most classifiers, hence becoming most likely to be prohibitive for embedded systems, in particular if big data are involved. As shown in [101] k-NN classifier is a particularly appealing solution since its training phase is immediate and reduces to the insertion of supervised couples in a table representing the KB of the classifier.

We recall that the k-NN classifier provides a label to a new instance to be classified as the label majority of the k closest instances. The figure of merit evaluating the affinity between two instances—mostly based on an euclidean distance in the input space—the inspection of instances in the KB and score ranking to identify the k closest neighbors are the main computationally demanding parts of the algorithm. If the cardinality of the KB is N the k-NN classifier is consistent [103] provided that

$$\frac{k}{N} \to 0 \text{ as } k \to \infty, N \to \infty.$$

However, k-NN classifiers are memory eager solutions—this is the price we have to pay for a computational negligible training phase—since all N instances need to be stored in the memory, despite the fact that efficient solutions can be envisaged to keep under control the memory request, e.g., those based on condensing or editing techniques. Both condensing and editing techniques aim at reducing the cardinality of the training set yet preserving the maximum classification accuracy. In particular, condensing techniques, e.g., Condensed Nearest Neighbor (CNN) [105], aim at keeping in the training set only those samples fundamental to shape the decision boundary. Differently, editing techniques, e.g., the Wilson Editing Rule (WER) [106], intervene on the training set by removing particularly noisy samples and request the Bayes's decision boundary to be smooth.

A better solution would involve thresholding the cardinality of KB by keeping a maximum of N_M instances. If they are in stationary conditions and N increases, N will be buffered to N_M supervised couples, for instance by keeping the most recent N_M instances. A simple circular buffer would solve the problem.

We detail the JIT classifier based on a k-NN structure and the H-CDT; the reference is Algorithm 25.

The initial knowledge base of the k-NN classifier is $Z_0 = \{(x(t), y(t)),\ t \in I_0\}$ (line 1), while the value of k is set to k_{LOO}, estimated by means of the Leave-One-Out (LOO) technique applied to Z_0 (line 2). The H-CDT is configured on the initial training set O_{T_0} (line 3). After this configuration phase, the algorithm works online by classifying upcoming samples as they arrive and by introducing, whenever available (line 7), new supervised information $(x(t), y(t))$ into the knowledge base of the classifier KB. In this case, the algorithm stores in I_T the time instant t when the sample has been received (line 8), includes the pair $(x(t), y(t))$ in Z_T (line 9), and updates the parameter k so that consistency is granted. The reader should be aware that k cannot be freely chosen as N increase to satisfy the consistency conditions; an effective computational-aware method to estimate the appropriate k is given in [104] and relies on the LOO performance evaluation method.

In stationary conditions, the classification accuracy always increases by introducing additional supervised samples during the operational life [103] but when available $x(t)$ is not supervised, I_t and Z_t sets are not updated (lines 11-12).

When the H-CDT notifies concept drift in the subsequence containing $x(t)$ (line 13), the refinement procedure also provides $\overline{t}$ (line 15). The H-CDT is then reconfigured on features s associated with the new state of the process, i.e., those in time interval $[\overline{t}, \hat{T}]$ (line 16).

The $\overline{t}$ information allows the JIT for removing those training samples acquired before $\overline{t}$ both from I_t and Z_t (lines 17, 18). The new value of k_{LOO} is then estimated by means of the LOO procedure on the new knowledge-base (line 19) and k set to it. Finally, $x(t)$ is classified by relying on the updated knowledge-base Z_t, and the current value of k (line 20).

9.4.2.2 Example: JIT Learning in a Classification Systems

The experiment refers to a synthetic monodimensional classification problem with two equiprobable classes $\{\omega_1, \omega_2\}$ each of which ruled by a Gaussian distribution

Algorithm 25: H-CDT-based JIT Adaptive Classifier

$I_0 = \{1, \ldots, T_0\}$, $Z_0 = \{(x(t), y(t)),\ t \in I_0\}$,$O_{T_0}$;
Estimate k_{LOO} by means of LOO on Z_0 and set $k = k_{\text{LOO}}$;
Configure the ICI-CDT part of H-CDT using O_{T_0};
$Z_t = Z_0$,$I_t = I_0$,$t = T_0 + 1$;
while *(1)* **do**
    Acquire $x(t)$ at time t;
    if *(supervised information $y(t)$ on $x(t)$ is available)* **then**
        $I_t = I_{t-1} \cup \{t\}$;
        $Z_t = Z_{t-1} \cup \{(x(t), y(t))\}$;
        update k as in [104];
    else
        $I_t = I_{t-1}$;
        $Z_t = Z_{t-1}$;
    end
    if *(H-CDT detects concept drift on the sequence containing $x(t)$)* **then**
        Let $\hat{T}$ be the concept-drift detection time;
        Extract $\bar{t}$ from H-CDT (Algorithm 23);
        Configure ICI-CDT on $[\bar{t}, \hat{T}]$ and Hotelling on feature sequence $s|t > \bar{t}$;
        $I_t = \left\{t \in T_t, t > \bar{t}\right\}$;
        $Z_t = \{(x(t), y(t)), t \in I_t\}$;
        Estimate k_{LOO} by means of LOO on Z_t and set $k = k_{\text{LOO}}$;
    end
    Classify $x(t)$ as $k - NN(x(t), k, Z_t)$;
    $t = t + 1$;
end

$p(x|\omega_1) = \mathcal{N}(0, 4)$ and $p(x|\omega_2) = \mathcal{N}(2.5, 4)$. The experiment is composed of $N = 10,000$ scalar observations. An abrupt concept drift affects both classes at time $T^{\text{o}} = 5000$ by modifying the pdfs as $p(x|\omega_1) = \mathcal{N}(2, 4)$ and $p(x|\omega_2) = \mathcal{N}(4.5, 4)$. Figure 9.10a shows the data instances for the two classes over time.

The following adaptive classification frameworks have been considered for comparison:

- the proposed JIT classifier (green dashed line with a square marker).
- A classifier trained on all available data every time a new supervised couple is provided (dotted black line). This classifier guarantees the best performance in stationary conditions.
- A short memory classifier trained on a sliding window open over the latest 40 supervised samples (solid red line with circle marker).

All the considered adaptive classification frameworks rely on the k-NN classifier as a base classifier.

A supervised sample out of $m = 5$ observations is provided to the classifiers.

The classification accuracy on unsupervised samples is the figure of merit used to assess the performance of the considered adaptive classification framework. In

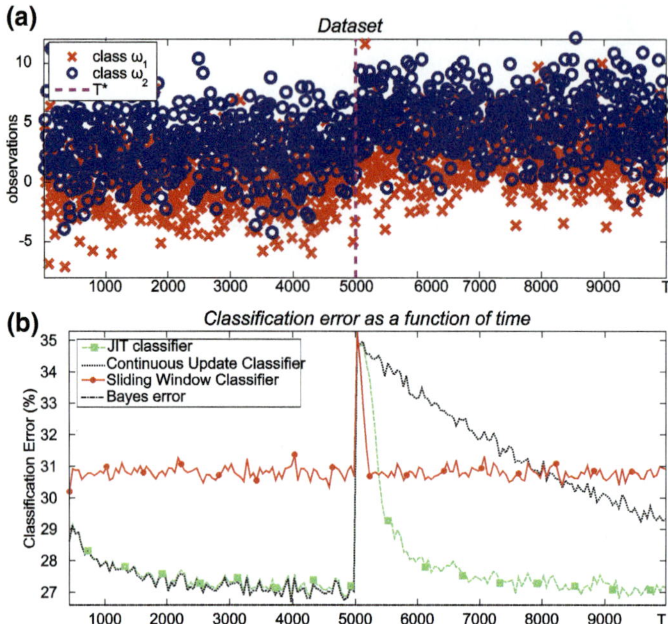

Fig. 9.10 An example of the just-in-time learning mechanism applied to a classifier. Data instances associated with the two classes are subject to a concept drift of abrupt type at T^{o} =5,000 that affects the mean of the distribution of class ω_2 (*upper plot*). The classification performance of the classifier are then compared with a short memory classifier implementing a batch online passive learning mechanism, the JIT classifier and the optimal Bays classifier (*lower plot*)

particular, Fig. 9.10b shows, at each time instant, the percentage of misclassified samples in 2,000 runs, averaged over a sliding window of 40 samples.

The JIT classifier tends to the Bayes error both before and after the change thanks to its ability to integrate fresh supervised samples during the operational life and to remove obsolete samples after a detected change. In fact, before T^{o} =5,000, the JIT classifier guarantees performance in line with those provided by the classifier trained on all available data (i.e., the black line) that, in stationary conditions, is able to guarantee the best performance.

After the change, the JIT classifier is able to promptly react to the change and adapt to the new working conditions thanks to its active detection/reaction approach. On the contrary, the classifier trained on all available data requires much more samples to adapt to the new working conditions since it is not endowed with a mechanism to remove obsolete samples.

Interestingly, the short memory classifier guarantees the best performance after the change since it is naturally able to remove obsolete samples through the windowing mechanism. Unfortunately, it is not able to improve its accuracy in stationary conditions since the sliding window is open over a fixed amount of samples (and this does not allow the base classifier to achieve the Bayes error).

9.4.3 Gradual Concept Drift

The proposed JIT grants asymptotic optimality when the process generating the data is affected by a sequence of abrupt concept drift [84] in the sense that, after concept drift is detected, the classifier's performance increases during operational life with provided additional supervised samples. The classic example is that of a quality inspection process where a supervisor is invoked time by time to provide an external quality evaluation which is the fresh information (supervised couple) that the JIT benefits to recover automatically from concept drift. Clearly, if the process is characterized by a sequence of concept drift that are too close in time, then the performance of the JIT might stay low even though the JIT classifier does its best to keep the highest accuracy possible compatible with the circumstances. In this case a passive-based classifier where k-NN is trained solely on the last N fixed data might provide better performance and be simpler from the complexity point of view.

This situation might also arise with gradual concept drift, obviously seen as a sequence of abrupt type of concept drift. Again, a passive solution might be preferable to the adaptive one if concept drift is fast, in the sense that its gradient over time is relevant (high developing concept drift). To address the gradual concept drift [97] proposes an extended JIT classifier introducing

- a modification of the ICI-CDT outlined in Sect. 9.3.2 that makes the CDT able to deal with a process whose expectation follows a polynomial trend.
- an adaptive classifier able to handle gradual concept drift affecting the process expectation. The classifier integrates an index estimating the evolution dynamics to improve classification accuracy.

Intuitively, the proposed extended classifier copes with gradual concept by estimating the concept drift trend, detrending the data and consider now the process as exhibiting a stationary state.

We model the gradual concept drift according to the formulation of equation (9.22). In particular, we focus on gradual concept drift that is represented by a possibly slow-time-varying stochastic process, whose expectation $\mathrm{E}[p(x|t)]$ follows a piecewise polynomial function $f_\theta(t)$. The parametric description of $f_\theta(t)$ is given by $\{(\theta_i, U_i)\}$ where θ_i is a parameter vector characterizing the polynomial $f_{\theta_i}(t)$ defined on the i-th time interval U_i (i.e., a subsequence of consecutive time instants). The expectations of the conditional probability distributions can be expressed as

$$\mathrm{E}[p(x|\omega_1, t)] = f_{\theta_i}(t) + q_{1,i} \tag{9.23}$$

$$\mathrm{E}[p(x|\omega_2, t)] = f_{\theta_i}(t) + q_{2,i} \tag{9.24}$$

where $t \in U_i$ and $q_{1,i}$ and $q_{2,i}$ are the expectations of the two classes ω_1 and ω_2 in stationary conditions. The process generating observations $x(t)$ at time t becomes

$$x(t) = \begin{cases} f_{\theta_i}(t) + \phi_{1,i}, & \text{if } y(t) = \omega_1 \\ f_{\theta_2}(t) + \phi_{2,i}, & \text{otherwise} \end{cases} \tag{9.25}$$

where $\phi_{1,i}$ and $\phi_{2,i}$ are random variables ruled by the pdfs characterizing the distributions of their respective classes ω_1 and ω_2 in stationary conditions with $\mathrm{E}[\phi_{1,i}] = q_{1,i}$ and with $\mathrm{E}[\phi_{2,i}] = q_{2,i}$.

We further assume that the probabilities $p(x|\omega_1, t)$ and $p(x|\omega_2, t)$ do not change within each interval defining the piecewise polynomial function, thus, the pdf of $x(t)$ is

$$p(x|t) = p_i(\omega_1)p(x|\omega_1, t) + p_i(\omega_2)p(x|\omega_2, t), \ t \in I_i.$$

The pdf of the inputs, the conditional distributions and the output distributions are unknown. The piecewise-polynomial function within each interval U_i, i.e., $f_{\theta_i}(t)$, is also unknown, but common between the two classes, as expressed in (9.24). We comment that the considered framework is an extension of the traditional one assuming constant $f_{\theta_i}(t)$.

9.4.4 JIT for Gradual Concept Drift

The key point of the proposed approach is to extend the observation model traditionally assumed in classification problems by allowing the expectation of the conditional probability density functions to evolve over time as a piecewise polynomial function, as expressed in (9.24). Under such a hypothesis, we can develop a CDT to assess variations in the (polynomial) trend of the process under monitoring, rather than in the value of its expectation. If the test does not detect variations, we perform a polynomial regression of the input samples and use the regression coefficients to modify online the knowledge base of an adaptive classifier. Differently, when a change is detected, the obsolete samples are removed from the knowledge base and the change detection test is restarted.

Since the ICI-CDT is natively able to deal with polynomial trends in the process under monitoring, it can be applied with minor modification at the feature level w.r.t. what is presented in Sect. 9.3.2. A detailed description can be found in [97].

Differently, the k-NN classifier has to be slightly modified to coherently adapt to gradual concept drift. Algorithm 26 presents the k-NN classifier able to deal with

Algorithm 26: Adaptive k-NN classifier for Gradual Concept drift

$N = |Z_T|$;
$i = 1$;
while $(i < N)$ **do**
  $d_i = \left(\left(x(t) - f_{\hat{\theta}(t)}(t)\right) - \left(x(t_i) - f_{\hat{\theta}(t)}(t_i)\right)\right)$;
  $i = i + 1$;
end
Identify the nearest k training samples according to the distances $\{d_i\}_{i=1,\ldots,N}$;
Classify $x(t)$ as the majority of labels in the k nearest training samples;

gradual concept drift. It is easy to see that the only difference w.r.t. the traditional k-NN classifier is the computation of the distance between the input sample and the training samples (line 4). Here, the parameter vector $\hat{\theta}(t)$ represents the coefficients of the best polynomial fit for the data during gradual concept drift. These coefficients can be estimated from the observations using any regression technique.

The polynomial fit represents the gradual concept drift, and is used to correct each term as function of the distance between the data and the fitted polynomial. In particular, the distance between the current sample $x(t)$ and the training sample $x(t_i)$ is computed after subtracting the values of the (estimated) polynomial having coefficients $\hat{\theta}(t)$ in their corresponding time instants (i.e., $f_{\hat{\theta}(t)}(t)$ and $f_{\hat{\theta}(t)}(t_i)$).

By replacing the CDT and the k-NN classifier it is then possible to formulate the JIT classifier for gradual concept drift following a similar scheme of Algorithm 25. Note that the proposed JIT for gradual concept drift is indeed an extension of Algorithm 25 as in absence of gradual concept drift, the higher order coefficients of the polynomial approach zero and the whole JIT operates as in stationary conditions.

9.4.5 Amygdala—VM-PFC—LPAC- ACC: The JIT Approach

The just-in-time learning framework is an example of a complex mechanism that founds its psychological roots in Piaget's theory of childhood learning. Detecting concept drift and reacting to it is aligned with Piagets psychological theory of human cognition [157], where learning is described as a constant effort to maintain or achieve balance between prior and new knowledge. As pointed out in [77], when new knowledge cannot be accommodated under existing schema because of severe conflict (i.e., nonstationarity), the need is to restructure the application to create new schemata that supplement or replace the prior knowledge base. While the former detection issue is addressed by the Amygdala—VM-PFC mechanism the need to supplement or replace the prior knowledge base (reaction process) is carried out by the LPAC-ACC layers.

Chapter 10
Fault Diagnosis Systems

The emergence of sensor-based networked embedded systems has made possible the real-time collection of a huge amount of data. However, not rarely, collected data are incomplete or do not make sense for various reasons, thus compromising the correctness of decisions made out of data by inducing possible dramatic outcomes. Even worse, most of research does not pay attention to the issue and implicitly assumes that data are correct and "true" by definition. It comes out that, in mission-critical applications or those applications significantly impacting on our lives, we should address this not negligible aspect.

An example is a rock-fall collapse or a landslide type of scenario. Assume that we have deployed the monitoring system that, once operational, provides the information we need to assess/infer a potential environmental risk, e.g., see [142, 143].

With reference to the Rialba towers application described in Sect. 8.6.5, if the strain gauge is detecting an enlargement of a fracture of 3.75 mm as it happens in Fig. 10.1 with deformometer 4 should we take actions? We might answer by saying that it depends on the introduced safety threshold. True. However, that implies that if the threshold is set to 2 mm we should alarm the population of the village living nearby since the potential risk appears to be real. And what about if the measurement is wrong since a permanent abrupt type of fault, i.e., a fault biasing the readout value with a constant component, affected my sensor? You could reply that we should look at the other strain gauges (deformometers 2 and 3 in the figure) which, in the considered case, do not show any enlargement in correspondence with the perceived event. Since the rock might introduce a rigid shift in its movement not perceivable by the other two strain gauges, we cannot came back with a definitive risk assessment. We immediately perceive that the issue cannot be underestimated. The above comment also brings us to the problem of sensors deployment. Sensors should be placed having well in mind the phenomenon and its expected evolution. Let us try to avoid covering the environment with unnecessary sensors and, even worse, deploy them in areas where the phenomenon is hardly observable due to sensitivity aspects (changes in the environment badly propagate to the sensor for physical reasons).

C. Alippi, *Intelligence for Embedded Systems*, DOI: 10.1007/978-3-319-05278-6_10,

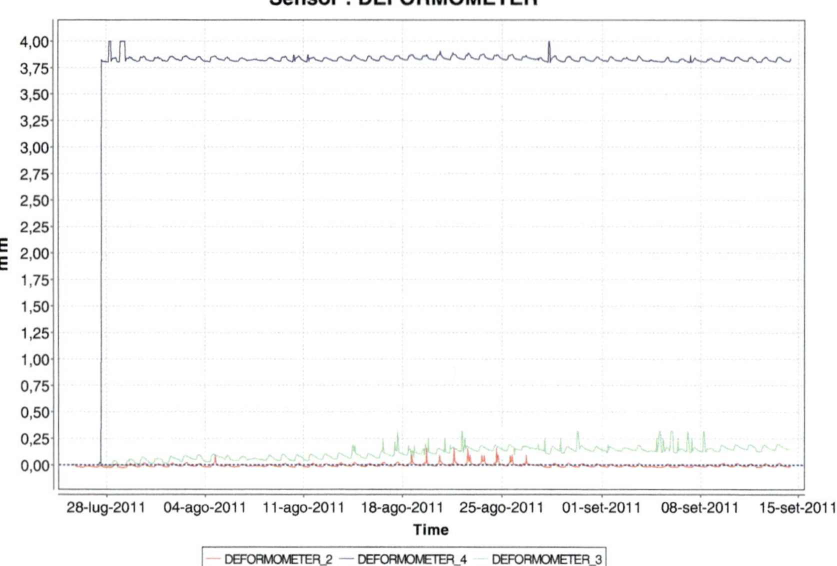

Fig. 10.1 The plot shows three measurements associated with three strain gauges (deformometers) mounted on the Rialba towers (see Sect. 8.6.5). Strain gauge 4 presents an enlargement of about 3.75 mm which might represent a potential problem for the towers stability. We see that all strain gauges are sensitive to the temperature which introduces a parasitic effect on the acquired data despite of the signal conditioning and compensation: the day–night seasonality emerges clearly. Small existing peaks are associated with transient faults in the readout mechanism. This is reality: faults, thermal drifts, missing, and erroneous data are the normality more then the exception not only in harsh environments

In general, unwished unpredictable situations are the result of faults affecting the sensor/actuator system or represent an abnormality in the monitored environment and may be either permanent or temporary, developing abruptly or incipiently. The problem becomes more pronounced as sensing/actuation systems get older since the sensors (along with the electronic chain up to the ADC) and the actuators are no more able to provide the correct functionality (and not always a calibration phase can solve the issue).

It is of paramount relevance for all applications involving a decision-making process to design methods able to analyze and interpret incoming data streams so that faults (from now on the term fault refers also to aging effects and thermal drifts) are detected, isolated, and identified as soon as possible and, possibly, accommodated for before decisions or actions are taken on the basis of carried information.

Despite the fact that hardware solutions can be envisaged to partly mitigate the problems, e.g., those based on modular redundancy by replicating the acquired hardware, they are not always able to deal with all types of faults that the sensor might encounter. Whereas an abrupt type of fault affecting a specific sensor can be easily detected by setting suitable thresholds, a drift type of fault would affect all sensors, hence making impossible to detect it with a strict hardware replication schema. A modular redundancy also implies an increment in cost that, by scaling linearly with the number of elements, might be acceptable for integrated sensors but not necessarily for more accurate and expensive traditional nonsilicon-based sensors. It should also be stressed that silicon integrated sensors suffer even more of aging than the others.

Fault Diagnosis Systems (FDSs) are software applications designed to detect potential insurgence of faults (fault detection), identify them (i.e., determine their type and magnitude), isolate faults (i.e., localize them within the system) and, possibly, mitigate their effects through ad-hoc actions (mitigation step).

The complexity of the FDS depends both on the functions requested by the application (e.g., we might be interested only in detecting the fault) and the computational power made available by the (embedded) processing system. In simplest embedded systems, we should consider a simple FDS, mostly based on the fault detection aspect (e.g., detection of faults affecting the sensors) and, possibly, adoption of some mitigation policies (e.g., the sensor is disabled and another one either physical or virtual is enabled or a software agent tries to recalibrate it). When embedded systems can cooperate, e.g., within a sensor network, more sophisticated, even distributed FDSs can be designed. In some other cases, the data stream inspection is centralized in a high performing remote processing system with outcomes enabling actions sent to the remote embedded units (e.g., disable the sensor, change the parameters of the filter in the conditioning stage, and modify the calibration curve w.r.t. the temperature, etc).

Most of FDSs operate by assuming hypotheses about the plant/system under inspection. For instance, we might assume that a model for the process generating the data is available and design FDS strategies on the base of that, e.g., by observing over time the discrepancy between the output of the available model and the acquired values and applying some CDTs to detect occurring changes. The more the a priori information we have the better the overall performance of the FDS, e.g., the smaller the latency in detecting a fault the easier the isolation step. For a general, deep treatment of problems behind traditional FDSs the interested reader can refer to [144, 145].

In the chapter we focus on more advanced FDSs based on a cognitive approach that, by exploiting computational intelligence techniques, address the FDS issue by assuming none—or little—a priori knowledge. As such, they are particularly suitable techniques for intelligent embedded systems. In fact, cognitive FDSs do not assume strong assumptions—if any—about the plant or the environment under investigation and learn the needed characteristics and the hidden system behavior directly from incoming data. Clearly, this approach requires a learning phase—which might even be executed online—to configure the FDS as well as implement those mechanisms able to identify and react to changes in the environment (not to be confused with faults which represent a particular instance of a more general class of changes, e.g., those including changes in the environment and model bias).

Most of cognitive methods for fault diagnosis aim at characterizing the expected behavior of the system (nominal fault-free state) based on the knowledge of mathematical description of the system (system model) with some parameters to be learned, or by using machine learning techniques under limited or none a priori knowledge availability. We classify the existing cognitive fault diagnosis approaches based on the available information about the system model. We have two extreme cases, the first one where a description for the system model is available (e.g., the equations describing a plant), the second case investigates the diagnosis problem when the system model in unavailable and the unique information is associated with acquired data.

10.1 Model-Based Fault Detection and Isolation

Even if we are assuming availability for the system model, mostly satisfying some canonical forms (e.g., a continuous or discrete state based description), there are unknown uncertainties affecting it, e.g., those that stem from the partial knowledge of the system parameters, unmodeled dynamics, linearizations of some nonlinear parts of the system, unknown disturbances, and measurement noise.

When a description for the system model is available, the cognitive aspect is mostly related to the characterization, through learning, of the uncertainties affecting the system and their dynamics as well as learning the thresholds allowing the FDS to detect, identify, and isolate faults.

Here, uncertainties are mostly regarded as unknown but bound and their characterization represents a fundamental step for designing an effective fault diagnosis system. In particular, several methods need to be applied depending on the available information

- filtering techniques for attenuating the effects of disturbances and measurement noise [148, 149];
- set-membership identification techniques for estimating bounds of parametric uncertainty [150, 152] when the bounds are not known;
- adaptive approximation techniques, that learn the modeling uncertainty [146, 147].

Fault detection is then carried out by checking for bounds violation, i.e., we inspect if new data are associated to features which are within or outside the identified bounds characterizing the fault-free nominal behavior. In the case of bound uncertainty and under nominal fault-free conditions, adaptive thresholds can be designed to bound residuals, i.e., the discrepancy between data obtained with the system model and those coming from measured sensor data [149, 153], convex sets are designed for outer-bounding the feasible parameter sets [151], interval constraints are determined that impose relations to be satisfied by certain variables and domains of these variables [152], predicted intervals are computed to bound values that output data can assume [150].

Model-based cognitive fault diagnosis methods following the bound approach grant robustness with respect to uncertainty (ensuring zero false positives when bounds represent a reliable information) but might suffer from the occurrence of false negatives (the existing fault is not detected by the method).

The specifications for the cognitive algorithm (e.g., system configuration, known/trained models, bounded/stochastic uncertainty, etc) for conducting fault detection are also used for fault isolation. It is important to mention that pursuing fault isolation in large-scale, complex systems in unstructured, open-ended environments that can be affected by multiple faults is a hard problem that might require additional a prior knowledge or knowledge to be learned to solve the fault diagnosis problem.

10.2 Model-Free Fault Detection and Isolation

Availability of a model for the system is a strong assumption in many applications, in particular those where the unique information about the system is constituted by acquired data. In these cases, fully cognitive mechanisms need to be created to address the fault diagnosis. The performance of the FDS to be designed depends on the complexity of the system, the type of expected faults, the chosen FDS strategy, and availability of a prior information about the system and the fault classes. For instance, we might know that the data stream has been generated by a time-invariant process (or time invariance is a good approximation for some time), that different data streams are mutually dependent, and that noise affects data streams with an additive or a multiplicative model (we recall these concepts are given in Chap. 5 where faults represent a particular type of perturbations). Each of these assumptions can help the FDS to improve its performance. The leitmotif is always the same: the more the available information the better the performance; we shall expect by a well-designed FDS. Different solutions might, in fact, take advantage of prior information to improve detectability, reduce latency in detection and/or false positive/negative rates.

Some methods, e.g., [156] exploit the spatial and temporal relationships existing among sensor data streams (faults are expected to change those relationships), others build a fault dictionary (library) containing fault signatures either at design-time by exploiting available cases of fault instances [154] or online, without assuming such an availability [155] to detect/classify the occurred faults.

Following these approaches, the structure of the cognitive fault isolation algorithm, that includes a fault dictionary, is modified in terms of the incoming new knowledge, i.e., new faults are added to the fault dictionary whenever detected or the same improves based on the new fault instances (new fault signatures are added to the dictionary). When a fault is detected and the embedded system can continue to operate, possibly after a fault accommodation phase has been taken into account to mitigate/reduce the impact of the fault, a new training phase needs to be activated since the system is operating on a different state w.r.t. the previous one. As such, all modules composing the model-free FDS need to be reconfigured accordingly.

The general conceptual framework behind a model-free FDS is given in Fig. 10.2. The key elements composing the FDS are the *Nominal concept*, the *Change Detection*, and the *Cognitive fault analysis* modules. All modules receive information or

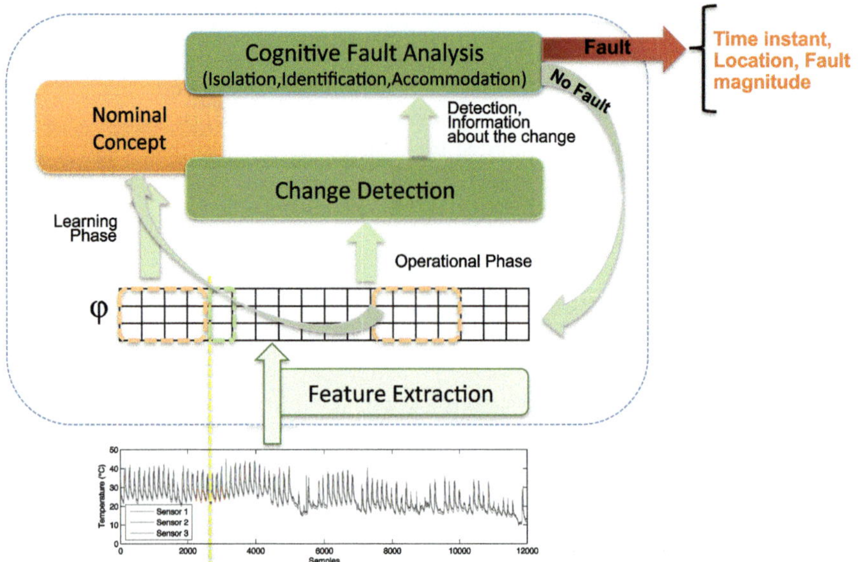

Fig. 10.2 A cognitive model-free FDS. Features extracted by sliding (possibly not overlapping) windows opened over input signals are used to generate features. In turn, the current available features are used to assess their coherence with the nominal fault-free state represented in the nominal concept module. If a change is detected, the cognitive level is activated and, after having assessed the change, carries out the fault isolation, identification, and accommodation procedures. If the change is not validated by the cognitive analysis level, the system continues in its operational modality, possibly by retraining the nominal concept module and the change detection one on new instances since a fault positive was detected

features φ obtained by processing raw data. Features provide a compact representation of the information for decision making and further processing. The nominal concept module contains a characterization of the nominal fault-free state built incrementally, directly from available features. Features extracted from the raw data up to a given instant of time constitute the training set (i.e., data up to the yellow vertical line in the figure) to be used by a learning phase to create the signatures/model representing the nominal conditions.

During the operational life features are extracted and the change detection module activated to inspect a sliding window opened in the feature data stream (the green box opened on the features). The change detection method verifies whether the current features are coherent with the model/signatures present in the nominal concept module where knowledge is stored and managed or not. If the answer is positive no changes are detected and the current features belong to the nominal state (and hence, no fault is perceived to have occurred in the raw signal that generated those features). Differently, if a change is detected, the cognitive fault analysis module is activated. The cognitive fault analysis module takes advantage of the time a change has been detected by the change detection module as well as it exploits additional information about the change and the knowledge associated with the nominal concept and carries out the validation of the change. If the change is validated and associated with a fault then, the isolation, identification, and accommodation phases are executed and provide the time instant the fault occurred, its location, and magnitude. It should be commented that identification of the type of fault might require an external assessment from a supervisor (e.g., the operator) that labels the type of fault after personal inspection. When this operation can be carried out a dictionary can be created online, over time, based on feature instance-label assigned by the operator. Clearly, if a fault dictionary is available the cognitive fault analysis module can take advantage of it and compare the current feature signatures with those present in the fault dictionary for fault type identification. If the change is not valid, the detection is associated to a false positive detection (e.g., due to the uncertainty in the characterization of the nominal concept) and the system keeps its operational modality, while the nominal concept module and the change detection can be possibly retrained on new instances of data.

Three fully cognitive methods for fault detection will be presented in the sequel. The first acting at sensor level, the second exploiting relationships existing between two sensors, the third taking advantage of relationships in space and time existing within a network of sensors. All these methods can be implemented in embedded systems. The first case is limited to a sensor mounted on an embedded system. The second one assumes that the embedded system has mounted at least two sensors and the third approach requires a full platform of sensors. Interestingly, the sensors can be attached to a single embedded system or being part of a more complex, distributed sensor network. The analysis is the same being methodological and, hence, technology independent.

10.2.1 FDS: The Sensor Level Case

Consider an unknown process generating the data $y = g(x, \eta)$ providing time series $y(1), y(2) \cdots, y(i), \cdots$ where $y(i) \in Y \subset \mathbb{R}$ is the datum acquired at time instance i affected by an unknown disturbance η and $x \in X \subset \mathbb{R}^l$ the input. FDS designed at the single sensor level are barely effective unless strong hypotheses are made about the process generating the signal.

We will summarize the most relevant cases under the common stated assumption that the process generating the data is time invariant (no environmental changes) and, hence, the fault is the only external cause inducing changes in the process.

- *i.i.d. assumption for raw data*. When measurements follow an i.i.d distribution CDTs can be considered to inspect changes in the data flow as proposed in Chap. 9. For instance, the assumption holds if the system model is $y = \bar{x} + \eta$, where $\bar{x}$ assumes a constant value and η is i.i.d. random noise. It also holds in quality analysis applications where, for each generated/produced item, a set of i.i.d. measurements are taken and describe the item (e.g., the weight and dimensions of an egg of class A).

- *i.i.d. feature assumption*. In general, we cannot assume sensor data to be independent. As such, the traditional way to construct a FDS requires a feature extraction step followed by the characterization of the nominal state in the feature space. Deviations from the nominal state are then seen as symptomatic situations to be further investigated. In order to be effective, the features associated with the nominal state must constitute a neighborhood in the feature space (or a finite set of neighborhoods) and features associated with faulty situations must be scattered apart, possibly grouped together to constitute a neighborhood different from the nominal one. If that is the case, namely, nominal and fault states are not overlapping (or weakly overlapping if perfect decoupling cannot be granted), fault detection is carried out by inspecting the features and discovering that current feature instances do not belong to the nominal state (either deterministically or in probability). Fault identification and isolation follow by identifying the cluster features belong to.
- *The complete model assumption and the abrupt type of faults*. An interesting case following the feature approach is that where features are the parameters of a model approximating the data in a running window open over the signal. Each parameter vector $\hat{\theta}$ learned/identified by taking advantage of data in the data window provides model $f(\hat{\theta}, x)$. Here, under the assumption that faults are of abrupt type and models are Linear and Time Invariant (LTI), we have that parameter vectors associated with models modeling, respectively, faulty data and fault-free data cover different areas of the parameter space $citeBasseville. However, the implicit strong assumption behind this method is that no model bias is present (complete model assumption) namely, $g(x, \eta) = f(\theta^o, x) + \eta$ within an additive signal plus noise model (other models might be considered). For instance, if LTI models are used to describe the data stream then also the system model must be generated by a LTI system of the same family used for model approximation.
- *The complete model assumption: predictive form*. Under the complete model assumption $g(x, \eta) = f(\theta^o, x) + \eta$, we can build a predictive model approximating the next time instance $y(t)$. Since the approximating model provides the estimate $\hat{y}(t) = f(\hat{\theta}, x)$,being x the regressor vector (e.g., in case of autoregressive models $x = [y(t-1), \cdots, y(t-\tau)]$ for a τ long window), we can compute the residual sequence $\varepsilon(t) = y(t) - \hat{y}(t)$. If $\hat{\theta}$ is a good estimate of θ^o (see Sects. 3.4.1 and 3.4.5) then the residual is i.i.d and CDTs can be applied to inspect changes induced by faults. In this case, time variance induced by the fault is detected by methods inspecting for changes in stationarity.

It must be commented that all the above methods suffer from the false positive and negatives issue. Not much can be done here, unless further hypotheses are assumed. It should be absolutely clear that, for each of the above methods, if an assumption is not met it is not possible to distinguish among:

- Changes in the environment. The stationarity/time-invariant assumption is not met;
- Existence of an approximation risk (model bias). If the model family is not complete since it does not contain the system model, then the model bias, if not negligible, will likely raise a false positive;
- False positives intrinsic with the chosen method.

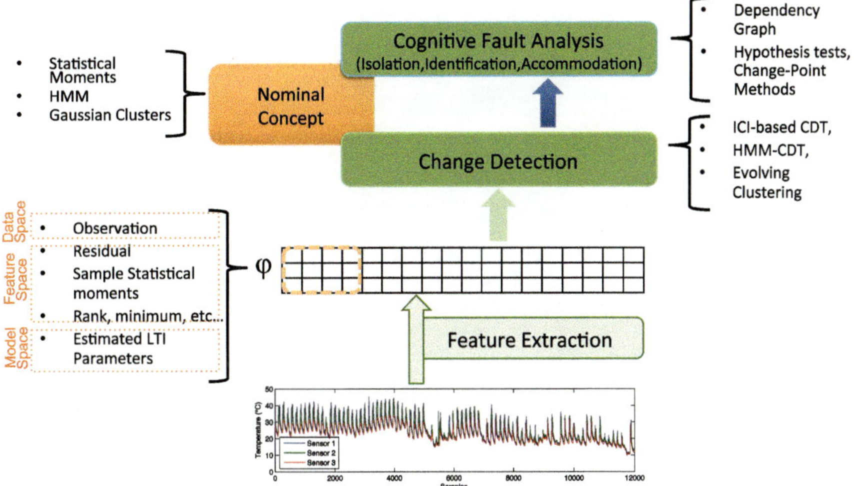

Fig. 10.3 A cognitive FDS detailing that of Fig. 10.2. Some possible instances for the elements composing the FDS are given

We strongly advise the reader to pay attention to the above issues since easy conceptual mistakes and wrong statements can be made.

Figure 10.3 details a bit more the general framework depicted in Fig. 10.2. In particular, the three basic elements associated with the framework are specialized to ease the understanding and cover some interesting cases itemized above. The detailed description follows:

- The feature extraction phase. As mentioned before the role of features is to represent the information carried by the signal (or by a window opened on it) in a compact way. Features can be intended at different abstraction levels. Depending on the application and our expertize, we select different features for different problems. Features at the data space level are the raw observations, possibly calibrated and compensated. When we move to a feature space, the set of input data present into an assigned temporal window are transformed according to a function into the feature vector x. Example of features are the prediction/reconstruction residuals, the sample statistical moments, the coefficients of a Fast Fourier Transform (FFT), the rank of the cross-correlation matrix, the minimum values assumed within each sensor window, etc. At the model space level, features are, for instance, the coefficients of LTI dynamic models (e.g., ARX, ARMAX, etc) used to predict the next expected sample. Over time, we generate a sequence of feature vectors x coinciding with the coefficients of the obtained local model approximating the signal in the considered window.
- The nominal concept module. The nominal concept describing the nominal fault-free state of the system can be modeled as a set of feature vectors associated with the nominal state. Statistical moments for the features, gaussian clusters, or HMMs

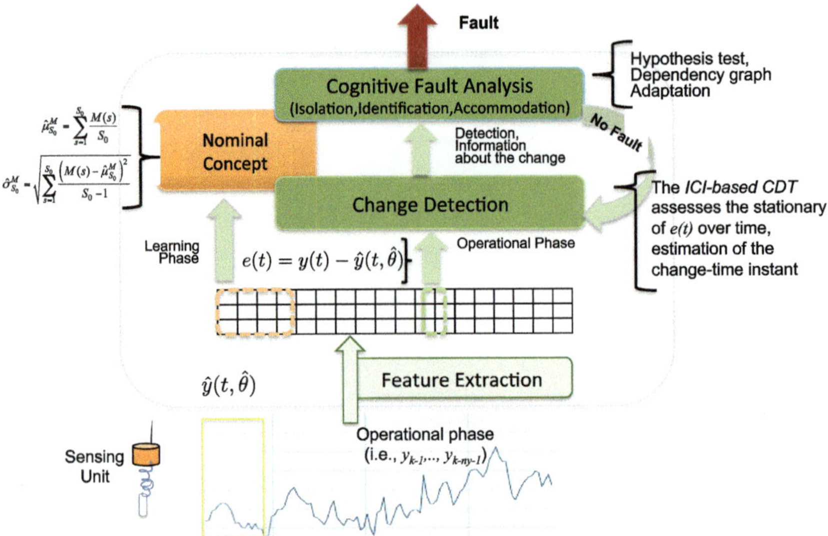

Fig. 10.4 An FDS inspecting the datastream associated with a specific sensor. The feature is the residual approximation error, constituting, over time, a sequence. The CDT operates by extracting further features, e.g., the sample mean and variance of windowed residuals for carrying out the change detection phase

describing the nominal fault-free state are other description mechanisms. It is clear that, here, the goal is to somehow describe the nominal state of the system. As such, any consistent representation works fine.

- The change detection module. The change detection aims at evaluating the affinity between the current acquired feature vector and the fault-free nominal concept. A k-NN classifier (or any classifier) can be built to evaluate the closest signature to the current feature vector. If the distance between the two is below a threshold then the current feature is assigned to the nominal state (and the system is said to operate correctly). In some other cases, the approach is statistical and a CDT test, e.g., based on the CUSUM, ICI rule, or an HMM-CDT presented in Chap. 9 can be used to assess changes in stationarity/time variance.
- Cognitive fault analysis. At the cognitive level, different methods can be considered to validate the change in stationarity and decide whether the change is associated with a fault, a model bias or a time variance in the environment. Methods based on the dependency graph or hypothesis tests can be used to solve the problem.

10.2.1.1 Example: A FDS-Based on a Residual Inspection

The example assumes that a given signal $y(t)$ can be approximated with a LTI predictive model and that the complete model assumption holds (i.e., the discrepancy between the model and the real data can be represented as a i.i.d. signal, here a white noise). The structure of the FDS we will discuss in the sequel is given in Fig. 10.4.

As we have seen above, the complete model framework assumes that the unknown process generating the data follows the system model, here rewritten as $y(t) = f(\theta^o, x(t)) + \eta$ where x is some uncontrollable unknown input driving the nature and η is an unknown white Gaussian noise. The signal is modeled with the LTI predictive model $\hat{y}(t, \theta)$, $x(t)$ being the regressor feature vector $x(t = k) = [y(k - 1), \cdots, y(k - \tau)]$ for a particular τ time lag and θ is the model parameter vector. The particular model $f(\hat{\theta}, x(t))$ is obtained after a learning procedure that leads to the parameter vector $\hat{\theta}$.

To instance the predictive model assumes that an Autoregressive (AR) function family grants perfect predictability, in the sense that the prediction residual $e(t) = y(t) - \hat{y}(t)$ is a white noise. Given the model linearity, $\hat{\theta}$ can be identified with a classic least squares procedure (learning phase, in the figure) and the predictive model assumes the form

$$\hat{y}(t, \hat{\theta}) = \hat{\theta}^T x(t), \; x(t = k) = [y(k - 1), \cdots, y(k - \tau)]$$

leading to the residual sequence $\cdots, e(t-2), e(t-1), e(t)$. Consider not overlapping windows of size S_O opened over the residual sequence. The generic instance of the window $M_w(s), s = 1, \cdots, S_O$ is relocated over $e(t)$ so that $M_w(s)$ for the w-th window satisfies $M_w(s) = e((w - 1)S_O + s), s = 1, \cdots, S_O$ and $w = 1, 2, \ldots$.

The feature vector for change detection $\varphi_w = [\hat{\mu}_w^M, \hat{\sigma}_w^M]$ extracted from the residual sequence are the sample mean and sample standard deviation estimated over the w-th window as

$$\hat{\mu}_w^M = \frac{1}{S_O} \sum_{s=1}^{S_O} M_w(s) \tag{10.1}$$

$$\hat{\sigma}_w^M = \sqrt{\frac{1}{S_O - 1} \sum_{s=1}^{S_O} \left(M_w(s) - \hat{\mu}_w^M\right)^2} \tag{10.2}$$

An ICI-based CDT can now be considered, e.g., the simple one presented in Chap. 9. If no changes are detected, the procedure is iterated: time passes and new features are generated and inspected for a potential change. Conversely, if a change is detected at a given instant of time, the cognitive fault analysis module is activated whose role is to validate/reject the change recommendation made by the change detection level and discover whether the change is associated with a change or not. Assume that the cognitive fault analysis level is implemented with an hypothesis test, e.g., based on Hotelling. When this is the case, the change detection and the cognitive fault analysis modules behave as an hierarchical CDT. Refer to Chap. 9 for further details.

We recall that even if the change is validated by the cognitive fault analysis module, by inspecting a single sensor we cannot claim that the change is associated with a fault since changes in the environment and false positives might occur. However, as pointed out above, we can take advantage of extra information made available by correlated sensors, e.g., as those present in multiple sensors embedded systems or sensor networks. There, by inspecting the dependency graph associated with existing sensors we can classify the change that, eventually, becomes a fault. These aspects will be addressed in next sections.

10.2.2 FDS: Changes in a Sensor–Sensor Relationship

It is not rare to find situations where there exists a dependency between two sensors. Such a dependency can either be *direct* or *indirect*. We say to have a direct dependency when the sensors are observing the same phenomenon but open different views of it. For instance, two thermal sensors deployed at some distance are likely to be dependent and the functional dependency can be expressed as a transfer function linking the two. However, we might —as we do in real life—experience an indirect relationship among two sensors that are apparently unrelated. For instance, if we have two strain gauges mounted onto two different bridges, we expect them to be unrelated. However, functional independence is only associated with ideal sensors. As we have seen in Chap. 2 not rarely temperature affects sensors. As such, if the strain gauge sensors provide measurements $y_1(t)$ and $y_2(t)$ and $T(t)$ is the temperature value, then there it exists a functional constraint

$$f(y_1(t - \tau_1),\ y_2(t - \tau_2),\ T(t)) = 0$$

for some τ_1 and τ_2 accounting for a possibly existing delay in propagating the effect of temperature on the sensors readout.

Such a constraint introduces an indirect relationship which can be exploited by any cognitive FDSs, e.g., by the cognitive fault diagnosis level of Fig. 10.4 that might integrate constraints existing among the set of sensors. Two different approaches for designing cognitive FDSs are given in the sequel.

10.2.2.1 An Evolving Fully Cognitive Approach

An interesting solution has been proposed in [155] under the hypotheses that the environment is initially time invariant and faults are of abrupt type (a sudden change in the affected instance that moves from a value to another one). More specifically, a model-free algorithm is proposed for fault diagnosis of nonlinear dynamic systems working in the parameter space of LTI predictive approximating models (the problem requires identification of time variance associated with a fault). It is shown that, under reasonable assumptions mainly about the dynamics of the process generating the data, the distribution of parameters of the LTI models is Gaussian in the parameter space. As such, features coincide with the LTI model parameters and the nominal concept can be described by a cluster induced by a Gaussian distribution.

More in detail, and following the general framework of Sect. 10.3, as data come over time an LTI model is constructed over independent N data windows synchronously opened on sensor data streams $y_1(t)$ and $y_2(t)$. Consider, for instance, the Autoregressive eXogenous (ARX) model as the reference LTI predictive family model that, applied to sequences of not overlapping windows, provides the sequence of parameters-features

$$\hat{\theta}_0, \hat{\theta}_1, \cdots, \hat{\theta}_w, \ldots$$

which spot the parameter space and must be inspected for fault detection, where w is the index of the window. The method operates online: as sensor data come and complete a full N instances window, the corresponding parameter vector $\hat{\theta}_t$ is generated for the w-th window of data. The change detection and fault analysis modules classify it as belonging to one of the following classes.

- Nominal class. Parameter vector $\hat{\theta}_w$ belongs to the nominal fault-free class at a given confidence level. In other words, parameter vector $\hat{\theta}_w$ is inspected by the change detection module which decides whether it belongs to the nominal concept or not. Since the nominal concept is based on a Gaussian distribution, it is checked whether $\hat{\theta}_w$ belongs to such a distribution once a confidence level has been set.
- Fault class. Instance $\hat{\theta}_w$ belongs to a fault class at a given confidence level. The change is detected and the cognitive fault analysis module assigns to the parameter the label of the class it belongs to.
- Outlier class. Instance $\hat{\theta}_w$ neither belongs to the nominal class nor to a fault class at a given confidence level.

The algorithm requires a training phase during which data streams are assumed to be time invariant. The set of parameter vectors generated during the training phase are used to generate the nominal (Gaussian) state for the system, suitably inserted in the nominal concept module. Since the nominal state is a cluster (or a set of clusters if the sensor-to-sensor relationship can operate in different nominal states, each of which described by an independent Gaussian cluster) the mean vector and the covariance matrix fully describe the state.

Note that no assumptions are made about the linearity of the relationship between the two data streams: the Gaussian distribution for the parameter vector holds even if the relationship is nonlinear. When a deviation from the nominal condition is detected, in the sense that the current instance does not belong to the nominal class, the parameter vector is moved to the outliers set. The outliers set is regularly inspected to see whether there are instances that, grouped together, provide sufficient statistical evidence to generate a new class different from the nominal one. If that is the case a first fault state is generated and, possibly labeled, e.g., by exploiting the information provided by the user or an automatic decision system.

An example is given in Fig. 10.5 which presents three clusters, a nominal one and two faulty ones.

Fault clusters constitute, de facto, a fault dictionary. At the beginning, no fault dictionary is given and the algorithm automatically builds it over time by following

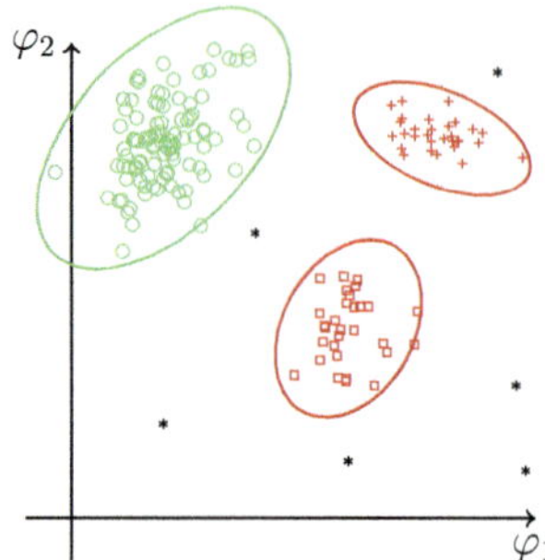

Fig. 10.5 The knowledge associated with the nominal concept module in a bi-dimensional parameter space. Three clusters are present. The first, is the nominal one. Instances associated with the nominal error-free one are *green circles*. Then, we have two "*red*" clusters associated with two fault classes (squared and cross instances are in there). Asterisks refer to the elements present in the outliers set

an evolving mechanism as faults occur. From now on, outliers are treated as separate instances until enough confidence is made available to either integrate some of them in existing classes or promote to a new fault class.

The method, is purely cognitive since all needed structures are built directly from sensors as data streams are provided.

Example: Changes at the Sensor-Sensor Relationship

We present a synthetic example describing the evolving FDS approach. Assume that the relationship between the two sensors is nonlinear as described by the system model

$$y_1(t) = \sin\left(\theta^T x\right) + \eta \tag{10.3}$$

where $x = [y_1(t-1), y_1(t-2), y_2(t-1)]$. Both θ and x are column vectors with $\theta = [0.1, 0.2, -0.1]$, $\eta \sim \mathcal{N}(0, 10^{-4})$ and the input sensor (behaving here as the exogenous input for the second sensor) is generated by the random walk

$$y_2(t) = 0.4 y_2(t-1) + \varepsilon(t) \tag{10.4}$$

$\varepsilon(t) \sim \mathcal{N}(0, 1)$. A data set of 80405 samples was generated by using the above system model and the FDS was trained on the first 16086 fault-free samples. Two faults were then injected in the remaining data, whose effect induces an abrupt change δ in the parameter vector. The multiplicative perturbation model is adopted so that the new configuration for parameters becomes $\theta_\delta = (1+\delta)\theta$.

The first fault is characterized by $\delta = 0.05$ and affects the system in the sample interval [32166, 48246]. The second fault, injected in sample interval [64326, 80405], is characterized by $\delta = -0.5$. At first, the nominal concept module is composed by

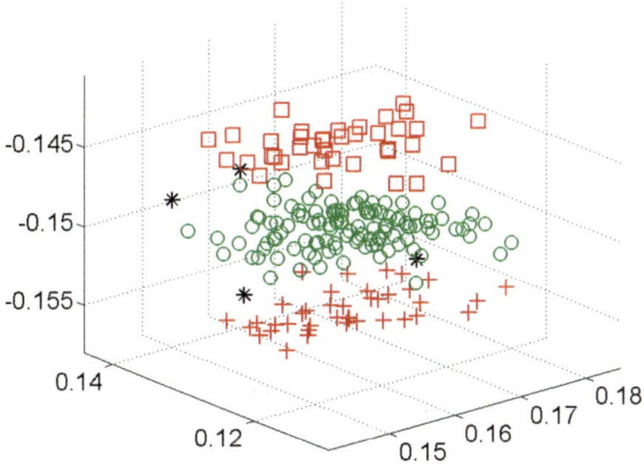

Fig. 10.6 *Green circles* represent estimated parameter vectors $\hat{\theta}_w$ belonging to the nominal state, red squares and + represent those parameter vectors associated with the two fault states. Black asterisks are outliers and compose the outliers set. The axis represents the three components of the approximating LTI parameter vector $\hat{\theta}$

the nominal error-free state only. At the end of the data set, the nominal concept contains the error-free state and two different fault states complete the available knowledge. To model the relationship between $y_1(t)$ and $y_2(t)$, ARX models have been considered where the order of the autoregressive and exogenous parts are 2 and 1, respectively.

The configuration of the parameters in the parameter space at the end of the execution of the evolving algorithm is given in Fig. 10.6. The green cluster represents the nominal concept. Instances belonging to the nominal set are generated during the training phase and updated during the operational life of the system whenever instances belonging to the nominal state are found. The two red clusters are associated with the two faulty conditions.

10.2.2.2 An HMM-Based Cognitive Approach

Changes at the sensor-to-sensor relationship can be also tackled by considering different mechanisms. For instance, [156] has addressed the problem by modeling the relationship with Hidden Markov Models (HMM), stochastic finite state machine where the number of states is learned as with the transition matrix. In this way, seasonalities present in the relationship can be modeled by the learning machine which is trained on the sequence $\hat{\theta}$ generated over either overlapping or non-overlapping windows of data. The rationale behind the machine is that the sensor-to-sensor relationship can be modeled by a probabilistic machine operating in the model space. The cognitive FDS framework is that depicted in Fig. 10.7.

Within this framework, the nominal concept is described by an HMM, whose number of states and transition matrix are learned during the training phase. Whenever the current model $\hat{\theta}_w$ and the previous ones cannot be explained anymore by the learned

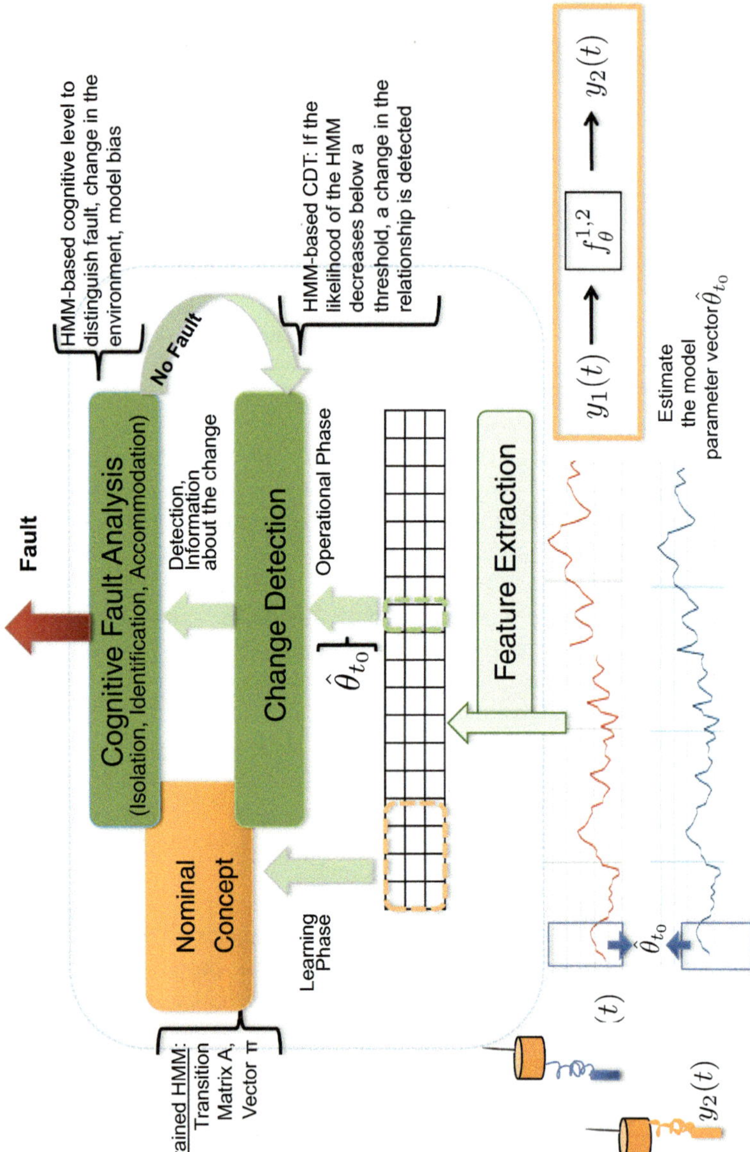

Fig. 10.7 An HMM-based cognitive FDS. The relationships between sensors are modeled as HMMs that model the functional dependencies as probabilistic finite state machines working in the model space

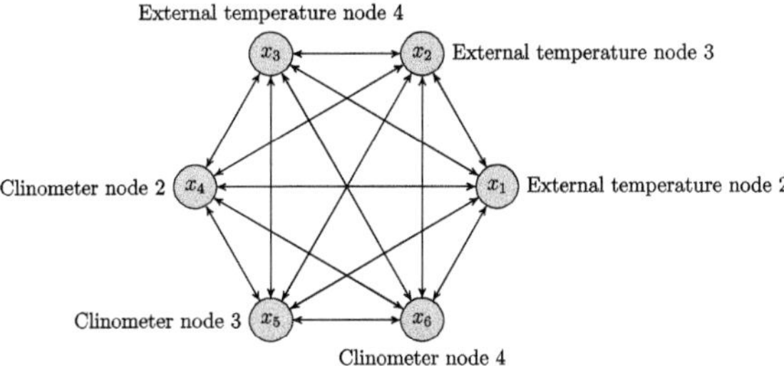

Fig. 10.8 The functional dependency graph for the Rialba towers deployment. The six sensors are clinometers and temperature sensors internal to the clinometers. All sensors are correlated, either directly because they measure the same entity (although in a different place) or indirectly, by means of the parasitic effect introduced by the temperature

HMM machine, a change is detected by the change detection module that inspects for drops in the likelihood function. When the likelihood is below a threshold, implying that the current models are believed not to belong to the learned nominal fault-free situation with high probability, a change is detected. Needed thresholds are learned during the training phase.

It should be pointed out that the above methods (the evolving one and the HMM-based) cannot distinguish among time variance in the system, model bias or the occurrence of faults: not enough a priori knowledge or information is in fact available to solve this problem whose solution requires a more structured algorithm as proposed in the next subsection.

10.2.3 FDS: The Multi Sensors Case

We have seen that, by modeling a sensor-to-sensor couple we introduce a relationship, i.e., a functional constraint that links the two sensor data streams. If a multisensor platform is attached to the embedded system or a network of sensor embedded systems (sensor networks) is available, the process can be iterated for all couples. The starting point of the whole procedure is the construction of a dependency graph where each node refers to an existing sensor and arcs represent the relationships (either direct or indirect) among sensors. Arcs are oriented since causality between data must hold.

An example of a dependency graph for the Rialba tower deployment presented in Sect. 8.6.5 is given in Fig. 10.8. The six sensors are three high resolution clinometer and three external temperature sensors, measuring the temperature of the clinometer (and not the external environmental temperature). Causality holds for all relationships and, as a consequence, relationships are bidirectional: if we consider the couple of sensors associated with data streams $y_1(t)$ and $y_2(t)$, there are the two relationships $f_\theta^{1,2}$ and $f_\theta^{2,1}$.

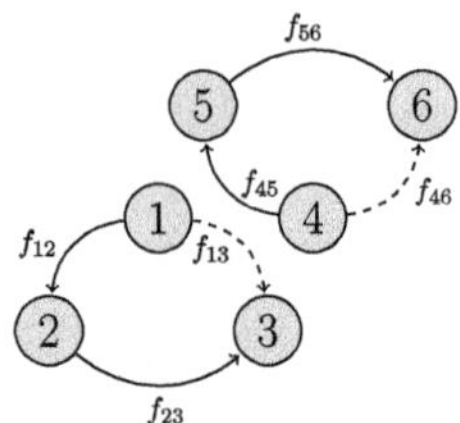

Fig. 10.9 Representation of the functional relationships for the synthetic network. The functional graph is disjoint and composed of two independent subgraphs. Relationships f_{13} and f_{46} are indirect relationships, and the others are associated with direct relationships

Since each relationship provides a constraint, we can use such—possibly—redundant information to build a cognitive FDS taking advantage of the existing dependency graph.

The class equivalence among *time variance of the environment, fault, model bias* is solved by taking advantage of the available information at the cognitive fault analysis module. More in detail, a CDT, e.g., based on the HMM likelihood inspects the stream of model parameters for changes. When a change is detected at the change detection module an alarm is raised, the cognitive level is activated and the current likelihoods associate with the dependency graph arcs provided to the cognitive fault analysis module. The cognitive level accesses the dependency graph as a whole and, based both on the functional topology and the set likelihoods, makes a decision that disambiguates the equivalence issue. If the dependency graph is poor, in the sense that not enough functional constraints can be created with the available sensor platform, then the equivalence issue cannot be solved.

Example: A Cognitive Fault Analysis Module. The Synthetic Multi Sensor Platform Case

Consider a multisensor platform for the embedded system (or the sensor network) composed of six sensors, whose functional graph is shown in Fig. 10.9. A solid arrow represents a functional relationship existing between a generic couple of sensors, modeled with an HMM based on ARX predictive models as described in Sect. 10.2.2.2. Dashed arcs refer to indirect functional relationships. Data streams generated by units 1 and 4 are coming from the random walk stochastic process similar to that used in experiment (10.4). The other data streams are created with a system model identical to that given in (10.3) with random generated θ vectors.

For each couple of sensors, data streams composed of 8165 samples were generated according to the above mechanism. The estimated parameters of the ARX models on the first 4085 fault-free data have been used to configure the HMMs.

A fault was then injected in the measurements provided by sensor 2 in the interval [6126, 8166], by generating an abrupt perturbation of the parameter vector θ_{12} relative to the functional relationship f_{12} between sensors 1 and 2 so that $\theta'_{12} = (1+\delta)\theta_{12}, \delta = 0.1$.

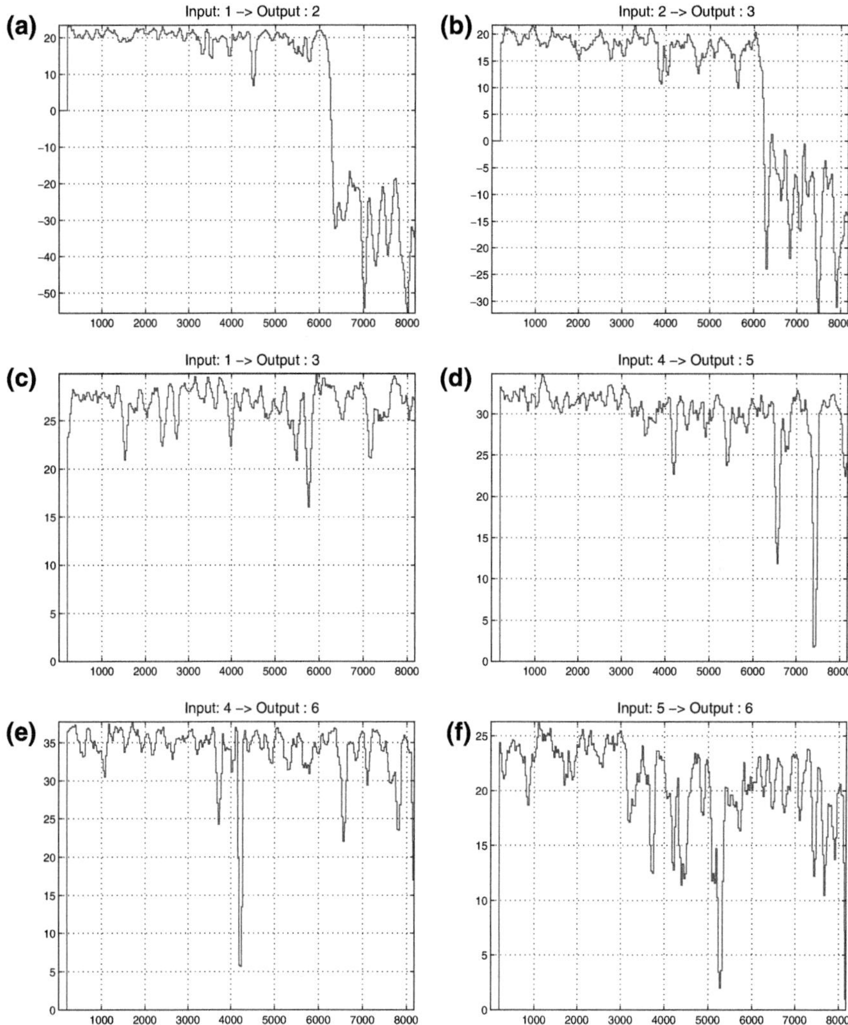

Fig. 10.10 The loglikelihoods for the selected relationships. **a** Likelihood f_{12}, **b** Likelihood f_{23}, **c** Likelihood f_{13}, **d** Likelihood f_{45}, **e** Likelihood f_{46}, **f** Likelihood f_{56}

Figure 10.10 presents the evolution over time of the loglikelihoods associated with the estimated ARX parameters modeling relationships f_{ij} of the synthetic example. We see in plots Fig. 10.10a and b the influence of the fault injected on measurements taken by sensor 2: the likelihoods drop and assume low values for the whole dataset; other relationships are not affected. The comments we can make, and will lead to the final decision taken from the cognitive FDS, are the following:

1. The only relationships affected by the drop in the likelihoods are those associated with f_{12} and f_{23}. The intersection between the sensors indexes (1, 2) and (2, 3)

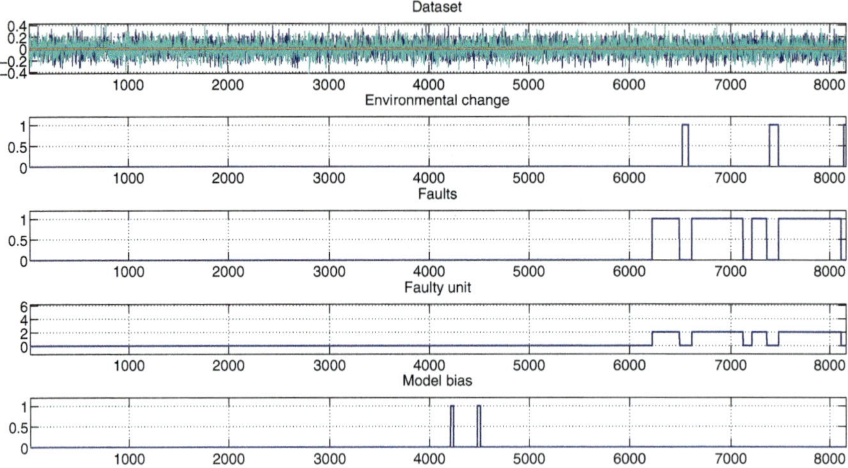

Fig. 10.11 Cognitive level for the synthetic network experiment. Top to bottom: the data streams of the six sensors, detection of environmental changes, detection of sensor faults, index of the faulty sensor (-1 stands for n.a.), and detection of model bias. When detections assume value 1, the specific detection happened

provides index 2. As a consequence, we can claim that sensor 2 is affected by concept drift.

2. All other relationships are not affected by the drop in the likelihood, hence stating that received parameters are consistent with the HMM-trained machine.
3. The change in sensor 2 cannot be associate with a change in the environment. In fact, the relationship f_{13} is not detecting a change: we expect that a change in the environment should be perceived by all related relationships.

If we assume, as a rule, that a model bias/false positive would affect one relationship at a time then events in correspondence with relationships f_{45} around time 7400, f_{46} around time 4100 and f_{56} around time 5200 are either true false positives or model bias.

The example has given an intuitive rationale of the rules behind the fault analysis module. However, in real more complex cases, the classification of the change carried out by inspecting the likelihoods associated with different relationships is a complex operation. The final decision made by the cognitive FDS depends on values provided by the likelihood integrated with some topological information through a more complex cognitive level that extends and formalizes the rules we intuitively presented in this simple synthetic example.

Details about the logic for partitioning the dependency graph and discriminate among model bias, sensor fault, and environmental change based on the likelihood values is given in [156].

By running the cognitive FDS on this experiment we obtain results plotted in Fig. 10.11. By inspecting the figure, we can make the following comments:

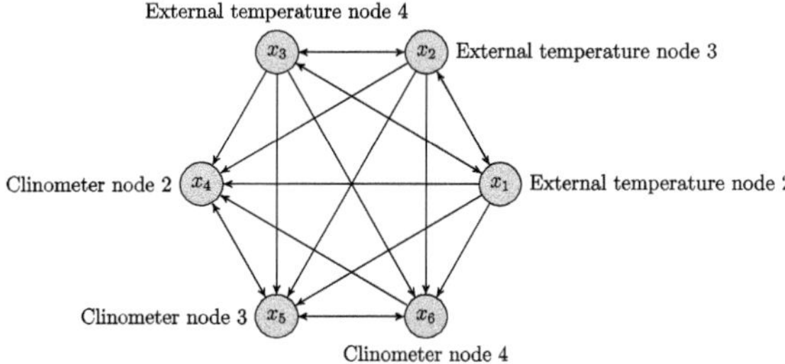

Fig. 10.12 The reduced graph of the functional dependency graph of Fig. 10.8. Only relationships between couples of sensors characterized by a linear correlation index above 0.9 are kept. The reduced dependency graph reduces the computational complexity load requested by the FDS to carry out its task

- The cognitive layer detects a model bias in interval $[4211, 4240]$ and interval $[4481, 4510]$, associated with the relationship f_{56}. If we wish to reduce the occurrence of false positives, we should introduce a more "strict" threshold. This modification would increase the detection delay;
- The cognitive level detects a fault affecting sensor 2 starting from sample 6221: as expected the system is able to correctly detect the induced fault at the cost of a reasonable detection delay;
- The cognitive level detects the presence of an environmental change in correspondence with the intervals $[6521, 6580]$, $[7391, 7480]$, and $[8141, 8165]$.

Example: A Cognitive Fault Analysis Module. The Rialba Towers Sensor Network Case

We now run the whole FDS algorithm based on the HMM methodology on data streams coming from the Rialba towers sensors whose dependency graph is given in Fig. 10.8. In particular, to keep under control the computational load requested by the FDS algorithm, the dependency graph of Fig. 10.8 was pruned, by keeping only those relationships between sensors couples showing at least a linear correlation above 0.9 (in this way identical existing bidirectional relationships are likely to be pruned). This hard threshold on correlation allows us to keep the most relevant arcs, and yielded the reduced dependency graph given in Fig. 10.12. However, we should keep the whole dependency graph if computation is not an issue.

Final results of the cognitive FDS based on the HMM method are associated with the evolution of the indexes given in Fig. 10.13. We observe that

- there it is an environmental change perceived in the time interval $[1321, 1365]$. A more specific analysis showed that the detected event was not associated with environmental change but a communication fault affecting all units of the network at the same time. Given the designed rules, the event was perceived as an environmental change since it was affecting all sensor-to-sensor relationships. However,

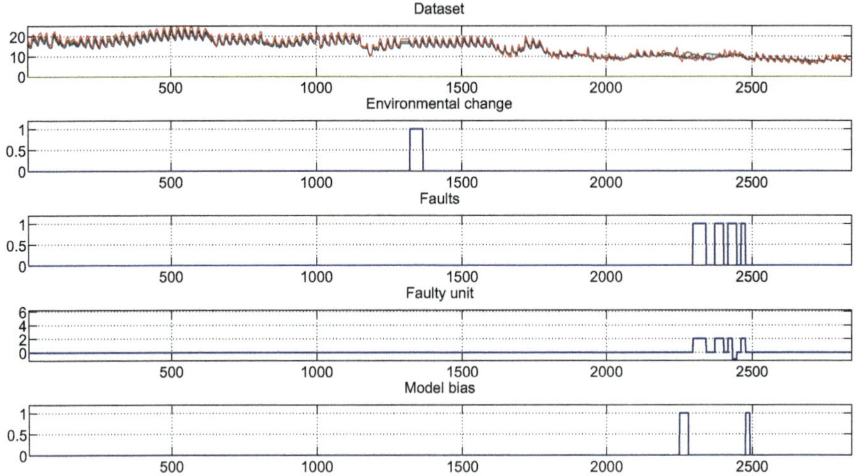

Fig. 10.13 Cognitive level for the Rialba towers scenario for the dependency graph given in Fig. 10.12

this situation can be simply detected by checking if the remote communication link is suitably working or not;

- a model bias and occurrence of faults affecting the second sensor are perceived in the interval [2250, 2491].

10.3 Amygdala and VM-PFC: FDS at the Multi Sensor Level

The behavior of a cognitive FDS resembles the cognitive mechanisms involving the Amygdala and VM-PFC. In this chapter we have seen that after processing the input data stream, a potential fault is detected (this operation resembles that carried out by the Amygdala) and the cognitive layer is activated (as it happens with the VM-PFC) which, by inspecting the dependency graph and the associated likelihoods, provides a refined more accurate control action. If a false positive is detected then the lower level undergoes an online learning so that structural parameters are modified to asymptotically improve the efficiency of the detection level.

References

1. Chernoff H (1952) A measure of asymptotic efficiency for tests of a hypothesis based on the sum of observations. Ann Math Stat 23(4):493507
2. Tempo R, Calafiore G, Dabbene F (2005) Randomized algorithms for analysis and control of uncertain systems. Springer, Berlin
3. Tempo R, Calafiore G, Dabbene F (1998) Uniform sample generation in l_p balls for probabilistic robustness analysis. In: Proceedings of the IEEE conference on decision and control, Tampa, Florida
4. Vidyasagar M (1997) A theory of learning and generalization, Springer, Berlin
5. Marwedel P (2011) Embedded systems design, Springer, Dordrecht
6. Tempo R, Dabbene F (2001) Perspectives in robust control, vol 268. Springer, Berlin/Heidelberg, pp 347–362
7. Chen X, Zhou R (1998) Order statistics and probabilistic robust control. Syst Control Lett 35:175–182
8. Metropolis N, Ulam SM (1949) The Monte Carlo method. J Am Stat Assoc 44:335–341
9. Bai EW, Tempo R, Fu M (1998) Worst-Case properties of the uniform distribution and randomized algorithms for Robustness analysis. Math Control, Signal Syst 11:183–196
10. Barmish BR, Lagoa CM (1997) The uniform distribution: A Rigorous justification for its use in Robustness analysis. Math Control, Signals Systems 10:203–222
11. Tempo R, Dabbene F (1999) Probabilistic Robustness analysis and design of uncertain systems, in dynamical systems. Control, Coding, Computer Vision, Birkhauser, pp 263–282
12. Liu JS (2001) Monte Carlo strategies in scientific computing dynamical systems. Springer, New York
13. Fishman G (1995) Monte Carlo. Springer, New York
14. Sen PK, Singer JM (1993) Large sample methods in statistics. Chapman Hall, New York
15. Ata MY (2007) A convergence criterion for the Monte Carlo estimates. Simul Model Pract Theory 15:237–246
16. Hastings WK (1970) Monte Carlo sampling methods using Markov chains and their applications. Biometrika 57:97–109
17. Giles MB (2007) Improved multilevel Monte Carlo convergence using the Milstein scheme, in Monte Carlo and Quasi-Monte Carlo Methods 2006. Springer, New York
18. Hoeffding W (1963) Probability inequalities for sums of bounded random variables. J Am Stat Assoc 58(301):1330
19. Spall JC (2003) Introduction to stochastic search and optimization estimation, simulation, and control. Wiley, Hoboken NJ

C. Alippi, *Intelligence for Embedded Systems*, DOI: 10.1007/978-3-319-05278-6,

20. Rudin W (1974) Real and complex analysis. McGrawHill, NewYork
21. Jech T (1978) Set theory. Academic, New York
22. Niederreiter H (1992) Random number generation nad quasi Monte Carlo methods. SIAM, Philadelphia
23. Cobham A (1965) The Intrinsic Computational difficulty of function. In: Proceedings of methodology and philosophy of science, vol 2. North-Holland, Amsterdam, p 2430
24. Knuth D (1998) The art of computer programming. Addison-Wesley, Reading
25. Garey MIR, Johnson DS (1990) Computers and intractability; a guide to the theory of NP-completeness. W. H. Freeman & Co., New York, NY, USA
26. Grinstead CM, Snell JL (2006) Introduction to probability. American Mathematical Society, Providence
27. Mallet Y, Coomans D, Kautsky J, De Vel O (1997) Classification using adaptive wavelets for feature extraction. IEEE Trans Pattern Anal Mach Intell 19:10581066
28. Dawid H, Meyr H (1996) The differential CORDIC algorithm: constant scale factor redundant implementation without correcting iterations. IEEE Trans Comput 45:307318
29. Hen Y, Hu H, Chern HM (1996) A novel implementation of CORDIC algorithm using backward angle recoding (BAR). IEEE Trans Comput 45:13701378
30. Boashash B (2003) Time-frequency signal analysis and processing a comprehensive reference. Elsevier Science, Oxford
31. Mallat S (1999) A wavelet tour of signal processing, 2nd edn. Academic Press, San Diego
32. Harris FJ (2004) Multirate signal processing for communication systems. Prentice Hall PTR, Upper Saddle River, NJ
33. Farrell JA, Polycarpou MM (2006) Adaptive approximation based control. Wiley, New York
34. Alippi C, Piuri V (1999) Neural modeling of dynamic systems with nonmeasurable state variables. IEEE-Trans Instrum Meas 48(6):1073–1080 Piscataway (NJ), USA
35. Billingsley P (1995) Probability and measure. Wiley, New York
36. Alippi C (1999) FPE-based criteria to dimension feedforward neural networks. IEEE Trans Circuits Syst: Part I, Fund Theory Appl 46:8:962–973
37. Alippi C, Briozzo L (1998) Accuracy versus precision in digital VLSI architectures for signal processing. IEEE Trans Comput 47(4):472–477
38. Bishop CM (1995) Neural networks for pattern recognition. Clarendon, London, U.K.
39. Hastie T, Tibshirani R, Friedman J (2009) The elements of statistical learning. Springer, New York
40. Wang Z, Bovik AC (2009) Mean squared error: love it or leave it? IEEE Signal Mag 1:98–117
41. Pappas TN, Safranek RJ, Chen J (2005) Perceptual criteria for image quality evaluation. In: Bovik AC (ed) Handbook of image and video processing, 2nd edn. Academic Press, New York
42. Wang Z, Bovik AC (2006) Modern image quality assessment. Morgan and Claypool, San Rafael, CA
43. Kullback S, Leibler RA (1951) On information and sufficiency. Ann Math Stat 22(1):7986
44. Kullback S (1959) Information theory and statistics. Wiley, New York
45. Mallela S, Dhillon IS, Kumar R (2003) A divisive information-theoretic feature clustering algorithm for text classification. J Mach Learn Res 3:12651287
46. Weinstein E, Feder M, Oppenheim A (1990) Sequential algorithms for parameter estimation based on the Kullback-Leibler information measure. IEEE Trans Acoust Speech Signal Process 38(9):1652–1654
47. Moreno PJ, Ho PP, Vasconcelos N (2004) A Kullback-Leibler divergence based kernel for SVM classification in multimedia applications. Hewlett-Packard Company Report
48. Prez-Cruz F (2008) Kullback-Leibler divergence estimation of continuous distributions. In: Proceedings of the IEEE international symposium on information theory
49. Gray RM (1990) Entropy and information theory. Springer, New York
50. Alippi C (2002) Selecting accurate, robust, and minimal feedforward neural networks. IEEE Trans Neural Netw 49(12):1799–1810
51. Stewart GW, Sun J (1990) Matrix perturbation theory. Academic, New York

52. Rauth DA, Randal VT (2005) Analog-to-digital conversion. part 5. IEEE Instrum Meas Mag 8(4):44–54
53. Kim N, Choi S, Cha H (2008) Automated sensor-specific power management for wireless sensor networks. In: Proceedings on IEEE MASS, 305–314
54. Anastasi G, Conti M, Di Francesco M, Passarella A (2009) Energy conservation in wireless sensor networks. Ad Hoc Netw 7:537568
55. Alippi C, Anastasi G, Di Francesco M, Roveri M (2009) Energy management in wireless sensor networks with energy-hungry sensors. IEEE-Instrum Meas Mag 2(2):16–23
56. Singh A, Budzik D, Chen W, Batalin MA, Stealey M, Borgstrom H, Kaiser WJ (2006) Multiscale sensing: a new paradigm for actuated sensing of high frequency dynamic phenomena. In: Proceedings on IEEE/RSJ IROS, pp 328–335
57. Tseng Y-C, Wang YC, Cheng K-Y, Hsieh Y-Y (2007) iMouse: an integrated mobile surveillance and wireless sensor system. IEEE Comput 40:60–66
58. Okabe A, Boots B, Sugihara K, Nok Chiu S (2000) Spatial Tessellations: concepts and applications of Voronoi diagrams, Wiley series in probability and statistics
59. Alippi C, Anastasi G, Galperti C, Mancini F, Roveri M (2007) Adaptive sampling for energy conservation in wireless sensor networks for snow monitoring applications. In: Proceedings of the IEEE MASS, pp 1–6
60. Alippi C, Anastasi G, Di Francesco M, Roveri M (2010) An adaptive sampling algorithm for effective energy management in wireless sensor networks with energy-hungry sensors. IEEE-Trans Instrum Meas 59(2):335–344
61. Zhou J, De Roure D (2007) FloodNet: coupling adaptive sampling with energy aware routing in a flood warning system. J Comput Sci Technol 22(1):121–130
62. Willett R, Martin A, Nowak R (2004) Backcasting: adaptive sampling for sensor networks. In: Proceedings of the IPSN, pp 124–133
63. Vuran MC, Akyildiz IF (2006) Spatial correlation-based collaborative medium access control in wireless sensor networks. IEEE/ACM Trans Netw 14(4):316–329
64. Rahimi M, Hansen M, Kaiser WJ, Sukhatme GS, Estrin D (2005) Adaptive sampling for environmental field estimation using robotic sensors. In: Proceedings of the IEEE/RSJ IROS, pp 3692–3698
65. Rahimi M, Baer R, Iroezi O, Garcia J, Warrior J, Estrin D, Srivastava MB (2005) Cyclops. In Situ Image Sensing and Interpretation. In: Proceedings of the SenSys 2005, pp 192–204
66. Kansal A, Hsu J, Zahedi S, Srivastava M (2007) Power management in energy harvesting sensor networks. ACM Trans Embed Comput Syst 43(4):1–38
67. Vigorito C, Ganesan D, Barto A (2007) Adaptive Control of duty-cycling in energy-harvesting wireless sensor networks. In: Proceedings of the IEEE SECON, pp 21–30
68. Deshpande A, Guestrin C, Madden S, Hellerstein JM, Hong W (2004) Model-driven data acquisition in sensor networks. In: Proceedings of the VLDB, pp 588–599
69. Ottman GK, Hofmann HF, Bhatt AC, Lesieutre GA (2002) Adaptive piezoelectric energy harvesting circuit for wireless remote power supply. IEEE Trans Power Electron 17(5):669776
70. Joseph AD (2005) Energy harvesting projects. IEEE Pervasive Comput 4(1):6971
71. Paradiso JA, Starner T (1827) Energy scaveging for mobile and wireless electronics. IEEE Pervasive Comput 4(1):2005
72. Roundy S, Leland ES, Baker J, Carleton E, Reilly E, Lai E, Otis B, Rabaey JM, Wright PK, Sundararajan V (2005) Improving power output for vibration-based energy scavengers. IEEE Pervasive Comput 4(1):2836
73. Raghunathan V, Kansal A, Hsu J, Friedman J, Srivastava M (2005) Design considerations for solar energy harvesting wireless embedded systems. In: Proceedings of IEEE international conference on information processing for sensor, networks, p 457462
74. Markvart T (1994) Solar electricity. Wiley, New York
75. Lim YH, Hamill DC (2001) Synthesis, simulation and experimental verification of a maximum power point tracker from nonlinear dynamics. In: Proceedings of the IEEE 32nd annual power electron. Specialists Conference, p 199204

76. Alippi C, Galperti C (2008) An adaptive system for optimal solar energy harvesting in wireless sensor network nodes. IEEE Trans Circuits Syst I: Regul Pap 55(6):1742–1750
77. Elwell R, Polikar R (2011) Incremental learning of concept drift in nonstationary environments. IEEE Trans Neural Netw 22(10):15171531
78. Manly BFJ, MacKenzie DI (2000) A cumulative sum type of method for environmental monitoring. Environmetrics 11:151–166
79. Kendall M (1975) Rank correlation methods, 4th edn. Griffin, London, U.K.
80. Chernoff H, Lehmann EL (1954) The use of maximum likelihood estimates in chi-square tests for goodness of fit. Ann Math Stat 25(3):579586
81. Gordon L, Pollak M (1994) An efficient sequential nonparametric scheme for detecting a change of distribution. Ann Stat 22(2):763–804
82. Alippi C, Boracchi G, Roveri M (2011) A hierarchical, nonparametric sequential change-detection test. In: Proceedings of IJCNN 2011, the international joint conference on neural networks
83. Alippi C, Boracchi G, Roveri M (2010) Change detection tests using the ICI rule. In: Proceedings of IEEE IJCNN 2010—International joint conference on neural networks
84. Alippi C, Roveri M (2008) Just-in-time adaptive classifiers. Part I. detecting non-stationary changes. IEEE Trans Neural Netw 19(7):1145–1153
85. Tartakovsky AG, Rozovskii BL, Blazek RB, Kim H (2006) Detection of intrusions in information systems by sequential change-point methods. Stat Method 3(3):252–293
86. McGilchrist CA, Woodyer KD (1975) Note on a distribution-free CUSUM technique. Technometrics 17(3):321–325
87. Basseville M, Nikiforov IV (1993) Detection of abrupt changes: theory and application. Prentice-Hall, Englewood Cliffs, N.J.
88. Wald A (1945) Sequential tests of statistical hypotheses. Ann Math Stat 16(2):117186
89. Mood AM, Graybill FA, Boes DC (1974) Introduction to the theory of statistics, 3rd edn. McGraw-Hill, New York
90. Massey FJ Jr (1951) The Kolmogorov-Smirnov test for goodness of fit. J Am Stat Assoc 46(253):68–78
91. Kruskal WH, Wallis WA (1952) Use of ranks in one-criterion variance analysis. J Am Stat Assoc 47(260):583–621
92. Johnson RA (1998) Applied multivariate statistical analysis. In: Wichern DW (ed). Prentice Hall, Englewood Cliffs
93. Ross GJ, Asoulis DKT, Adams NM (2011) Nonparametric monitoring of data streams for changes in location and scale. Technometrics 53(4):379389
94. Alippi C, Boracchi G, Roveri M (2011) A just-in-time adaptive classification system based on the intersection of confidence intervals rule. Neural Netw 24:791–800
95. Mudholkar GS, Trivedi MC (1981) A Gaussian approximation to the distribution of the sample variance for nonnormal populations. J Am Stat Assoc 76(374):479–485
96. Goldenshluger A, Nemirovski A (1997) On spatial adaptive estimation of nonparametric regression. Math Meth Stat 6:135–170
97. Alippi C, Boracchi G, Roveri M (2011) An effective just-in-time adaptive classifier for gral concept drifts. In: Proceedings of the 2011 International Joint Conference on Neural Networks (IJCNN), pp 1675–1682
98. Alippi C, Boracchi G, Ditzler G, Polikar R, Roveri M (2013) Adaptive Classifiers for Nonstationary Environments. IEEE/Wiley Press Book Series, Contemporary Issues in Systems Science and Engineering
99. Stenwart I (2005) Consistency of support vector machines and other regularized kernel classifier. IEEE Trans Inf Theory 51(1):128–142
100. Hastie T, Tibshirani R, Friedman J (2009) The elements of statistical learning. Springer, New York
101. Alippi C, Bu DL, Zhao D (2013) Detecting and reacting to changes in sensing units: the active classifier case. IEEE Trans Syst Man Cybern—Part A: Syst Hum (To appear).
102. http://www.i-sense.org/open_library.html

103. Fukunaga K (1972) Introduction to statistical pattern recognition. Academic, New York
104. Alippi C, Roveri M (2008) Just-in-time adaptive classifiers. Part II. Designing the classifier. IEEE Trans Neural Netw 19(12):2053–2064
105. Gates GW (1972) The reduced nearest neighbor rule. IEEE Trans Inf Theory 18(3):431–433
106. Wilson D (1972) Asymptotic properties of nearest neighbor rules using edited data. IEEE Trans Syst Man Cybern 2(3):408–421
107. Rauth DA, Randal VT (2005) Analog-to-digital conversion. IEEE Instrum Meas Mag 8:4454
108. Fowler KR, Schmalzel JL (2004) Sensors: the first stage in the measurement chain. IEEE Instrum Meas Mag 7:60–66
109. Ferrero A, Salicone S (2006) Measurement uncertainty. IEEE Instrum Meas Mag 9(3):44–51
110. Schmalzel JL, Rauth DA (2005) Sensors and signal conditioning. IEEE Instrum Meas Mag 8(2):48–53
111. Fowler KR (2003) Analog-to-digital conversion in real-time systems. IEEE Instrum Meas Mag 6:58–64
112. Lindley D (2008) Uncertainty: Einstein. Bohr, and the struggle for the soul of science, Anchor, Heisenberg
113. Stewart CV (1997) Bias in robust estimation caused by disconti- nuities and multiple structures. IEEE Trans Pattern Anal Mach Intell 19(8):818–833
114. Dutt S, Assaad FT (1996) Mantissa-preserving operations and robust algorithm-based fault tolerance for matrix computations. IEEE Trans Comput 45(4):408–424
115. Saab YG (1995) A fast and robust network bisection algorithm. IEEE Trans Comput 44(7):903–913
116. Gu D, Petkov PH, Konstantinov MM (2013) Robust control design with MATLAB. Springer, London
117. Holt J, Hwang J (1993) Finite precision error analysis of neural network hardware implementations. IEEE Trans Comput 42:281290
118. Dundar G, Rose K (1995) The effects of quantization on multilayer neural networks. IEEE Trans. Neural Networks 6(11):14461451
119. Edwards PJ, Murray AF (1998) Fault tolerance via weight noise in analog VLSI implementations of MLPsA case study with EPSILON. IEEE Trans Circuits Syst II 45(9):12551262
120. Buckingham MJ (1983) Noise in electronic devices and systems. Ellis Horwood, Chichester
121. Seber G Wild C (1989) Nonlinear regression, Wiley, New york
122. Murata N, Yoshizawa S, Amari S (1994) Network information criterion determining the number of hidden units for an artificial neural network model. IEEE Trans Neural Networks 5(11):865872
123. Alippi C (1999) FPE-based criteria to dimension feedforward neural topologies. IEEE Trans Circuits Syst I 46(8):962973
124. Levin AU, Leen TK, Moody JE (1994) Fast pruning using principal components, In: Paper presented at the Proceedings of the 6th NIPS, vol 6, p 3542
125. Nocedal J, Wright SJ (2006) Numerical optimization, 2nd edn. Springer, New York
126. Dugunji J (1966) Topology. Allyn and Bacon, Boston
127. Alippi C (2002) Randomised algorithms: a system-level. Poly-time analysis of robust computation. IEEE Trans Comput 51(7):740–749
128. Alippi C (2002) Application-level robustness and redundancy in linearsystems. IEEE Trans Circuits Syst Part I Fundam Theory Appl 49(7):1024–1027
129. Saltelli A, Ratto M, Andres T, Campolongo F, Cariboni J, Gatelli D, Saisana M, Tarantola S (2008) Global sensitivity analysis: the primer. John Wiley and Sons, New Jersey
130. Ljung L (1999) System identification. Wiley Online Library, New York
131. Cybenko G (1989) Approximation by superpositions of a sigmoidal function. Math Contr Signals Syst 2:303314
132. Vapnik V (1995) The nature of statistical learning theory. Springer- Verlag, New York
133. Vapnik V (1998) Statistical learning theory. Wiley, New York
134. Vapnik V (1999) An overview of statistical learning theory. IEEE Trans Neural Networks 10(5):988999

135. Alippi C, Braione P (2006) Classification methods and inductive learning rules: what we may learn from theory. IEEE Trans Syst Man Cybern Part C Appl Rev 36(5):649–655
136. Ljung L, Caines PE (1978) Asymptotic normality of prediction error estimators for approximate system models. In: paper presented at the IEEE conference on decision and control symposium on adaptive processes, vol 17, p 927932
137. Ljung L (1978) Convergence analysis of parametric identification methods. IEEE Trans Autom Control 23(5):770783
138. Gupta Rupa, Koscik Timothy R, Bechara Antoine, Tranel Daniel (2011) The amygdala and decision making. Neuropsychologia 49(4):760766
139. Gupta Rupa, Duff Melissa C, Denburg Natalie L, Cohen Neal J, Bechara Antoine, Tranel Daniel (2009) Declarative memory is critical for sustained advantageous complex decision-making. Neuropsychologia 47(7):1686–1693
140. Ochsner K, Barret L (in press) A multiprocess perspective on the neuroscience of emotion. In: Mayne T, Bonnano G (eds) Emotion: current issues and future directions, Guilford Press, New York
141. Bechara A, Damasio H, Damasio A R, Lee GP (1999) Different contributions of the human amygdala and ventromedial prefrontal cortex to decision-making. J Neurosci 19(13):54735481
142. Alippi C, Camplani R, Galperti C, Marullo A, Roveri M (2013) A high frequency sampling monitoring system for environmental and structural applications. Part A ACM Trans Sens Netw (accepted for publication)
143. Alippi C, Camplani R, Roveri M. Viscardi G (2012) NetBrick: a high-performance, low-power hardware platform for wireless and hybrid sensor network. In: paper presented at the 9th IEEE international conference on mobile Ad hoc and sensor systems, Las Vegas, USA, 8–11 October 2012 (accepted for publication at IEEE MASS 2012)
144. Isermann R (2006) Fault-diagnosis systems: an introduction from fault detection to fault tolerance. Springer Verlag, Berlin
145. Basseville M, Nikiforov I et al (1993) Detection of abrupt changes: theory and application, vol. 15. Prentice Hall Englewood Cliffs, New Jersey
146. Reppa V, Polycarpou M, Panayiotou CG (2012) A distributed architecture for sensor fault detection and isolation using adaptive approximation, In: paper presented at Proceedings of the IEEE world congress on computational intelligence, 2012, Brisbane, Australia, p 23402347
147. Reppa V, Polycarpou M, Panayiotou CG (2014) Adaptive approximation for multiple sensor fault detection and isolation of nonlinear uncertain systems. IIEEE Trans Neural Networks and Learn Syst 25(1):1–10
148. Keliris C, Polycarpou M, Parisini T (2012) A distributed fault detection filtering approach for a class of interconnected input-output nonlinear systems. In: paper presented at the European control conference (ECC2012), Zurich, Switzerland
149. Keliris C, Polycarpou M, Parisini T (2013) A distributed fault detection filtering approach for a class of interconnected continuous-time nonlinear systems, IEEE Trans Autom Control 58(8):2032–2047
150. Blesa J, Puig V, Saludes J (2012) Robust identification and fault diagnosis based on uncertain multiple inputmultiple output linear parameter varying parity equations and zonotopes. J Process Control (article in press)
151. Blesa J, Puig V, Saludes J (2012) Robust fault detection using polytope-based set-membership consistency test. IET Control Theory Appl 111(10):104910
152. Tornil-Sin S, Ocampo-Martinez C, Puig V, Escobet T (2013) Robust fault diagnosis of nonlinear systems using interval constraint satisfaction and analytical redundancy relations. IEEE Trans Syst Man Cybern Part A Syst Hum 44(1):18–29 (accepted)
153. Puig V, Oca S, Blesa J (2012) Adaptive threshold generation in robust fault detection using interval models: time-domain and frequency-domain approaches. Int J Adapt Control Signal Process 26(3):258–283
154. Chen H, Tino P, Yao X, Rodan A (2014) Learning in the model space for fault diagnosis. IEEE Trans Neural Networks Learn Syst 25(1):124–136

155. Alippi C, Roveri M, Trovo F (2012) A learning from models cognitive fault diagnosis system. In: paper presented at the artificial neural networks and machine learning (ICANN 2012), p 305313
156. Alippi C, Ntalampiras S, Roveri M (2013) A cognitive fault diagnosis system for sensor networks. IEEE Trans Neural Networks Learn Syst 24(8):1213–1226
157. Flavell H (1996) Piagets legacy. Psychol. Sci. 7(4):200203
158. Esmaeilzadeh H, Sampson A, Ceze L, Burger D (2013) Neural acceleration for general purpose approximate programs. In: paper presented at 46th annual IEEE/ACM international symposium on microarchitecture (MICRO 2013), pp 16–21
159. Mudge T (2001) Power: a first-class architectural design constraint. Computer 34(4):52–58
160. Pillai P, Shin KG (2001) Real-time dynamic voltage scaling for low-power embedded operating systems. In: paper presented at proceedings of the eighteenth ACM symposium on operating systems principles, pp 89–102
161. Shirako J et al (2006) Compiler control power saving scheme for multi core processors. In: paper presented at languages and compilers for parallel computing—lecture notes in computer science, Springer
162. Hsu CH, Kremer U (2003) The design, implementation, and evaluation of a compiler algorithm for CPU energy reduction. SIGPLAN 38(5):38–48
163. Talpes E, Marculescu D (2005) Toward a multiple clock/voltage island design style for power-aware processors. IEEE Trans Very Large Scale Integr VLSI Syst 13(5):539–552
164. Wu Q, Juang P, Martonosi M, Clark DW (2004) Formal online methods for voltage/frequency control in multiple clock domain microprocessors. SIGOPS Oper Syst Rev 38(5):248–259
165. Magklis G, Chaparro P, Gonzalez J, Gonzalez A (2006) Independent front-end and back-end dynamic voltage scaling for a GALS microarchitecture. In: Proceedings of the 2006 international symposium on low power electronics and design, 2006. ISLPED'06, pp 49–54
166. Moeng M, Melhem R (2010) Applying statistical machine learning to multicore voltage and frequency scaling. In: Proceedings of the 7th ACM international conference on computing frontiers (CF '10)
167. Lamport L (1978) Time, clocks, and the ordering of events in a distributed system. Commun ACM 21(7):558–565
168. Vasanthavada N, Marinos PN (1988) Synchronization of fault-tolerant clocks in the presence of malicious failures. IEEE Trans Comput 37(4):440–448
169. Olson A, Shin KG (1994) Probabilistic clock synchronization in large distributed systems. IEEE Trans Comput 43(9):1106–1112
170. van Greunen J, Rabaey J (2003) Lightweight time synchronization for sensor networks. In: Proceedings of the 2nd ACM international conference on wireless sensor networks and applications (WSNA '03), pp 11–19
171. Elson J, Girod L, Estrin D (2002) Fine-grained network time synchronization using reference broadcasts. In: Proceedings of the 5th symposium on operating systems design and implementation
172. Marti M, Kusy B, Simon G, Ldeczi A (2004) The flooding time synchronization protocol. In: Proceedings of the 2nd international conference on embedded networked sensor systems (SenSys '04), pp 39–49
173. Wu YC, Chaudhari Q, Serpedin E (2011) Clock synchronization of wireless sensor networks. Sig Process Mag IEEE 28(1):124–138
174. IEEE Standard for a Precision Clock Synchronization Protocol for Networked Measurement and Control Systems (2008) IEEE Std 1588–2008 (Revision of IEEE Std 1588–2002). doi:10.1109/IEEESTD.2008.4579760
175. Noh K, Chaudhari QM, Serpedin E, Suter BW (2007) Novel clock phase offset and skew estimation using two-way timing message exchanges for wireless sensor networks. IEEE Trans Commun 55(4):766–777
176. Mills DL (1991) Internet time synchronization: the network time protocol. IEEE Trans Commun 39(10):1482–1493

177. Kim H, Ma X, Hamilton BR (2012) Tracking low-precision clocks with time-varying drifts using Kalman filtering. IEEE/ACM Trans Netw 20(1):257–270
178. Bletsas A (2003) Evaluation of Kalman filtering for network time keeping. In: Proceedings of the first IEEE international conference on pervasive computing and communications (PERCOM '03), pp 1452–1460
179. Auler LF, D'Amore R (2007) Adaptive Kalman filter for time synchronization over packet-switched networks: an heuristic approach. In: 2nd international conference on communication systems software and middleware COMSWARE 2007, pp 7–12
180. ISO guide 99
181. ISO guide 98
182. Doebelin E (1990) Measurement systems: application and design. McGraw-Hill, New York
183. Hawkins DM, Qiu P, Kang CW (2003) The changepoint model for statistical process control. J Qual Technol 35(4):355–366
184. Ross GJ Sequential change detection in R: the cpm package. J Stat Softw (Forthcoming)
185. Ross GJ, Tasoulis DK, Adams NM (2011) Nonparametric monitoring of data streams for changes in location and scale. Technometrics 53(4):379–389
186. Mann HB, Whitney DR (1947) On a test of whether one of two random variables is stochastically larger than the other. Ann Math Stat 18(1):50–60
187. Mood AM (1954) On the asymptotic efficiency of certain nonparametric two-sample tests. Ann Math Stat 25(3):514–522
188. Lepage Y (1974) A combination of Wilcoxon's and Ansari-Bradley's statistics. Biometrika 58(1):213–217
189. Pettitt AN (1979) A non-parametric approach to the change-point problem. Appl Stat 28(2):126–135
190. Ross G, Adams NM (2012) Two nonparametric control charts for detecting arbitrary distribution changes. J Qual Technol 44(22):102–116
191. Anderson TW (1962) On the distribution of the two-sample Cramer-von Mises criterion. Ann Math Stat 33(3):1148–1159
192. Zamba KD, Hawkins DM (2006) A multivariate change-point model for statistical process control. Technometrics 48(4):539–549
193. Gustafsson F (2000) Adaptive filtering and change detection. Wiley, New York
194. Alippi C, Vanini G (2004) Wireless sensor networks and radio localization: a metrological analysis of the MICA2 received signal strenght indicator. In: Proceedings of the first IEEE workshop on embedded networked sensors
195. Alippi C, Vanini G (2006) A RSSI-based and calibrated centralized localization technique for wireless sensor networks. In: Proceedings of the second IEEE international workshop on sensor networks and systems for pervasive computing, pp 1–5
196. Morelli C, Nicoli M, Rampa V, Spagnolini U, Alippi C (2006) Particle filters for RSS-based localization in wireless sensor networks: an experimental study. In: Proceedings of ICASPS, pp 957–960
197. Patwari N, Ash JN, Kyperountas S, Hero AO, Moses RL, Correal NS (2005) Locating the nodes: cooperative localization in wireless sensor networks. Sig Process Mag IEEE 22(4):54–69
198. Mao G, Fidan B, Anderson BDO (2007) Wireless sensor network localization techniques. Comput Netw 51(10):2529–2553
199. Maung Maung NA, Kawai M (2012) Hybrid RSS-SOM localization scheme for wireless ad hoc and sensor networks. In: 2012 international conference on indoor positioning and indoor navigation (IPIN), pp 1–7
200. Kaune R (2012) Accuracy studies for TDOA and TOA localization. In: 2012 15th international conference on information fusion (FUSION), pp 408–415
201. Dardari D, Conti A, Ferner U, Giorgetti A, Win MZ (2009) Ranging with ultrawide bandwidth signals in multipath environments. Proc IEEE 97(2):404–426
202. Galler S, Gerok W, Schroeder J, Kyamakya K, Kaiser T (2007) Combined AOA/TOA UWB localization. In: International symposium on communications and information technologies, ISCIT '07, pp 1049–1053

203. Römer K (2003) The lighthouse location system for smart dust. In: Proceedings of MobiSys 2003 (ACM/USENIX conference on mobile systems, applications, and services), pp 15–30
204. Alippi C, Boracchi G, Roveri M (2013) Ensembles of change-point methods to estimate the change point in residual sequences. Soft Comput 17(11):1971–1981
205. Hawkins DM, Zamba KD (2005) A change-point model for a shift in variance. J Qual Technol 37(1):21–31
206. Hamacher V, Vranesic Z, Zaky S (1997) Computer organization. Mc Graw Hill, New York
207. Piche SW (1995) The selection of weights accuracies for madalines. IEEE Trans Neural Netw 6(2):432–445
208. Salivahanan S, Vallavaraj A (2010) Digital signal processing, 2nd edn. Mc Graw Hill, New York
209. Cormen TH, Leiserson CE, Rivest R, Stein C (2001) Introduction to algorithms. Mc Graw Hill, New York
210. Lieberherr KJ, Palsberg J (1993) Engineering adaptive software. Project Proposal. ftp://ftp.ccs.neu.edu/pub/people/lieber/proposal.ps
211. Kephart JO, Chess DM (2003) The vision of autonomic computing. Computer 36(1):41–50
212. Murphy KP (2012) Machine learning a probabilistic perspective. MIT press, Cambridge
213. Walsh WE, Tesauro G, Kephart JO, Das R, Utility functions in autonomic systems. In: Proceedings of the international conference on autonomic computing, pp 70–77
214. Tesauro G (2007) Reinforcement learning in autonomic computing: a manifesto and case studies. IEEE Internet Comput 11(1):22–30
215. Sutton RS, Barto AG (1998) Reinforcement learning: an introduction. MIT press, Cambridge
216. Antola A, Marullo A, Mezzalira L, Roveri M (2014) GINGER: a novel reprogramming paradigm for distributed sensor networks, in PerCom 2014. Internal report Politecnico di Milano
217. Schwartz A (1993) A reinforcement learning method for maximizing undiscounted rewards. In: Proceedings of the tenth international conference on machine learning, pp 298–305
218. Alippi C, Camplani R, Roveri M, Viscardi G (2012) NetBrick: a high-performance, low-power hardware platform for wireless and hybrid sensor network. In: 9th IEEE international conference on mobile ad hoc and sensor systems (IEEE MASS 2012)
219. Alippi C, Camplani R, Galperti C, Marullo A, Roveri M (2013) A high-frequency sampling monitoring system for environmental and structural applications. ACM Trans Sen Netw 9(4):41:1–41:32
220. Kandel ER, Schwartz JH, Jessell TM, Siegelbaum SA, Hudspeth AJ (2013) Principles of neural science, 5th edn. Mc Graw-Hill, New York
221. Nunez R, Freeman WJ (eds) (2000) Reclaiming cognition: the primacy of action, intention and emotion. Imprint Academic, Thorverton
222. Salehie M, Tahvildari L (2009) Self-adaptive software: landscape and research challenges. ACM Trans Auton Adapt Syst 4(2):14:1–14:42
223. An X, Rutten E, Diguet J, Le Griguer N, Gamatie A (2013) Autonomic management of reconfigurable embedded systems using discrete control: application to FPGA, INRIA research report RR-8308
224. Sironi F, Triverio M, Hoffmann H, Maggio M, Santambrogio MD (2010) Self-aware adaptation in FPGA-based systems. In: International conference on field programmable logic and applications (FPL), pp 187–192
225. Majer M, Teich J, Ahmadinia A, Bobda C (2007) The Erlangen slot machine: a dynamically reconfigurable FPGA-based computer. J VLSI Signal Process Syst 47(1):15–31
226. Allan DW, Ashby N, Hodge CC (1997) The science of timekeeping. Hewlett Packard Application Note 1289
227. Ramus R (1989) Alpha quartz. IEEE Potentials 8(4):9
228. ENEA (1999) http://clisun.casaccia.enea.it/pagine/tabelleradiazione.htm. In Tabella della Radiazione So- lare, ENEA—Italian national agency for new technologies, Energy and sustainable economic development

229. Ongaro F, Saggini S, Giro S, Mattavelli P (2010) Two-dimensional MPPT for photovoltaic energy harvesting systems. In: IEEE 12th workshop on control and modeling for power electronics (COMPEL), pp 1–5
230. Alippi C, Camplani R, Galperti C, Marullo A, Roveri M (2013) A high frequency sampling monitoring system for environmental and structural applications. ACM Trans Sen Netw 9(4):1–32 (Article 41)
231. Cassandras CG, Lafortune S (2008) Introduction to discrete event systems. Springer, New York
232. Morelli C, Nicoli M, Rampa V, Spagnolini U, Alippi C (2006) Particle filters for RSS-based localization in wireless sensor networks: an experimental study. In: Proceedings of ICASPS, Toulouse, pp IV-957–IV-960 14–19 May 2006
233. Thrun S, Burgard W, Fox D (2005) Probabilistic robotics. MIT Press, Cambridge
234. Di Francesco M, Anastasi G, Conti M, Das S, Neri V (2011) Reliability and energy efficiency in IEEE 802.15.4/ZigBee sensor networks: a cross-layer approach. In: IEEE J Sel Areas Commun 29(8):1508–1524
235. Brienza S, De Guglielmo D, Alippi C, Anastasi G, Roveri M (2013) A Learning-based algorithm for optimal MAC parameters setting in IEEE 802.15.4 wireless sensor networks. In: Proceedings of ACM international symposium on performance evaluation of wireless ad hoc, sensor, and ubiquitous networks ACM PE-WASUN, Barcelona, pp 3–7
236. Efron B, Tibshirani RJ (1998) An introduction to the bootstrap. Chapman and Hall/CRC, Boca Raton
237. Bickel PJ, Gotze F, van Zwet W (1997) Resampling fewer than n observations: gains, losses, and remedies for losses. Statistica Sinica 7:1–31
238. Bickel PJ, Sakov A (2002) Extrapolation and the bootstrap. Sankhya Indian J Stat 64:640–652
239. Kleiner A, Talwalkar A, Sarkar P, Jordan M (2012) The big data bootstrap. In: Proceedings of the 29th international conference on machine learning, Edinburgh, pp 1–8
240. Zhou ZH (2012) Ensemble methods: foundations and algorithms. Chapman and Hall/CRC, Boca Raton
241. Johnson RA, Wichern DW (2002) Applied multivariate statistical analysis. Prentice Hall, New Jersey
242. Bocca M, Eriksson LM, Mahmood A, Jantti R, Kullaa J (2011) A synchronized wireless sensor network for experimental modal analysis in structural health monitoring. Comput Aided Civil Infrastruct Eng 26(7):483–499
243. Rappaport TS (1996) Wireless communications: principles and practice. Prentice-Hall Inc., New Jersey
244. Vallet J, Kaltiokallio O, Myrsky M, Saarinen J, Bocca M (2012) Simultaneous RSS-based localization and model calibration in wireless networks with a mobile robot. In: International workshop on cooperative robots and sensor networks (RoboSense 2012), pp 1106–1113
245. Vallet J, Kaltiokallio O, Myrsky M, Saarinen J, Bocca M (2013) On the sensitivity of RSS based localization using the log-normal model: an empirical study. In: 10th workshop on positioning, navigation and communication 2013 (WPNC'13), Dresden, 20–21 March 2013
246. Ganeriwal S, Kumar R, Srivastava MB (2003) Timing-sync protocol for sensor networks. In: First international conference on embedded networked sensor systems, pp 138–149
247. Graybiel AM (1990) Neurotransmitters and neuromodulators in the basal ganglia. Trends Neurosci 13(7):244–254
248. Graybiel AM et al (1994) The basal ganglia and adaptive motor control Science 265(5180):1826–1831
249. Graybiel AM (1998) The basal ganglia and chunking of action repertoires. Neurobiol Learn Mem 70(1):119–136
250. Squire LR (1992) Memory and the hippocampus: a synthesis from findings with rats, monkeys, and humans. Psychol Rev 99(2):195–231

Index

C. Alippi, *Intelligence for Embedded Systems*, DOI: 10.1007/978-3-319-05278-6,

R

S

T

U

V

W